Table of the Elements

Element	Symbol	Atomic Number	Atomic Weight	Element	Symbol	Atomic Number	Atomic Weight
actinium	Ac	89	227.03	mendelevium	Md	101	258.10
aluminum	Al	13	26.98	mercury	Hg	80	200.59
americium	Am	95	243.06	molybdenum	Mo	42	95.94
antimony	Sb	51	121.75	neodymium	Nd	60	144.24
argon	Ar	18	39.95	neon	Ne	10	20.18
arsenic	As	33	74.92	neptunium	Np	93	237.05
astatine	At	85	209.99	nickel	Ni	28	58.69
barium	Ba	56	137.33	niobium	Nb	41	92.91
berkelium	Bk	97	247.07	nitrogen	N	7	14.01
beryllium	Be	4	9.012	nobelium	No	102	259.10
bismuth	Bi	83	208.98	osmium	Os	76	190.23
bohrium	Bh	107	(264.12)*	oxygen	O	8	16.00
boron	B	5	10.81	palladium	Pd	46	106.42
bromine	Br	35	79.90	phosphorus	P	15	30.97
cadmium	Cd	48	112.41	platinum	Pt	78	195.08
calcium	Ca	20	40.08	plutonium	Pu	94	244.06
californium	Cf	98	251.08	polonium	Po	84	208.98
carbon	C	6	12.01	potassium	K	19	39.10
cerium	Ce	58	140.12	praseodymium	Pr	59	140.91
chlorine	Cl	17	35.45	promethium	Pm	61	144.91
chromium	Cr	24	52.00	proactinium	Pa	91	231.04
cobalt	Co	27	58.93	radium	Ra	88	266.03
copper	Cu	29	63.55	radon	Rn	86	222.02
curium	Cm	96	247.07	rhenium	Re	75	186.21
dubnium	Db	105	262.11	rhodium	Rh	45	102.91
dysprosium	Dy	66	162.50	rubidium	Rb	37	85.47
einsteinium	Es	99	252.06	ruthenium	Ru	44	101.07
erbium	Er	68	167.26	rutherfordium	Rf	104	261.11
europium	Eu	63	151.96	samarium	Sm	62	150.36
fermium	Fm	100	257.09	scandium	Sc	21	44.96
fluorine	F	9	19.00	seaborgium	Sg	106	263.12
francium	Fr	87	223.02	selenium	Se	34	78.96
gadolinium	Gd	64	157.25	silicon	Si	14	28.09
gallium	Ga	31	69.72	silver	Ag	47	107.87
germanium	Ge	32	72.61	sodium	Na	11	22.99
gold	Au	79	196.97	strontium	Sr	38	87.62
hafnium	Hf	72	178.49	sulfur	S	16	32.07
hassium	Hs	108	265.13	tantalum	Ta	73	180.95
helium	He	2	4.003	technetium	Tc	43	98.81
holmium	Ho	67	164.93	tellurium	Te	52	127.60
hydrogen	H	1	1.008	terbium	Tb	65	158.93
indium	In	49	114.82	thallium	Tl	81	204.38
iodine	I	53	126.90	thorium	Th	90	232.04
iridium	Ir	77	192.22	thulium	Tm	69	168.93
iron	Fe	26	55.85	tin	Sn	50	118.71
krypton	Kr	36	83.80	titanium	Ti	22	47.88
lanthanum	La	57	138.91	tungsten	W	74	183.85
lawrencium	Lr	103	260.11	uranium	U	92	238.03
lead	Pb	82	207.2	vanadium	V	23	50.94
lithium	Li	3	6.941	xenon	Xe	54	131.29
lutetium	Lu	71	174.97	ytterbium	Yb	70	173.04
magnesium	Mg	12	24.31	yttrium	Y	39	88.91
manganese	Mn	25	54.94	zinc	Zn	30	65.39
meitnerium	Mt	109	(268)*	zirconium	Zr	40	91.22

* Atomic weights in parentheses have an uncertainty greater than ±1 in the last figure.

Applications of

ENVIRONMENTAL CHEMISTRY

A Practical Guide for Environmental Professionals

Eugene R. Weiner, Ph.D.

LEWIS PUBLISHERS

Boca Raton London New York Washington, D.C.

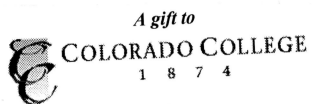

A gift to

COLORADO COLLEGE
1 8 7 4

by

FORESTRY SCHOOL FUND

Library of Congress Cataloging-in-Publication Data

Weiner, Eugene R.
  Applications of environmental chemistry: a practical guide for environmental
professionals / Eugene R. Weiner.
      p.   cm.
  Includes index.
  ISBN 1-56670-354-9 (alk. paper)
  1. Environmental chemistry.   I. Title
TD193.W45 2000
577'.194—dc21   99-087370
                                                                               CIP

© 2000 by CRC Press LLC
Lewis Publishers is an imprint of CRC Press LLC

No claim to original U.S. Government works
International Standard Book Number 1-56670-354-9
Library of Congress Card Number 99-087370
Printed in the United States of America   1 2 3 4 5 6 7 8 9 0
Printed on acid-free paper

Preface

"By sensible definition, any by-product of a chemical operation for which there is no profitable use is a waste. The most convenient, least expensive way of disposing of said waste — up the chimney or down the river — is the best."

From *American Chemical Industry — A History*, by W. Haynes, Van Nostrand Publishers, 1954.

The quotation above describes the usual approach to waste disposal as it was practiced in the first half of the 1900s. Current disposal and cleanup regulations are aimed at correcting problems caused by such misguided advice and go further toward maintaining a nondegrading environment. Regulations, such as federal and state Clean Water Acts, have set in motion a great effort to identify the chemical components and other characteristics that influence the quality of surface and groundwaters and the soils through which they flow. The number of drinking water contaminants regulated by the U.S. government has increased from about 5 in 1940 to more than 150 in 1999.

There are two distinct spheres of interest for an environmental professional: the ever-changing constructed sphere of regulations and the comparatively stable sphere of the natural environment. Much of the regulatory sphere is bound by classifications and numerical standards for waters, soils, and wastes. The environmental sphere is bound by the innate behavior of chemicals of concern. While this book focuses on the environmental sphere, it makes an excursion into a small part of the regulatory sphere in Chapter 1 where the rationale for stream classifications and standards and the regulatory definition of water quality are discussed.

This book is intended to serve as a guide and reference for professionals and students. It is structured to be especially useful for those who must use the concepts of environmental chemistry but are not chemists and do not have the time and/or the inclination to learn all the relevant background material. Chemistry topics that are most important in environmental applications are succinctly summarized with a genuine effort to walk the middle ground between too much and too little information. Frequently used reference materials are also included, such as water solubilities, partition coefficients, natural abundance of trace metals in soil, and federal drinking water standards. Particularly useful are the frequent "rules of thumb" lists which conveniently offer ways to quickly estimate important aspects of the topic being discussed.

Although it is often true that "a little knowledge can be dangerous," it is also true that a little chemical knowledge of the "right sort" can be of great help to the busy nonchemist. Although no "practical guide" will please everyone with its choice of inclusions and omissions, I have based my choices on the most frequently asked questions from my colleagues and on the material I find myself looking up frequently. The main goal of this book is to offer nonchemist readers enough chemical insight to help them contend with those environmental chemistry problems that seem to arise most frequently in the work of an environmental professional. Environmental chemists and students of environmental chemistry should also find the book valuable as a "general purpose" reference.

Chapter 1 outlines part of the administrative regulatory structure with which the reader, presumably, must interact. Chapter 2 offers some elementary theoretical background for those who may need it or find it interesting. Professionals with little time to spare will find Chapters 3–7 and the appendices of greatest interest, which is where pollutant properties and environmental applications are described.

About the Author

Eugene R. Weiner, Ph.D., is professor emeritus of chemistry at the University of Denver, Colorado. He joined the University of Denver's faculty in 1965. From 1967 to 1992, Dr. Weiner was a consultant with the U.S. Geological Survey, Water Resources Division in Denver, and has consulted on environmental issues for many other private, state, and federal entities. After 27 years of research and teaching environmental and physical chemistry, he joined Wright Water Engineers Inc., an environmental and water resources engineering firm in Denver, as senior scientist.

Dr. Weiner received a B.S. degree in mathematics from Ohio University, an M.S. degree in physics from the University of Illinois, and a Ph.D. degree in chemistry from Johns Hopkins University. He has authored and coauthored approximately 200 research articles, books, and technical reports. In recent years, he conducted 16 short courses, dealing with the movement and fate of contaminants in the environment, at major cities around the U.S. for the continuing education program of the American Society of Civil Engineers.

Table of Contents

Chapter 3   Major Water Quality Parameters

1 Water Quality

CONTENTS

1.1 DEFINING WATER QUALITY

In most parts of the world, the days are long gone when rivers, lakes, springs, and wells from which one can directly drink, could readily meet almost all needs for high quality water. Where such water remains — mostly in high mountain regions untouched by mining, grazing, or industrial fallout — it must be protected by strict regulations. In the U.S., many states seek to preserve high quality waters with antidegradation policies. But most of the water that is used for drinking water supplies, irrigation, and industry, not to mention supplying a supporting habitat for natural flora and fauna, is much-reused water that often needs treatment to become acceptable.

Whenever it is recognized that water treatment is required, new issues arise concerning the level of quality sought, the costs involved, and, perhaps, restrictions imposed on the uses of the water. Since it is economically impossible to make all waters suitable for all purposes, it becomes necessary to designate which uses various waters are suitable for.

In this context, a practical evaluation of water quality depends on how the water is used, as well as its chemical makeup. The quality of water in a stream might be considered good if the water is used for irrigation but poor if it is used as a drinking water supply. To determine water quality, one must first identify the ways in which the water will be used and only then determine appropriate numerical standards for important parameters of the water that will support and protect the designated water uses.

WATER USE CLASSIFICATIONS AND WATER QUALITY STANDARDS

Many impurities in water are beneficial. For example, carbonate (CO_3^{2-}) and bicarbonate (HCO_3^-) make water less sensitive to acid rain and acid mine drainage; hardness and alkalinity decrease the solubility and toxicity of metals; nutrients, dissolved carbon dioxide (CO_2), and dissolved oxygen (O_2) are essential for aquatic life. Outside a chemical laboratory, truly pure water generally is not desirable. Pure water is more corrosive (aggressive) to metal than water containing a measure of hardness, cannot sustain aquatic life, and certainly does not taste as good as natural water saturated with dissolved oxygen and containing a healthy mix of minerals.

The quality of water is not judged by its purity but rather by its suitability for the different uses intended for it. The water contaminant nitrate (NO_3^-) illustrates this point. In drinking water supplies, nitrate concentrations greater than 10 mg/L are considered a potential health hazard, particularly to young children. On the other hand, nitrate is a beneficial plant nutrient in agricultural water and is added as a fertilizer. Water containing more than 10 mg/L of nitrate is of poor quality if it is used for potable water but may be of good quality for agricultural use.

Thus, water uses must be identified before water quality can be judged. The following preliminary steps, taken by a state or federal agency, are a common approach to evaluating water quality:

1. Define the basic purposes for which natural waters will potentially be used (water supply, aquatic life, recreation, agriculture, etc.). These will be the categories used for classifying existing bodies of water.
2. Set numerical water quality standards for physical and chemical characteristics that will support and protect the different water use categories.
3. Compare the water quality standards with field measurements of existing bodies of water, then assign appropriate use classifications to the water bodies according to whether their present or potential quality is suitable for the assigned water uses.

After a body of water is classified for one or more uses, compile an appropriate set of numerical standards to protect its assigned use classifications. Where different assigned classifications have different standards for the same parameter, the more stringent standard applies.

It is clear that measuring the chemical composition of a water sample collected in the field is just one step in determining water quality. The sample data must then be compared with the standards assigned to that water body. If no standards are exceeded, the water quality is defined as good within its classified uses. As new information is collected about environmental and health effects of individual water constituents, it may be necessary to revise the standards for different water uses. Federal and state regulations require that water quality standards be reviewed periodically and modified when appropriate.

TYPICAL WATER USE CLASSIFICATIONS

All states classify surface waters and groundwater according to their current and intended uses. Typical classifications are

1. Recreational:
   a. **Class 1 — primary contact:** Surface waters that are suitable or intended to become suitable for prolonged and intimate contact with the body, or for recreational activities where the ingestion of small quantities of water is likely to occur, e.g., swimming, rafting, kayaking, water skiing, etc.
   b. **Class 2 — secondary contact:** Surface waters that are suitable or intended to become suitable for recreation in or around the water, which are not included in the primary contact subcategory, e.g., shore fishing, motor yachting, etc.
2. Aquatic Life: Surface waters that are suitable or intended to become suitable for the protection and maintenance of vigorous communities of aquatic organisms and populations of significant aquatic species. Separate standards should be applied to protect:
   a. **Class 1 — cold water aquatic life:** These are waters where conditions of physical habitat, water flows and levels, and chemical quality are (1) currently capable of sustaining a wide variety of cold water biota (considered to be the inhabitants, including sensitive species, of water in which temperatures do not normally exceed 20°C), or (2) could sustain such biota if correctable water quality conditions were improved.

b. **Class 1 — warm water aquatic life:** These are waters where conditions of physical habitat, water flows and levels, and chemical quality are (1) currently capable of sustaining a wide variety of warm water biota (considered to be the inhabitants, including sensitive species, of water in which temperatures normally exceed 20°C), or (2) could sustain such biota if correctable water quality conditions were improved.

c. **Class 2 — cold and warm water aquatic life:** These are waters that are not capable of sustaining a wide variety of cold or warm water biota, including sensitive species, due to conditions of physical habitat, water flows and levels, or uncorrectable water quality that result in substantial impairment of the abundance and diversity of species.

3. Agriculture: Surface waters that are suitable or intended to become suitable for irrigation of crops and that are not hazardous as drinking water for livestock.

4. Domestic water supply: Surface waters that are suitable or intended to become suitable for potable water supplies. After receiving standard treatment — defined as coagulation, flocculation, sedimentation, filtration, and disinfection with chlorine or its equivalent — these waters will meet federal and state drinking water standards.

5. Wetlands: Surface water and groundwater that supply wetlands. Wetlands may be defined as areas that are inundated or saturated by surface or groundwater at a frequency and duration sufficient to support, and under normal circumstances do support, a prevalence of vegetation and organisms typically adapted for life under saturated soil conditions.

6. Groundwater: Subsurface waters in a zone of saturation that are or can be brought to the surface of the ground or to surface waters through wells, springs, seeps, or other discharge areas. Separate standards are applied to groundwater used for:

a. **Domestic use:** Groundwaters that are used or are suitable for a potable water supply.

b. **Agricultural use:** Groundwaters that are used or are suitable for irrigating crops and livestock water supply.

c. **Surface water quality protection:** This classification is used for groundwaters that feed surface waters. It places restrictions on proposed or existing activities that could impact groundwaters in a way that water quality standards of classified surface water bodies could be exceeded.

d. **Potentially usable:** Groundwaters that are not used for domestic or agricultural purposes, where background levels are not known or do not meet human health and agricultural standards, where total dissolved solids (TDS) levels are less than 10,000 mg/L, and where domestic or agricultural use can be reasonably expected in the future.

e. **Limited use:** Groundwaters where TDS levels are equal to or greater than 10,000 mg/L, where the groundwater has been specifically exempted by regulations of the state, or where the criteria for any of the above classifications are not met.

SETTING NUMERICAL WATER QUALITY STANDARDS

Numerical water quality standards are chosen to protect the current and intended uses for the water. The water quality standards for each water body are based on all the uses for which it is classified. In addition, site-specific standards may be established where special conditions exist, such as where aquatic life has become acclimated to high levels of dissolved metals. Each state has tables of water quality standards for each classified water body. In addition to standards for environmental waters, there are standards for treated drinking water as delivered from a water treatment plant or, for some parameters such as lead and copper, as delivered at the tap.

The U.S. Environmental Protection Agency (EPA) sets baseline standards for different use classifications that serve as minimum requirements for the state standards. Water quality standards are defined in terms of

- Chemical composition: concentrations of metals, organic compounds, chlorine, nitrates, ammonia, phosphorus, sulfate, etc.
- General physical and chemical properties: temperature, alkalinity, conductivity, pH, dissolved oxygen, hardness, total dissolved solids, chemical oxygen demand, etc.
- Biological characteristics: biological oxygen demand, fecal coliforms, whole effluent toxicity, etc.
- Radionuclides: radium-226, radium-228, uranium, radon, gross alpha and gross beta emissions, etc.

Rule of Thumb

Generally, the most stringent standards are for drinking water and aquatic life classifications.

STAYING UP-TO-DATE WITH STANDARDS AND OTHER REGULATIONS

This is a daunting challenge and, in the opinion of some, an impossible one. Not only are the federal regulations constantly changing, individual states may also promulgate different rules because of local needs. The usual approach is to obtain the latest regulatory information as the need arises, always recognizing that your current understanding may be outdated. Part of the problem is that few environmental professionals can find time to regularly read the *Federal Register*, where the EPA first publishes all proposed and final regulations.

Fortunately, most trade magazines and professional journals highlight important changes in standards and regulations that are of interest to their readers. If you stay abreast of this literature, you will be aware of the regulatory changes and their implications. For the greatest level of security, one has to often contact state and federal information centers to ensure working with the regulations that are currently being enforced. Among the most useful sources for staying abreast of the latest information is the EPA Web site on the Internet (www.epa.gov/). The website has links to information hotlines, laws and regulations, databases and software, available publications, and other information sources.

1.2 SOURCES OF WATER IMPURITIES

A water impurity is any substance other than water (H_2O) that is found in the water sample. Thus, calcium carbonate ($CaCO_3$) is a water impurity even though it is not considered hazardous and is not regulated. Impurities can be divided into two classes: (1) unregulated impurities not considered harmful, and (2) regulated impurities (pollutants) considered harmful.

In water quality analysis, unregulated as well as regulated impurities are measured. For example, hardness is a water quality parameter that results mainly from the presence of dissolved calcium and magnesium ions, which are unregulated impurities. However, high hardness levels can partially mitigate the toxicity of many dissolved metals to aquatic life. Hence, it is important to measure water hardness in order to evaluate the hazards of dissolved metals.

Data concerning unregulated impurities are also helpful for anticipating certain non-health-related potential problems, such as pipe and boiler deposits, corrosivity, and low soil permeability. Unregulated impurities can also help to identify the recharge sources of wells and springs, learn about the mineral formations through which surface water or groundwaters pass, and age-date water samples.

NATURAL SOURCES

Snow and rain water contain dissolved and particulate minerals collected from atmospheric particulate matter, and small amounts of gases dissolved from atmospheric gases. Snow and rain have virtually no bacterial content until they reach the surface of the earth.

After precipitation reaches the surface of the earth and flows over and through the soil, there are innumerable opportunities for the introduction of mineral, organic, and biological substances. Water can dissolve at least a little of nearly anything it contacts. Because of its relatively high density, water can also carry suspended solids. Even under pristine conditions, surface and groundwaters will usually contain various dissolved and suspended chemical substances.

HUMAN-CAUSED SOURCES

Many human activities cause additional possibilities for water contamination. Some important sources are

- Construction and mining where freshly exposed soils and minerals can contact flowing water
- Industrial waste discharges and spills
- Petroleum discharges from leaking storage tanks, pipelines, tankers, and trucks
- Agricultural applications of chemical fertilizers, herbicides, and pesticides
- Urban storm water runoff, which contains all the debris of a city, including spilled fuels, animal feces, dissolved metals, organic scraps, road salt, tire and brake particles, construction rubble, etc.
- Effluents from industries and waste treatment plants
- Leachate from landfills, septic tanks, treatment lagoons, and mine tailings
- Fallout from atmospheric pollution

The environmental professional must remain alert to the possibility that natural impurity sources may be contributing to problems that at first appear to be solely the result of human-caused sources. Whenever possible, one should obtain background measurements that demonstrate what impurities are present in the absence of known human-caused contaminant sources. For instance, groundwater in an area impacted by mining often contains relatively high concentrations of dissolved metals. Before any remediation programs are initiated, it is important to determine what the groundwater quality would have been if the mines had not been there. This generally requires finding a location upgradient of the area influenced by mining, where the groundwater encounters subsurface mineral structures similar to those in the mined area.

1.3 MEASURING IMPURITIES

There are four characteristics of water impurities that are important for an initial assessment of water quality:

1. What impurities are present? Are they regulated compounds?
2. How much of each impurity is present? Are any standards exceeded for the water body being sampled?
3. How do the impurities influence water quality? Are they hazardous? Beneficial? Unaesthetic? Corrosive?
4. What is the fate of the impurities? How will their location, quantity, and chemical form change with time?

WHAT IMPURITIES ARE PRESENT?

The chemical content of a water sample is found by *qualitative* chemical analysis of collected environmental samples, which identifies the chemical species present. Some of the analytical methods used are gas and ion chromatography, mass spectroscopy, optical emission and absorption spectroscopy, electrochemical probes, and immunoassay testing.

HOW MUCH OF EACH IMPURITY IS PRESENT?

The amount of impurity is found by *quantitative* chemical analysis of the water sample. The amount of impurity can be expressed in terms of *total mass*, (e.g., "There are 15 tons of nitrate in the lake.") or in terms of *concentration*. (e.g., "Nitrate is present at a concentration of 12 mg/L.") Concentration is usually the measure of interest for predicting the effect of an impurity on the environment. It is used for defining environmental standards, and is reported in most laboratory analyses. An additional limitation of total mass is applied to some rivers in the form of total maximum daily loads (TMDLs).

WORKING WITH CONCENTRATIONS

Unfortunately, there is not one all-purpose method for expressing concentration. The best choice of concentration units depends in part on the medium (liquid or solid), and in part on the purpose of the measurement.

For regulatory purposes, concentration is usually expressed as *mass* of impurity per unit volume or unit mass of sample.

Water samples: Constituent concentrations are typically reported as milligrams of impurity per liter of sample (mg/L), or micrograms of impurity per liter (µg/L) of sample.

$$1 \text{ mg/L} = 1 \text{ part per million (ppm).}$$

$$1 \text{ µg/L} = 0.001 \text{ mg/L} = 1 \text{ part per billion (ppb).}$$

Soil samples: Constituent concentrations are typically reported as milligrams of impurity per kilogram of sample (mg/kg) or micrograms of impurity per kilogram (µg/kg) of sample.

$$1 \text{ mg/kg} = 1 \text{ part per million (ppm).}$$

$$1 \text{ µg/kg} = 0.001 \text{ mg/kg} = 1 \text{ part per billion (ppb).}$$

For chemical calculations, concentration is usually expressed either as *moles* of impurity per liter of sample (mol/L), moles of impurity per kilogram of sample (mol/kg), or as *equivalents* of impurity per liter (eq/L) or kilogram (eq/kg) of sample.

Moles per liter (mol/L): Are related to the *number* of impurity molecules, rather than the *mass* of impurity molecules, present in a liter of sample. This is more useful for chemical calculations because chemical reactions involve one-on-one molecular interactions, regardless of the mass of the reacting molecules. A common chemical notation for expressing a concentration as mol/L is to enclose the constituent in square brackets. Thus, writing $[Na^+] = 16.4$, is the same as writing $Na^+ = 16.4$ mol/L.

To convert mg/L to mol/L, divide by 1000 and multiply by the molecular weight of the impurity. Obtain the molecular weight by adding the atomic weights of all the atoms in the molecule. Look up the atomic weights in the periodic table inside the front cover of this book.

Example 1.1: Converting mg/L to moles/L

Benzene in a water sample was reported as 0.017 mg/L. Express this concentration as mol/L.

Answer: The chemical formula for benzene is C_6H_6. Therefore, its molecular weight is $(6 \times 12 + 6 \times 1) = 78$ g/mol. The concentration of benzene in the sample can be expressed as

$$\frac{0.017 \text{ mg/L}}{(1000 \text{ mg/g})(78 \text{ g/mol})} = 2.18 \times 10^{-7} \text{ mol/L.}$$

Equivalents per liter (eq/L): Express the *moles of ionic charge per liter of sample*. This is useful for chemical calculations involving ions, because ionic reactions must always balance electrically, for example, with respect to ionic charge.

The *equivalent weight* of a substance is its molecular or atomic weight *divided* by the magnitude of charge (without regard for the sign of the charge) for ionic species or, for non-ionic species, what the charge would be if they were dissolved (also called the oxidation number). Thus, the equivalent weight of Ca^{2+} is 1/2 its atomic weight, because each calcium ion carries two positive charges, and a 1/2 mole of Ca^{2+} contains 1 mole of positive charge.

Equivalents per liter of an impurity are equal to the moles per liter *multiplied* by the ionic charge or oxidation number, because, for example, 1 mole of Ca^{2+} contains 2 moles of charge. That this is consistent with the fact that the *equivalent weight* of a substance is its molecular weight *divided* by the charge or oxidation number is shown by Example 1.2.

Example 1.2: Working with Equivalent Weights

The equivalent weight of Cr^{3+} is the mass that contains 1 mole of charge. Since each ion of Cr^{3+} contains 3 units of charge, the moles of charge in a given amount of chromium are 3 times the moles of ions. Thus, 1 mole of Cr^{3+} or 52 grams contains 3 moles of charge or 3 equivalent weights. Therefore,

$$\text{eq. wt. } Cr^{3+} = (\text{at. wt. } Cr^{3+})/3 = 52.0/3 = 17.3 \text{ g/eq.}$$

If a water sample contains one mol/L (52 g/L) of Cr^{3+}, it contains 3×17.3 g/L or 3 eq/L of Cr^{3+}.

Working with equivalents is useful for comparing the balance of positive and negative ions in a water sample or making cation exchange calculations. To convert mol/L to eq/L, multiply by the ionic charge or oxidation number of the impurity. Use the absolute value of the charge or oxidation number, i.e., multiply by +2 for a charge of either +2 or –2.

Example 1.3

Chromium III in a water sample is reported as 0.15 mg/L. Express the concentration as eq/L. (The Roman numeral III indicates that the oxidation number of chromium in the sample is +3. It also indicates that the dissolved ionic form would have a charge of +3.)

Answer: The atomic weight of chromium is 52.0 g/mol. Chromium III ionizes as Cr^{3+}, so its concentration in mol/L is multiplied by 3 to obtain its equivalent weight.

$$0.15 \text{ mg/L} = \frac{0.15 \text{ g/L}}{52.0 \text{ g/mol}} = 2.88 \times 10^{-3} \text{ mol/L or 2.88 mmol/L.}$$

$$0.15 \text{ mg/L} = 2.88 \times 10^{-3} \text{ mol/L} \times 3 \text{ eq/mol} = 8.65 \times 10^{-3} \text{ eq/L or 8.65 meq/L.}$$

Example 1.4

Alkalinity in a water sample is reported as 450 mg/L as $CaCO_3$. Convert this result to eq/L of $CaCO_3$. Alkalinity is a water quality parameter that results from more than one constituent. It is

TABLE 1.1
Molecular Weights and Equivalent Weights of Some Common Water Species

Species	atomic wt.	|charge|	equiv. wt.	Species	atomic wt.	|charge|	equiv. wt.
Na^+	23.0	1	23.0	Cl^-	35.4	1	35.4
K^+	39.1	1	39.1	F^-	19.0	1	19.0
Li^+	6.9	1	6.9	Br^-	79.9	1	79.9
Ca^{2+}	40.1	2	20.04	NO_3^-	62.0	1	62.0
Mg^{2+}	24.3	2	12.2	NO_2^-	46.0	1	46.0
Sr^{2+}	87.6	2	43.8	HCO_3^-	61.0	1	61.0
Ba^{2+}	137.3	2	68.7	CO_3^{2-}	60.0	2	30.0
Fe^{2+}	55.8	2	27.9	CrO_4^{2-}	116.0	2	58.0
Mn^{2+}	54.9	2	27.5	SO_4^{2-}	96.1	2	48.03
Zn^{2+}	65.4	2	32.7	S^{2-}	32.1	2	16.0
Al^{3+}	27.0	3	9.0	PO_4^{3-}	95.0	3	31.7
Cr^{3+}	52.0	3	17.3	$CaCO_3$	100.1	2	50.04

expressed as the amount of $CaCO_3$ that would produce the same analytical result as the actual sample (see Chapter 2).

Answer: The molecular weight of $CaCO_3$ is: $(1 \times 40 + 1 \times 12 + 3 \times 16) = 100$ g/mol. The dissolution reaction of $CaCO_3$ is

$$CaCO_3 \xrightarrow{H_2O} Ca^{2+} + CO_3^{2-}.$$

Since the absolute value of charge for either the positive or negative species equals 2, eq/L = mol/L × 2.

$$450 \text{ mg/L} = \frac{450 \text{ mg/L}}{(1000 \text{ mg/g})(100 \text{ g/mol})} = 4.5 \times 10^{-3} \text{ mol/L or 4.5 mmol/L.}$$

$$450 \text{ mg/L} = 4.5 \times 10^{-3} \text{ mol/L} \times 2 \text{ eq/mol} = 9.0 \times 10^{-3} \text{ eq/L or 9.0 meq/L.}$$

How Do Impurities Influence Water Quality?

The effects of different impurities on water quality are found by research and experience. For example, concentrations of arsenic in drinking water greater than 0.05 mg/L are deemed to be hazardous to human health. This judgment is based on research and epidemiological studies. Frequently, regulations have to be based on an interpretation of studies that are not rigorously conclusive. Such regulations may be controversial, but until they are revised due to the emergence of new information, they serve as the legal definition of the concentration above which an impurity is deemed to have a harmful effect on water quality.

The EPA has a policy of publishing newly proposed regulatory rules before the rules are finalized, and of explaining the rationale used to justify the rules in order to receive feedback from interested parties. During the time period dedicated for public comment, interested parties can support or take issue with the EPA's position. The public input is then added to the database used for establishing a final regulation. An example of such a regulation may be a numerical standard for a chemical not previously regulated, a revised standard for a chemical already regulated, or a new procedure for the analysis of a pollutant. The EPA has published extensive documentation for all their standards describing the data on which the numerical values are based.

2 Principles of Contaminant Behavior in the Environment

CONTENTS

2.1 THE BEHAVIOR OF CONTAMINANTS IN NATURAL WATERS

Every part of our world is continually changing, the unwelcomed contaminants as well as the essential ecosystems. Some changes occur imperceptibly on a geological time scale; others are rapid occurring within days, minutes, or less. Oil and coal are formed from animal and vegetable matter over millions of years. When oil and coal are burned, they can release their stored energy in fractions of a second. Control of environmental contamination depends on understanding how pollutants are affected by environmental conditions, and learning how to bring about desired changes. For example, metals that are dangerous to our health, such as lead, are often more soluble in water under acidic conditions than under basic conditions. Knowing this, one can plan to remove dissolved lead from drinking water by raising the pH and making the water basic. Under basic conditions, a large part of dissolved lead can be made to precipitate as a solid and can be removed from drinking water by settling out or filtering.

Contaminants in the environment are driven to change by

- *Physical forces* that move contaminants to new locations, often without significant change in their chemical properties. Contaminants released into the soil and water can move into regions far from their origin under the forces of wind, gravity, and water flow. An increase in temperature will cause an increase in the rate at which gases and volatile substances evaporate from water or soil into the atmosphere. Electrostatic attractions can cause dissolved substances and small particles to adsorb to solid surfaces, where they may leave the water flow and become immobilized in soils or filters.
- *Chemical changes* such as oxidation and reduction which break chemical bonds and allow atoms to rearrange into new compounds.
- *Biological activity* whereby microbes, in their constant search for survival energy, break down many kinds of contaminant molecules and return their atoms to the environmental cycles that circulate carbon, oxygen, nitrogen, sulfur, phosphorus, and other elements repeatedly through our ecosystems. Biological processes are a special kind of chemical change.

We are particularly interested in processes that move pollutants to less hazardous locations or change the nature of a pollutant to a less harmful form because these processes are the tools of environmental protection. The effectiveness of these processes depends on properties of the pollutant and its water and soil environment. Important properties of *pollutants* can usually be found in handbooks or chemistry references. However, the important properties of the *water* and *soil* in which the pollutant resides are always unique to the particular site and must be measured anew for every project.

IMPORTANT PROPERTIES OF POLLUTANTS

The six properties listed below are the most important for predicting the environmental behavior of a pollutant. They are often tabulated in handbooks and other chemistry references.

1. Solubility in water
2. Volatility
3. Density
4. Chemical reactivity
5. Biodegradability
6. Tendency to adsorb to solids

If not known, these properties often can be estimated from the chemical structure of the pollutant. Whenever possible, this book will offer "rules of thumb" for estimating pollutant properties.

IMPORTANT PROPERTIES OF WATER AND SOIL

The properties of water and soil that influence pollutant behavior can be expected to differ at every location and must be measured for each project. Since environmental conditions are so varied, it is difficult to generate a simple set of properties that is always the most important to measure. The lists below include the most commonly needed properties.

Water Properties

- Temperature
- Water quality (chemical composition, pH, oxidation-reduction potential, alkalinity, hardness, turbidity, dissolved oxygen, biological oxygen demand, fecal coliforms, etc.)
- Flow rate and flow pattern

Properties of Solids and Soils in Contact with Water

- Mineral composition
- Percentage of organic matter
- Sorption attractions for contaminants (sorption coefficients)
- Mobility of solids (colloid and particulate movement)
- Porosity
- Particle size distribution
- Hydraulic conductivity

The properties of environmental waters and soils are always site-specific and must be estimated or measured in the field.

2.2   WHAT ARE THE FATES OF DIFFERENT POLLUTANTS?

There are three possible naturally occurring fates of pollutants other than the results of engineered remediation processes:

1. All or a portion might remain unchanged in their present location.
2. All or a portion might be carried elsewhere by transport processes.
   a. Movement to other phases (air, water, or soil) by volatilization, dissolution, adsorption, and precipitation.
   b. Movement within a phase under gravity, diffusion, and advection.
3. All or a portion might be transformed into other chemical species by natural chemical and biological processes.
   a. *Biodegradation (aerobic and anaerobic)*: Pollutants are altered structurally by biological processes, mainly the metabolism of microorganisms present in aquatic and soil environments.
   b. *Bioaccumulation*: Pollutants accumulate in plant and animal tissues to higher concentrations than in their original environmental locations.
   c. *Weathering*: Pollutants undergo a series of environmental non-biological chemical changes by processes such as oxidation-reduction, acid-base, hydration, hydrolysis, complexation, and photolysis reactions.

2.3   PROCESSES THAT REMOVE POLLUTANTS FROM WATER

Transport Processes

Contaminants that are dissolved or suspended in water can move to other phases by the following processes:

- *Volatilization*: Dissolved contaminants move from water or soil into air, in the form of gases or vapors.
- *Sorption*: Dissolved contaminants become bound to solids by attractive chemical and electrostatic forces.
- *Precipitation*: Dissolved contaminants are caused to precipitate as solids by changes in pH or oxidation-reduction potential, or they react with other species in water to form compounds of low solubility. Precipitation often produces finely divided solids that will not settle out under gravity unless sedimentation processes occur.
- *Sedimentation*: Small suspended solids in water grow large enough to settle to the bottom under gravity. There are two stages to sedimentation:

   a. Coagulation: Suspended solids generally carry an electrostatic charge that keeps them apart. Chemicals may be added to lower the repulsive electrostatic energy barrier between the particles (destabilization), allowing them to coagulate.
   b. Flocculation: Lowering the repulsive energy barrier by coagulation allows suspended solids to collide and clump together to form a floc. When floc particles aggregate, they can become heavy enough to settle out.

ENVIRONMENTAL CHEMICAL REACTIONS

The following are brief descriptions of important environmental chemical reactions. More detailed discussions are given throughout this book.

- *Photolysis*: In molecules that absorb solar radiation, exposure to sunlight can break chemical bonds and start chemical breakdown. Many natural and synthetic organic compounds are susceptible to photolysis.
- *Complexation and chelation*: Polar or charged dissolved species (such as metal ions) bind to electron-donor ligands* to form *complex* or *coordination* compounds. Complex compounds are often soluble and resist removal by precipitation because the ligands must be displaced by other anions (such as sulfide) before an insoluble species can be formed. Common ligands include hydroxyl, carbonate, carboxylate, phosphate, and cyanide anions, as well as humic acids and synthetic chelating agents such as nitrilotriacetate (NTA) and ethylenediaminetetraacetate (EDTA).
- *Acid-base*: Protons (H^+ ions) are transferred between chemical species. Acid-base reactions are part of many environmental processes and influence the reactions of many pollutants.
- *Oxidation-reduction (OR, or redox)*: Electrons are transferred between chemical species, changing the oxidation states and the chemical properties of the electron donor and the electron acceptor. Water disinfection, electrochemical reactions such as metal corrosion, and most microbial reactions such as biodegradation are oxidation-reduction reactions.
- *Hydrolysis and hydration*: A compound forms chemical bonds to water molecules or hydroxyl anions. In water, all ions and polar compounds develop a hydration shell of water molecules. When the attraction to water is strong enough, a chemical bond can result. Many metal ions form hydroxides of low solubility because of hydrolysis reactions. In organic compounds, a water molecule may replace an atom or group, a step that often breaks the organic compound into smaller fragments. Hydration of dissolved carbon dioxide (CO_2) and sulfur dioxide (SO_2) forms carbonic acid, H_2CO_3 and sulfurous acid (H_2SO_3), respectively.
- *Precipitation*: Two or more dissolved species react to form an insoluble solid compound. Precipitation can occur if a solution of a salt becomes oversaturated, as in when the concentration of a salt becomes greater than its solubility limit. For example, the solubility of calcium carbonate, $CaCO_3$, at 25°C is about 10 mg/L. In a water solution containing 5 mg/L of $CaCO_3$, all the calcium carbonate will be dissolved. If more $CaCO_3$ is added or water is evaporated, the concentration of dissolved calcium carbonate can increase only to 10 mg/L. Any $CaCO_3$ in excess of the solubility limit will precipitate as solid $CaCO_3$.

   Precipitation can also occur if two soluble salts react to form a different salt of low solubility. For example, silver nitrate ($AgNO_3$) and sodium chloride (NaCl) are both highly soluble. They react in solution to form the insoluble salt silver chloride (AgCl) and the soluble salt sodium nitrate ($NaNO_3$). The silver chloride precipitates as a solid. Breaking the reaction into separate conceptual steps helps to visualize what happens. Refer to the solubility table inside the back cover, which gives qualitative solubilities for ionic compounds in water.

---

* Ligands are polyatomic chemical species that contain non-bonding electron pairs.

In the first step, silver nitrate and sodium chloride are added to water and dissolve as ions:

$$AgNO_3(s) \xrightarrow{\text{H}_2\text{O}} Ag^+(aq) + NO_3^-(aq). \qquad (2.1)$$

$$NaCl(s) \xrightarrow{\text{H}_2\text{O}} Na^+(aq) + Cl^-(aq). \qquad (2.2)$$

Immediately after the salts have dissolved, the solution contains Ag^+, Na^+, Cl^-, and NO_3^- ions.

In the second conceptual step, these ions can combine in all possible ways that pair a positive ion with a negative ion. Thus, besides the original $AgNO_3$ and $NaCl$ pairs, $AgCl$ and $NaNO_3$ are also possible. $NaNO_3$ is a soluble ionic compound, so the Na^+ and NO_3^- ions remain in solution. However, $AgCl$ is insoluble and will precipitate as a solid. The overall reaction is written:

$$AgNO_3(aq) + NaCl(aq) \rightarrow Na^+(aq) + NO_3^-(aq) + AgCl(s). \qquad (2.3)$$

BIOLOGICAL PROCESSES

Biodegradation

Microbes can degrade organic pollutants by facilitating oxidation-reduction reactions. During microbial metabolism (the biological reactions that convert organic compounds into energy and carbon for growth), there is a transfer of electrons from a pollutant molecule to other compounds present in the soil or water environment that serve as electron acceptors. The electron acceptors most commonly available in the environment are molecular oxygen (O_2), carbon dioxide (CO_2), nitrate (NO_3^-), sulfate (SO_4^{2-}), manganese (Mn^{2+}), and iron (Fe^{3+}). When O_2 is available, it is always the preferred electron acceptor and the process is called *aerobic* biodegradation. Otherwise it is called *anaerobic* biodegradation.

Organic pollutants are generally toxic because of their chemical structure. Changing their structure in any way will change their properties and may make them innocuous or, in a few cases, more toxic. Eventually, usually after many reaction steps in a process called mineralization, biodegradation converts organic pollutants into carbon dioxide, water, and mineral salts. Although these final products represent the destruction of the original pollutant, some of the intermediate steps may produce compounds that are also pollutants, sometimes more toxic than the original. Biodegradation is discussed in more detail in Chapter 4.

2.4 MAJOR CONTAMINANT GROUPS AND THEIR NATURAL PATHWAYS FOR REMOVAL FROM WATER

METALS

Dissolved metals such as iron, lead, copper, cadmium, mercury, etc., are removed from water mainly by sorption and precipitation processes. Some metals — particularly As, Cd, Hg, Ni, Pb, Se, Te, Sn, and Zn — can form volatile metal-organic compounds in the natural environment by microbial mediation. For these, volatilization can be an important removal mechanism. Bioaccumulation of metals in animals can lead to toxic effects but usually is not very significant as a removal process. Bioaccumulation in plants on the other hand, has been developed into a useful remediation technique called *phytoremediation*. Biotransformation of metals, by which some metals are caused to precipitate, has shown promise as a removal method.

CHLORINATED PESTICIDES

Chlorinated pesticides, such as atrazine, chlordane, DDT, dicamba, endrin, heptachlor, lindane, etc., are removed from water mainly by sorption, volatilization, and biotransformation. Chemical processes like oxidation, hydrolysis, and photolysis appear to play a usually minor role.

HALOGENATED ALIPHATIC HYDROCARBONS

Halogenated hydrocarbons mostly originate as industrial and household solvents. Compounds such as 1,2-dichloropropane, 1,1,2-trichlorethane, tetrachlorethylene, etc. are removed mainly by volatilization. Under natural conditions, biotransformation and biodegradation processes are usually very slow, with half-lives of tens or hundreds of years. However, engineered biodegradation procedures have been developed. These procedures have short enough half-lives to be useful remediation techniques.

FUEL HYDROCARBONS

Gasoline, diesel fuel, and heating oils are mixtures of hundreds of different organic hydrocarbons. The lighter weight compounds such as benzene, toluene, ethylbenzene, xylenes, naphthalene, trimethylbenzenes, and the smaller alkanes, etc. are removed mainly by sorption, volatilization, and biotransformation. The heavier compounds including polycyclic aromatic hydrocarbons (PAHs) such as fluorene, benzo(a)pyrene, anthracene, phenanthrene, etc. are not volatile and are removed mainly by sorption, sedimentation, and biodegradation.

INORGANIC NONMETAL SPECIES

These include ammonia, chloride, cyanide, fluoride, nitrite, nitrate, phosphate, sulfate, sulfide, etc. They are removed mainly by sorption, volatilization, chemical processes, and biotransformation.

It is important to note that many normally minor pathways such as photolysis can become important, or even dominant, in special circumstances.

2.5   CHEMICAL AND PHYSICAL REACTIONS IN THE WATER ENVIRONMENT

Chemical and physical reactions in water can be

- *Homogeneous* — occurring entirely among dissolved species.
- *Heterogeneous* — occurring at the liquid-solid-gas interfaces.

Most environmental water reactions are heterogeneous. Purely homogeneous reactions are relatively rare in natural waters and wastewaters. Among the most important reactions occurring at the liquid-solid-gas interfaces are those that move pollutants from one phase to another.

The following are processes by which a pollutant becomes distributed (or is *partitioned*) into all the phases it comes in contact with.

- *Volatilization*: At the liquid-air and solid-air interfaces, volatilization transfers volatile contaminants from water and solid surfaces into the atmosphere, and into air in soil pore spaces. Volatilization is most important for compounds with high vapor pressures. Contaminants in the vapor phase are the most mobile in the environment.
- *Dissolution*: At the solid-liquid and air-liquid interfaces, dissolution transfers contaminants from air and solids to water. It is most important for contaminants of high water solubility. The environmental mobility of contaminants dissolved in water is generally intermediate between volatilized and sorbed contaminants.
- *Sorption**: At the liquid-solid and air-solid interfaces, sorption transfers contaminants from water and air to soils and sediments. It is most important for compounds of low

---

* *Sorption* is a general term including both adsorption and absorption. *Adsorption* means binding to a particle surface. *Absorption* means becoming bound in pores and passages within a particle.

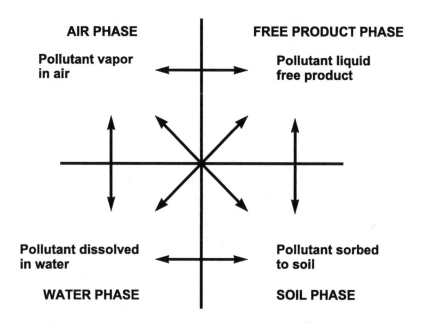

FIGURE 2.1   Partitioning of a pollutant among air, water, soil, and free product phases.

solubility and low volatility. Sorbed compounds undergo chemical and biological transformations at different rates and by different pathways than dissolved compounds. The
binding strength with which different contaminants become sorbed depends on the nature
of the solid surface (sand, clays, organic particles, etc.), and on the properties of the
contaminant. Contaminants sorbed to solids are the least mobile in the environment.

2.6   PARTITIONING BEHAVIOR OF POLLUTANTS

A pollutant in contact with water, soil, and air will partially dissolve into the water, partially
volatilize into the air, and partially sorb to the soil surfaces, as illustrated in Figure 2.1. The relative
amounts of pollutant that are found in each phase with which it is in contact, depends on intermolecular attractive forces existing between pollutant, water, and soil molecules. *The most important
factor for predicting the partitioning behavior of contaminants in the environment is an understanding of the intermolecular attractive forces between contaminants and the water and soil
materials in which they are found.*

Partitioning from a Diesel Oil Spill

Consider, for example, what happens when diesel oil is spilled at the soil's surface. Some of the
liquid diesel oil (commonly called *free product*) flows downward under gravity through the soil
toward the groundwater table. Before the spill, the soil pore spaces above the water table (called
the soil unsaturated zone) were filled with air and water, and the soil surfaces were partially covered
with adsorbed water. As diesel oil, which is a mixture of many different compounds, passes
downward through the soil, its different components become partitioned among the pore space air
and water, the soil particle surfaces, and the oil free product. After the spill, the pore spaces are
filled with air containing diesel vapors, water carrying dissolved diesel components, and diesel free
product that has changed in composition by losing some of its components to other phases. The
soil surfaces are partially covered with diesel free product and adsorbed water containing dissolved
diesel components.

Diesel oil is a mixture of hundreds of different compounds each having a unique partitioning, or distribution pattern. The pore space air will contain mainly the most volatile components, the pore space water will contain mainly the most soluble components, and the soil particles will sorb mainly the least volatile and soluble components. The quantity of the free product diminishes continually as it moves downward through the soil because a significant portion is lost to other phases. The composition of the free product also changes continually because the most volatile, soluble, and strongly sorbed compounds are lost preferentially. The chemical distributions attain quasi-equilibrium, with compounds continually passing back and forth across each phase interface, as indicated in Figure 2.1. As the remaining free product continues to change by losing components to other phases (part of the "weathering process"), it increasingly resists further change. Since the lightest weight components tend to be the most volatile and soluble, they are the first to be lost to other phases, and the remaining free product becomes increasingly more viscous and less mobile. Severely weathered free product is very resistant to further change, and can persist in the soil for decades. It only disappears by biodegradation or by actively engineered removal.

Depending on the amount of diesel oil spilled, it is possible that all of the diesel free product becomes "immobilized" in the soil before it can reach the water table. This occurs when the mass of free product diminishes and its viscosity increases to the point where capillary forces in the soil pore spaces can hold the remaining free product in place against the force of gravity. There is still pollutant movement, however, mainly in the non-free product phases. The volatile components in the vapor state usually diffuse rapidly through the soil, moving mostly upward toward the soil surface and along any high permeability pathways through the soil, such as a sewer line backfill. New water percolating downward, from precipitation or other sources, can dissolve additional diesel compounds from the sorbed phase and carry it downward. Percolating water can also displace some soil pore water already carrying dissolved pollutants, as well as free product held by capillary forces, forcing them to move farther downward. Although the diesel free product is not truly immobilized, its downward movement can become imperceptible.

However, if the spill is large enough, diesel free product may reach the water table before becoming immobilized. If this occurs, liquid free product being lighter than water, cannot enter the water-saturated zone but remains above it, effectively floating on top of the water table. There, the free product spreads horizontally on the groundwater surface, continuing to partition into ground-water, soil pore space air, and to the surfaces of soil particles. In other words, a portion of the free product will always become distributed among all the solid, liquid and gas phases that it comes in contact with. This behavior is governed by intermolecular forces that exist between molecules.

2.7   INTERMOLECULAR FORCES

Volatility, solubility, and sorption processes all result from the interplay between intermolecular forces. All molecules have attractive forces acting between them. The attractive forces are electro-static in nature, created by a nonuniform distribution of valence shell electrons around the positively charged nuclei of a molecule. When electrons are not uniformly distributed, the molecule will have regions that carry net positive and negative charges. A charged region on one molecule is attracted to oppositely charged regions on adjacent molecules, resulting in the so-called *polar attractive forces*. There can be momentary electrostatic repulsive forces as well. On average, however, molecular arrangements will favor the lower energy attractive positions, and the attractive forces always prevail. The most obvious demonstrations of intermolecular attractive forces are the phase changes of matter that inevitably accompany a sufficient lowering of temperature, where a cooling gas turns into a liquid and into a solid, when the temperature becomes low enough.

Temperature dependent phase changes: Attractive forces always work to bring order to molec-ular configurations, in opposition to thermal energy which always works to randomize configura-tions. Gases are always the higher temperature form of any substance and are the most randomized state of matter. If the temperature of a gas is lowered enough, every gas will condense to a liquid, a more ordered state. Condensation is a manifestation of intermolecular attractive forces. As the temperature falls, the thermal energy of the gas molecules decreases, eventually reaching a point where there is insufficient thermal kinetic energy to keep the molecules separated against the intermolecular attractive forces. The temperature at which condensation occurs is called the boiling point, and it is dependent on environmental pressure as well as temperature. If the temperature of the liquid is lowered further, it eventually freezes to a solid when the thermal energy becomes low enough for intermolecular attractions to pull the molecules into a rigid solid arrangement. Solids are the most highly ordered state of matter. Whenever lowering the temperature causes a change of phase, the decrease in thermal energy allows the always-present attractive forces to overcome molecular kinetic energy and to pull gas and liquid molecules closer together into more ordered liquid or solid phases.

Volatility, solubility, and sorption: The model of attractive forces working to bring increased order, against the randomizing effects of thermal energy, also explains the volatility, solubility, and sorption behavior of molecules. Molecules of volatile liquids have relatively weak attractions to one another. Thermal energy at ordinary environmental temperatures is sufficient to allow the most energetic of the weakly held molecules to escape from their liquid neighbors and fly into the gas phase. Molecules in water-soluble solids are attracted to water more strongly than they are attracted to themselves. If a water-soluble solid is placed in water, its surface molecules are drawn from the solid phase into the liquid phase by attractions to water molecules. Dissolved molecules that become sorbed to sediment surfaces are held to the sediment particle by attractive forces that pull them away from water molecules. Understanding intermolecular forces is the key to predicting how contaminants become distributed in the environment.

PREDICTING RELATIVE ATTRACTIVE FORCES

When you can predict relative attractive forces between molecules, you can predict their relative solubility, volatility, and sorption behavior. For example, the freezing and boiling temperatures of a substance (and, hence, its volatility) are related to the attractive forces between molecules of that substance. The water solubility of a compound is related to the strength of the attractive forces between molecules of water and molecules of the compound. The soil-water partition coefficient of a compound indicates the relative strengths of its attraction to water and soil. From these concepts, the following may be deduced:

- Boiling a liquid means that it is heated to the point where thermal energy is high enough to overcome the attractive forces and drive the molecules apart from one another into the gas phase. A higher boiling temperature indicates stronger intermolecular attractive forces between the liquid molecules. With stronger forces, the thermal energy has to be higher in order to overcome the attractions and allow liquid molecules to escape into the gas phase. Thus, the fact that water boils at a higher temperature than does methanol means that water molecules are attracted to one another more strongly than are methanol molecules.
- Freezing a liquid means that its thermal energy is reduced to the point where attractive forces can overcome the randomizing effects of thermal motion and pull freely-moving liquid molecules into fixed positions in a solid phase. A lower freezing point indicates weaker attractive forces. The thermal energy has to be reduced to lower values so that the weaker attractive forces can pull the molecules into fixed positions in a solid phase. The fact that methanol freezes at a lower temperature than water is another indicator that attractive forces are weaker between methanol molecules than between water molecules.
- Wax is solid at room temperature (20°C or 68°F), while diesel fuel is liquid. The freezing temperature of diesel fuel is well below room temperature. This indicates that the attractive forces between wax molecules are stronger than between molecules in diesel fuel. At the same temperature where diesel molecules can still move about randomly in

the liquid phase, wax molecules are held by their stronger forces in fixed positions in the solid phase.

- Compounds that are highly soluble in water have strong attractions to water molecules. Compounds that are found associated mostly with soils have stronger attractions to soil than to water. Compounds that volatilize readily from water and soil have weak attractions to water and soil.

2.8   PREDICTING BOND TYPE FROM ELECTRONEGATIVITIES

Intermolecular forces are electrostatic in nature. Molecules are composed of electrically charged particles (electrons and protons), and it is common for them to have regions that are predominantly charged positive or negative. Attractive forces between molecules arise when electrostatic forces attract positive regions on one molecule to negative regions on another. The strength of the attractions between molecules depends on the *polarities* of chemical bonds within the molecules and the *geometrical shapes* of the molecules.

Chemical bonds — ionic, nonpolar covalent, and polar covalent: At the simplest level, the chemical bonds that hold atoms together in a molecule are of two types:

1. *Ionic bonds:* occur when one atom attracts an electron away from another atom to form a positive and a negative ion. The ions are then bound together by electrostatic attraction. The electron transfer occurs because the electron-receiving atom has a much stronger attraction for electrons in its vicinity than does the electron-losing atom.
2. *Covalent bonds:* are formed when two atoms share electrons, called *bonding electrons*, in the space between their nuclei. The electron-attracting properties of covalent bonded atoms are not different enough to allow one atom to pull an electron entirely away from the other. However, unless both atoms attract bonding electrons equally, the average position of the bonding electrons will be closer to one of the atoms. The atoms are held together because their positive nuclei are attracted to the negative charge of the shared electrons in the space between them.

When two covalent bonded atoms are identical, as in Cl_2, the bonding electrons are always equally attracted to each atom and the electron charge is uniformly distributed between the atoms. Such a bond is called a *nonpolar covalent bond*, meaning that it has no polarity, i.e., no regions with net positive or negative charge.

When two covalent bonded atoms are of different kinds, as in HCl, one atom may attract the bonding electrons more strongly than the other. This results in a non-uniform distribution of electron charge between the atoms where one end of the bond is more negative than the other, resulting in a *polar bond*.

Figure 2.2 illustrates the electron distributions in nonpolar and polar covalent bonds. The strength with which an atom attracts bonding electrons to itself is indicated by a quantity called *electronegativity*. Electronegativities of the elements, shown in Table 2.1, are relative numbers with an arbitrary maximum value of 4.0 for fluorine, the most electronegative element. Electronegativity values are approximate, to be used primarily for predicting the relative polarities of covalent bonds.

The electronegativity difference between two atoms indicates what kind of bond they will form. The greater the difference in electronegativities of bonded atoms, the more strongly are the bonding electrons attracted to the more electronegative atom, and the more polar is the bond. The following "rules of thumb" usually apply, with very few exceptions.

Because electronegativity differences can vary continuously between zero and four, bond character also can vary continuously between nonpolar covalent and ionic, as illustrated in Figure 2.3.

Rules of Thumb (Use Table 2.1)

1. If the electronegativity difference between two bonded atoms is zero, they will form a nonpolar covalent bond. Examples are O_2, H_2, N_2, and NCl.
2. If the electronegativity difference between two atoms is between zero and 1.7, they will form a polar covalent bond. Examples are HCl, NO, and CO.
3. If the electronegativity difference between two atoms is greater than 1.7, they will form an ionic bond. Examples are NaCl, HF, and KBr.
4. Relative electronegativities of the elements can be predicted by an element's position in the Periodic Table. Ignoring the noble gases:
   a. The most electronegative element (F) is at the upper right corner of the Periodic Table.
   b. The least electronegative element (Fr) is at the lower left corner of the Periodic Table.
   c. In general, electronegativities increase diagonally up and to the right in the Periodic Table. Within a given Period (or row), electronegativities tend to increase in going from left to right; within a given Group (or column), electronegativities tend to increase in going from bottom to top.
   d. The farther apart two elements are in the Periodic Table the more different are their electronegativities, and the more polar will be a bond between them.

Cl ⬭ Cl	
Nonpolar covalent bond established by a uniform distribution of bonding electrons between identical atoms in Cl_2.	Polar bond established by a non-uniform distribution of bonding electrons between different atoms in HCl. Electron charge is concentrated toward the more electronegative atom, indicated by δ–. The less electronegative atom is indicated by δ+.
Nonpolar covalent bond indicated by a straight line joining the atoms.	Polar bond indicated by an arrow over the bond pointing to the more electronegative atom. A vertical cross line shows the more positive end of the bond. The length of the arrow indicates the magnitude of the bond's dipole moment.

FIGURE 2.2 Uniform and non-uniform electron distributions, resulting in nonpolar and polar covalent chemical bonds. The use of a delta (δ) in front of the + and – signs signifies that the charges are partial, arising from a non-uniform electron charge distribution rather than from the transfer of a complete electron.

DIPOLE MOMENTS

For polar bonds, we can define a quantity, called the *dipole moment,* which serves as a measure of the non-uniform charge separation. Hence, the dipole moment measures the degree of the bond polarity. The more polar the bond, the larger is its dipole moment. The dipole moment, μ, is equal to the magnitude of positive and negative charges at each end of the dipole multiplied by the distance, d, between the charges.

Polarity arrows, as shown in Figure 2.4, are vector quantities. They show both the magnitude and direction of the bond dipole moment. The length of the arrow indicates how large is the dipole moment, and the direction of the arrow points to the charge separation.

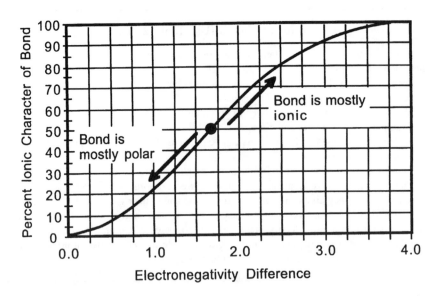

FIGURE 2.3   Bond character as a function of the electronegativity difference.

TABLE 2.1
Electronegativity Values of the Elements

1 1A												13 3A	14 4A	15 5A	16 6A	17 7A
1 **H** 2.1	2 2A															
3 **Li** 1.0	4 **Be** 1.5											5 **B** 2.0	6 **C** 2.5	7 **N** 3.0	8 **O** 3.5	9 **F** 4.0
11 **Na** 1.0	12 **Mg** 1.2	3 3B	4 4B	5 5B	6 6B	7 7B	8 8B	9 8B	10 8B	11 1B	12 2B	13 **Al** 1.5	14 **Si** 1.8	15 **P** 2.1	16 **S** 2.5	17 **Cl** 3.0
19 **K** 0.9	20 **Ca** 1.0	21 **Sc** 1.3	22 **Ti** 1.4	23 **V** 1.5	24 **Cr** 1.6	25 **Mn** 1.6	26 **Fe** 1.7	27 **Co** 1.7	28 **Ni** 1.8	29 **Cu** 1.8	30 **Zn** 1.6	31 **Ga** 1.7	32 **Ge** 1.9	33 **As** 2.1	34 **Se** 2.4	35 **Br** 2.8
37 **Rb** 0.9	38 **Sr** 1.0	39 **Y** 1.2	40 **Zr** 1.3	41 **Nb** 1.5	42 **Mo** 1.6	43 **Tc** 1.7	44 **Ru** 1.8	45 **Rh** 1.8	46 **Pd** 1.8	47 **Ag** 1.6	48 **Cd** 1.6	49 **In** 1.6	50 **Sn** 1.8	51 **Sb** 1.9	52 **Te** 2.1	53 **I** 2.5
55 **Cs** 0.8	56 **Ba** 1.0	57 **ˈLa** 1.1	72 **Hf** 1.3	73 **Ta** 1.4	74 **W** 1.5	75 **Re** 1.7	76 **Os** 1.9	77 **Ir** 1.9	78 **Pt** 1.8	79 **Au** 1.9	80 **Hg** 1.7	81 **Tl** 1.6	82 **Pb** 1.7	83 **Bi** 1.8	84 **Po** 1.9	85 **At** 2.1
87 **Fr** 0.8	88 **Ra** 1.0	89 **#Ac** 1.1	104 **Rf** ?	105 **Db** ?	106 **Sg** ?	107 **Bh** ?	108 **Hs** ?	109 **Mt** ?								

*Lanthanide series	58 **Ce** 1.1	59 **Pr** 1.1	60 **Nd** 1.1	61 **Pm** 1.1	62 **Sm** 1.2	63 **Eu** 1.1	64 **Gd** 1.2	65 **Tb** 1.1	66 **Dy** 1.2	67 **Ho** 1.2	68 **Er** 1.2	69 **Tm** 1.3	70 **Yb** 1.0	71 **Lu** 1.3
# Actinide series	90 **Th** 1.3	91 **Pa** 1.5	92 **U** 1.5	93 **Np** 1.3	94 **Pu** 1.3	95 **Am** 1.3	96 **Cm** 1.3	97 **Bk** 1.3	98 **Cf** 1.3	99 **Es** 1.3	100 **Fm** 1.3	101 **Md** 1.3	102 **No** 1.3	103 **Lr** 1.5

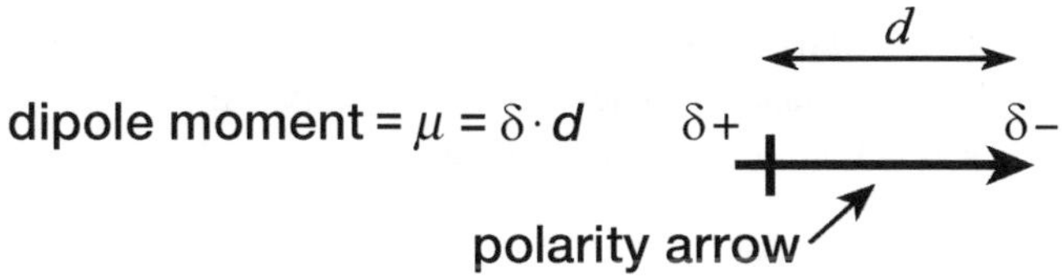

$$\text{dipole moment} = \mu = \delta \cdot d$$

FIGURE 2.4   Molecular dipole moment as indicated by a polarity arrow.

2.9   MOLECULAR GEOMETRY, MOLECULAR POLARITY, AND INTERMOLECULAR FORCES

Knowing whether a molecule is polar or not helps to predict its water solubility and other properties. The presence of polar bonds in a molecule may make the molecule polar also. A molecule is polar if the polarity vectors of all its bonds add up to give a net polarity vector to the molecule. Like polar bonds, a polar molecule has a negatively charged region where electron density is concentrated, and a positively charged region where electron density is diminished. The polarity of a molecule is the vector sum of all its bond polarity vectors. A polar molecule can be experimentally detected by observing whether an electric field exerts a force on it that makes it align its charged regions in the direction of the field. Polar molecules will point their negative ends toward the positive source of the field, and their positive ends toward the negative source.

To predict if a molecule is polar, we need to answer two questions:

1. Does the molecule contain polar bonds? If it does, then it *might* be polar; if it doesn't, it cannot be polar.
2. If the molecule contains polar bonds, do all the bond polarity vectors add to give a resultant molecular polarity? If the molecule is symmetrical in a way that the bond polarity vectors add to zero, then the molecule is nonpolar although it contains polar bonds. If the molecule is asymmetrical and the bond polarity vectors add to give a resultant polarity vector, the resultant vector indicates the molecular polarity.

EXAMPLES OF NONPOLAR MOLECULES

Nonpolar molecules invariably have low water solubility. A molecule with no polar bonds cannot be a polar molecule. Thus, all diatomic molecules where both atoms are the same, such as H_2, O_2, N_2, and Cl_2, are nonpolar because there is no electronegativity difference across the bond. On the other hand, a molecule with polar bonds whose dipole moments add to zero because of molecular symmetry is not a polar molecule. Carbon dioxide, carbon tetrachloride, hexachlorobenzene, *para*-dichlorobenzene, and boron tribromide are all symmetrical and nonpolar, although all contain polar bonds.

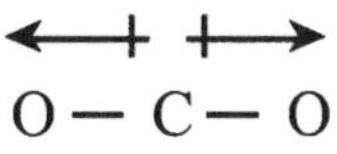

$$O - C - O$$

Carbon dioxide: Oxygen is more electronegative ($EN(O_2) = 3.5$) than carbon ($EN(C) = 2.5$). Each bond is polar, with the oxygen atom at the negative end of the dipole. Because CO_2 is linear with carbon in the center, the polarity vectors cancel each other and CO_2 is nonpolar.

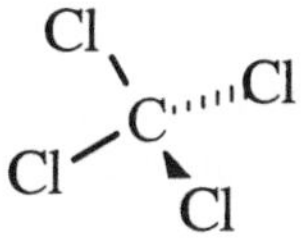

Carbon tetrachloride: $EN(C) = 2.5$, $EN(Cl) = 3.0$ $C \leftrightarrow Cl$. Although each bond is polar, the tetrahedral symmetry of the molecule results in no net dipole moment so that CCl_4 is nonpolar.

Hexachlorobenzene: The bond polarities are the same as in CCl_4 above. C_6Cl_6 is planar with hexagonal symmetry. All the bond polarities cancel one another and the molecule is nonpolar.

***Para*-dichlorobenzene:** This molecule also is planar. It has polar bonds of two magnitudes, the smaller polarity H+⟶C bond and the larger polarity C+⟶Cl bond. The H and Cl atoms are positioned so that all polarity vectors cancel and the molecule is nonpolar. Check the electronegativity values in Table 2.1.

Boron tribromide: EN(B) = 2.0, EN(Br) = 2.8 B+⟶Br.
BBr_3 has trigonal planar symmetry, with 120° between adjacent bonds. All the polarity vectors cancel and the molecule is nonpolar.

EXAMPLES OF POLAR MOLECULES

Polar molecules are generally more water-soluble than nonpolar molecules of similar molecular weight. Any molecule with polar bonds whose dipole moments do not add to zero is a polar molecule. Carbon monoxide, carbon trichloride, pentachlorobenzene, *ortho*-dichlorobenzene, boron dibromochloride, and water are all polar.

Carbon monoxide: Oxygen is more electronegative (EN(O_2) = 3.5) than carbon (EN(C) = 2.5). Every diatomic molecule with a polar bond must be a polar molecule.

Carbon trichloride: EN(C) = 2.5, EN(Cl) = 3.0, EN(H) = 2.1.
It has polar bonds of two magnitudes, the smaller polarity H+⟶C bond and the larger polarity C+⟶Cl bond. The asymmetry of the molecule results in a net dipole moment, so that $CHCl_3$ is polar.

Pentachlorobenzene: The bond polarities are the same as in $CHCl_3$ above. The bond polarities do not cancel one another and the molecule is polar.

***Ortho*-dichlorobenzene:** This molecule is planar and has two kinds of polar bonds: H+⟶C and C+⟶Cl. The bond polarity vectors do not cancel, making the molecule polar.

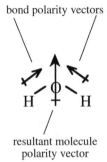

resultant molecule
polarity vector

Boron dibromochloride: EN(B) = 2.0, EN(Br) = 2.8, EN(Cl) = 3.0. In BBr$_2$Cl, the polarity vectors of the polar bonds, B$\longleftrightarrow$Br and B$\longleftrightarrow$Cl, do not quite cancel and the molecule is slightly polar.

Water: is a particularly important polar molecule. Its bond polarity vectors add to give the water molecule a high polarity (i.e., dipole moment). The dipole-dipole forces between water molecules are greatly strengthened by *hydrogen bonding* (see discussion below), which contributes to many of water's unique characteristics, such as relatively high boiling point and viscosity, low vapor pressure, and high heat capacity.

THE NATURE OF INTERMOLECULAR ATTRACTIONS

All molecules are attracted to one another because of electrostatic forces. Polar molecules are attracted to one another because the negative end of one molecule is attracted to the positive ends of other molecules, and vice versa. Attractions between polar molecules are called *dipole-dipole forces*. Similarly, positive ions are attracted to negative ions. Attractions between ions are called *ion-ion forces*. If ions and polar molecules are present together, as when sodium chloride is dissolved in water, there can be *ion-dipole forces*, where positive and negative ions (e.g., Na$^+$ and Cl$^-$) are attracted to the oppositely charged ends of polar molecules (e.g., H$_2$O).

However, nonpolar molecules also are attracted to one another although they do not have permanent charges or dipole moments. Evidence of attractions between nonpolar molecules is demonstrated by the fact that nonpolar gases such as methane (CH$_4$), oxygen (O$_2$), nitrogen (N$_2$), ethane (CH$_3$CH$_3$), and carbon tetrachloride (CCl$_4$) condense to liquids and solids when the temperature is lowered sufficiently. Knowing that positive and negative charges attract one another makes it easy to understand the existence of attractive forces among polar molecules and ions. But how can the attractions among nonpolar molecules be explained?

In nonpolar molecules, the valence electrons are distributed about the nuclei so that, on average, there is no net dipole moment. However, molecules are in constant motion, often colliding and approaching one another closely. When two molecules approach closely, their electron clouds interact by electrostatically repelling one another. These repulsive forces momentarily distort the electron distributions within the molecules and create transitory dipole moments in molecules that would be nonpolar if isolated from neighbors. A transitory dipole moment in one molecule induces electron charge distortions and transitory dipole moments in all nearby molecules. At any instant in an assemblage of molecules, nearly every molecule will have a non-uniform charge distribution and an instantaneous dipole moment. An instant later, these dipole moments would have changed direction or disappeared so that, averaged over time, nonpolar molecules have no net dipole moment. However, the effect of these transitory dipole moments is to create a net attraction among nonpolar molecules. Attractions between nonpolar molecules are called *dispersion forces* or *London forces* (after Professor Fritz London who gave a theoretical explanation for them in 1928).

Hydrogen bonding: An especially strong type of dipole-dipole attraction, called *hydrogen bonding*, occurs among molecules containing a hydrogen atom covalently bonded to a small, highly electronegative atom that contains at least one valence shell nonbonding electron pair. An examination of Table 2.1 shows that fluorine, oxygen, and nitrogen are the smallest and most electronegative elements that contain nonbonding valence electron pairs. Although chlorine and sulfur have similarly high electronegativities and contain nonbonding valence electron pairs, they are too large to consistently form hydrogen bonds (H-bonds). Because hydrogen bonds are both strong and common, they influence many substances in important ways.

Hydrogen bonds are very strong (10 to 40 kJ/mole) compared to other dipole-dipole forces (from less than 1 to 5 kJ/mole). The hydrogen atom's very small size makes hydrogen bonding so uniquely strong. Hydrogen has only one electron. When hydrogen is covalently bonded to a small, highly electronegative atom, the shift of bonding electrons toward the more electronegative atom leaves the hydrogen nucleus nearly bare. With no inner core electrons to shield it, the partially positive hydrogen can approach very closely to a nonbonding electron pair on nearby small polar molecules. The very close approach results in stronger attractions than with other dipole-dipole forces.

Because of the strong intermolecular attractions, hydrogen bonds have a strong effect on the properties of the substances in which they occur. Compared with nonhydrogen bonded compounds of similar size, hydrogen bonded substances have relatively high boiling and melting points, low volatilities, high heats of vaporization, and high specific heats. Molecules that can H-bond with water are highly soluble in water; thus, all the substances in Figure 2.5 are water-soluble.

COMPARATIVE STRENGTHS OF INTERMOLECULAR ATTRACTIONS

The strength of dipole-dipole forces depends on the magnitude of the dipole moments. The strength of ion-ion forces depends on the magnitude of the ionic charges. The strength of dispersion forces depends on the *polarizability* of the nonpolar molecules. Polarizability is a measure of how easily the electron distribution can be distorted by an electric field — that is, how easily a dipole moment can be induced in an atom or a molecule. Large atoms and molecules have more electrons and larger electron clouds than small ones. In large atoms and molecules, the outer shell electrons are farther from the nuclei and, consequently, are more loosely bound. The electron distributions can be more easily distorted by external charges. In small atoms and molecules, the outer electrons are closer to the nuclei and are more tightly held. Electron charge distributions in small atoms and molecules are less easily distorted.

Therefore, large atoms and molecules are more polarizable than small ones. Since atomic and molecular sizes are closely related to atomic and molecular weights, we can generalize that polarizability increases with increasing atomic and molecular weights. The greater the polarizability of atoms and molecules, the stronger are the intermolecular dispersion forces between them. Molecular shape also affects polarizability. Elongated molecules are more polarizable than compact molecules. Thus, a linear alkane is more polarizable than a branched alkane of the same molecular weight.

All atoms and molecules have some degree of polarizability. Therefore, all atoms and molecules experience attractive dispersion forces, whether or not they also have dipole moments, ionic charges, or can hydrogen-bond. Small polar molecules are dominated by dipole-dipole forces since the contribution to attractions from dispersion forces is small. However, dispersion forces may dominate in very large polar molecules.

Rules of Thumb

1. The higher the atomic or molecular weights of nonpolar molecules, the stronger are the attractive dispersion forces between them.
2. For different nonpolar molecules with the same molecular weight, molecules with a linear shape have stronger attractive dispersion forces than do branched, more compact molecules.
3. For polar and nonpolar molecules alike, the stronger the attractive forces, the higher the boiling point and freezing point, and the lower the volatility of the substance.

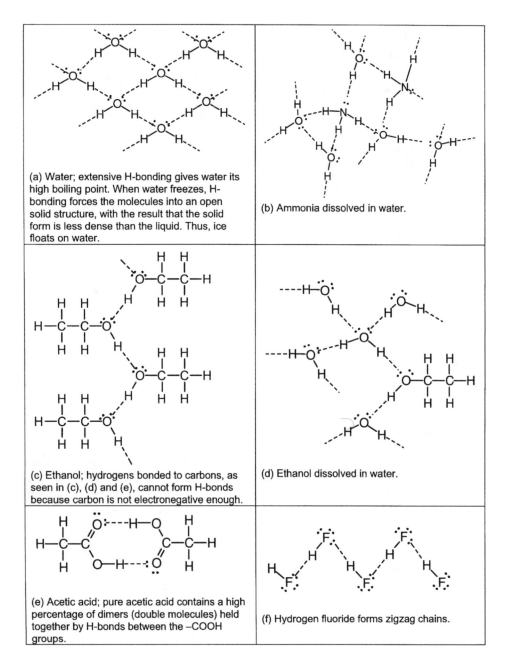

(a) Water; extensive H-bonding gives water its high boiling point. When water freezes, H-bonding forces the molecules into an open solid structure, with the result that the solid form is less dense than the liquid. Thus, ice floats on water.

(b) Ammonia dissolved in water.

(c) Ethanol; hydrogens bonded to carbons, as seen in (c), (d) and (e), cannot form H-bonds because carbon is not electronegative enough.

(d) Ethanol dissolved in water.

(e) Acetic acid; pure acetic acid contains a high percentage of dimers (double molecules) held together by H-bonds between the –COOH groups.

(f) Hydrogen fluoride forms zigzag chains.

FIGURE 2.5   Examples of hydrogen bonding among different molecules.

Examples

1. Consider the halogen gases fluorine (F_2, MW = 38), chlorine (Cl_2, MW = 71), bromine (Br_2, MW = 160), and iodine (I_2, MW = 254). All are nonpolar, with progressively greater molecular weights and correspondingly stronger attractive dispersion forces as you go from F_2 to I_2. Accordingly, their boiling and melting points increase with their molecular weights. At room temperature, F_2 is a gas (bp = –188°C), Cl_2 is also a gas but with a higher boiling point (bp = –34°C), Br_2 is a liquid (bp = 58.8°C), and I_2 is a solid (mp = 184°C).

TABLE 2.2
Some Properties of the First Twelve Straight-Chain Alkanes

Alkane	Formula	Molecular Weight	Melting Point[a] °C	Boiling Point °C
methane	CH_4	16	−183	−162
ethane	C_2H_6	30	−172	−89
propane	C_3H_8	44	−188	−42
n-butane	C_4H_{10}	58	−138	0
n-pentane	C_5H_{12}	72	−130	36
n-hexane	C_6H_{14}	86	−95	69
n-heptane	C_7H_{16}	100	−91	98
n-octane	C_8H_{18}	114	−57	126
n-nonane	C_9H_{20}	128	−51	151
n-decane	$C_{10}H_{22}$	142	−29	174
n-dodecane	$C_{12}H_{26}$	170	−10	216

[a] Deviations from the general trend in melting points occur because melting points for the smallest alkanes are more strongly influenced by differences in crystal structure and lattice energy of the solid.

2. Alkanes are compounds of carbon and hydrogen only. Although C—H bonds are slightly polar (electronegativity of C = 2.5; electronegativity of H = 2.1) all alkanes are nonpolar because of their bond geometry. In the straight-chain alkanes (called *normal*-alkanes), as the alkane carbon chain becomes longer, the molecular weights and, consequently, the attractive dispersion forces become greater. Consequently, melting points and boiling points become progressively higher. The physical properties of the *normal*-alkanes in Table 2.2 reflect this trend.

3. *Normal*-butane [n-C_5H_{12}] and dimethylpropane [$CH_3C(CH_3)_2CH_3$] are both nonpolar and have the same molecular weights (MW = 72). However, n-C_5H_{12} is a straight-chain alkane while $CH_3C(CH_3)_2CH_3$ is branched. Thus, n-C_5H_{12} has stronger dispersion attractive forces than $CH_3C(CH_3)_2CH_3$ and a correspondingly higher boiling point.

normal-pentane: bp = 36°C   Dimethylpropane: bp = 9.5°C

2.10   SOLUBILITY AND INTERMOLECULAR ATTRACTIONS

In liquids and gases, the molecules are in constant, random, thermal motion, colliding and intermingling with one another. Even in solids, the molecules are in constant, although more limited, motion. If different kinds of molecules are present, random movement tends to mix them uniformly. If there were no other considerations, random motion would cause all substances to dissolve completely into one another. Gases and liquids would dissolve more quickly and solids more slowly.

However, intermolecular attractions must also be considered. Strong attractions between molecules tend to hold them together. Consider two different substances A and B, where A molecules are attracted strongly to other A molecules, B molecules are attracted strongly to other B molecules, but A and B molecules are attracted weakly to one another. Then, A and B molecules tend to stay separated from each other. A molecules try to stay together and B molecules try to stay together, each excluding entry from the other. In this case, A and B are not soluble in one another.

As an example of this situation, let A be a nonpolar, straight-chain liquid hydrocarbon such as n-octane (C_8H_{18}) and let B be water (H_2O). Octane molecules are attracted to one another by strong dispersion forces, and water molecules are attracted strongly to one another by dipole-dipole forces and H-bonding. Dispersion attractions are weak between the small water molecules. Because the small water molecules have low polarizability, octane cannot induce a strong dispersion force attraction to water. Because octane is nonpolar, there are no dipole-dipole attractions to water. When water and octane are placed in the same container, they remain separate forming two layers with the less dense octane floating on top of the water.

However, if there were strong attractive forces between A and B molecules, it would help them to mix. The solubility of one substance (the solute) in another (the solvent) depends mostly on intermolecular forces and, to a much lesser extent, on conditions such as temperature and pressure. *Substances are more soluble in one another when intermolecular attractions between solute and solvent are similar in magnitude to the intermolecular attractions between the pure substances.* This principle is the origin of the rules of thumb that say "like dissolves like" or "oil and water don't mix." "Like" molecules have similar polar properties and, consequently, similar intermolecular attractions. Oil and water do not mix because water molecules are attracted strongly to one another, and oil molecules are attracted strongly to one another; but water molecules and oil molecules are attracted only weakly to one another.

Rules of Thumb

1. The more symmetrical the structure of a molecule containing polar bonds, the less polar and the less soluble it is in water.
2. Molecules with OH, NO, or NH groups can form hydrogen bonds to water molecules. They are the most water-soluble non-ionic compounds, even if they are nonpolar because of geometrical symmetry.
3. The next most water-soluble compounds contain O, N, and F atoms. All have high electronegativities and allow water molecules to H-bond with them.
4. Charged regions in ionic compounds (like sodium chloride) are attracted to polar water molecules. This makes them more soluble.
5. Most compounds in oil and gasoline mixtures are nonpolar. They are attracted to water very weakly and have very low solubilities.
6. All molecules, including nonpolar molecules, are attracted to one another by dispersion forces. The larger the molecule the stronger the dispersion force.
7. Nonpolar molecules, large or small, have low solubilities in water because the small-sized water molecules have weak dispersion forces, and nonpolar molecules have no dipole moments. Thus, there are neither dispersion nor polar attractions to encourage solubility.

Examples

1. Alcohols of low molecular weight are very soluble in water because of hydrogen bonding. However, their solubilities decrease as the number of carbons increase. The –OH group on alcohols is hydrophilic (attracted to water), while the hydrocarbon part is hydrophobic (repelled from water). If the hydrocarbon part of an alcohol is large enough, the hydrophobic behavior overcomes the hydrophilic behavior of the –OH group and the alcohol has low solubility. Solubilities for alcohols with increasingly larger hydrocarbon chains are given in Table 2.3.

TABLE 2.3
Solubilities and Boiling Points of Some Straight Chain Alcohols

Name	Formula	Molecular Weight	Melting Point[a] (°C)	Boiling Point (°C)	Aqueous solubility at 25°C (mol/L)
Methanol	CH_3OH	32	–98	65	∞ (miscible)
Ethanol	C_2H_5OH	46	–130	78	∞ (miscible)
1-propanol	C_3H_7OH	60	–127	97	∞ (miscible)
1-butanol	C_4H_9OH	74	–90	117	0.95
1-pentanol	$C_5H_{11}OH$	88	–79	138	0.25
1,5-pentanediol[b]	$C_5H_{10}(OH)_2$	104	–18	239	∞ (miscible)
1-hexanol	$C_6H_{13}OH$	102	–47	158	0.059
1-octanol	$C_8H_{17}OH$	130	–17	194	0.0085
1-nonanol	$C_9H_{19}OH$	144	–6	214	0.00074
1-decanol	$C_{10}H_{21}OH$	158	+6	233	0.00024
1-dodecanol	$C_{12}H_{25}OH$	186	+24	259	0.000019

[a] Deviations from the general trend in melting points occur because melting points for the smallest alcohols are more strongly influenced by differences in crystal structure and lattice energy of the solid.
[b] The properties of 1,5-pentanediol deviate from the trends of the other alcohols because it is a diol and has two –OH groups available for hydrogen bonding. See text.

2. For alcohols of comparable molecular weight, *the more hydrogen bonds a compound can form, the more water-soluble the compound, and the higher the boiling and melting points of the pure compound.* In Table 2.3, notice the effect of adding another –OH group to the molecule. The double alcohol 1,5-pentanediol is more water-soluble and has a higher boiling point than single alcohols of comparable molecular weight, as a result of its two –OH groups capable of hydrogen bonding. This effect is general. Double alcohols (diols) are more water-soluble and have higher boiling and melting points than single alcohols of comparable molecular weight. Triple alcohols (triols) are still more water-soluble and have higher boiling and melting points.

3 Major Water Quality Parameters

CONTENTS

3.1 INTERACTIONS AMONG WATER QUALITY PARAMETERS

This chapter deals with important water quality parameters which serve as controlling variables that strongly influence the behavior of many other constituents present in the water. The major controlling variables are pH, oxidation-reduction (redox) potential, alkalinity and acidity, temperature, and total dissolved solids. This chapter also discusses several other important parameters, such as ammonia, sulfide, carbonates, dissolved metals, and dissolved oxygen, that are strongly affected by changes in the controlling variables.

It is important to understand that chemical constituents in environmental water bodies react in an environment far more complicated than if they simply were surrounded by a large number of water molecules. The various impurities in water interact in ways that can affect their chemical behavior markedly. The water quality parameters defined above as controlling variables have an especially strong effect on water chemistry. For example, a pH change from pH 6 to pH 9 will lower the solubility of Cu^{2+} by five orders of magnitude. At pH 6 the solubility of Cu^{2+} is about 40 mg/L while at pH 9 it is about 4×10^{-3} mg/L. If, for example, a pH 6 water solution contained 20 mg/L of Cu^{2+} and the pH were raised to 9, all but 4×10^{-3} mg/L of the Cu^{2+} would precipitate as solid $Cu(OH)_2$.

As another example, consider a shallow lake with algae and other vegetation growing in it. Suspended and lake-bottom sediments contain high concentrations of decaying organic matter. The lake is fed by surface and groundwaters containing high levels of sulfate. During the day, photosynthesis can produce enough dissolved oxygen to maintain a positive oxidation-reduction potential in the water. At night, photosynthesis stops and biodegradation of suspended and lake-bottom organic sediments consumes nearly all of the dissolved oxygen in the lake. This causes the water to change from oxidizing (aerobic) to reducing (anaerobic) conditions and also causes the oxidation-reduction potential to change from positive to negative values. Under reducing conditions, dissolved sulfate in the lake is reduced to sulfide, producing hydrogen sulfide gas which smells like rotten eggs. Thus, there is an odor problem at night that generally dissipates during the day. A remedy for this problem entails finding a way to maintain a positive oxidation-reduction potential for longer periods of time.

Rule of Thumb

Because they strongly influence other water quality parameters, the controlling variables listed below are usually included among the parameters that are measured in water quality sampling programs.

- pH
- Temperature
- Alkalinity and/or acidity
- Total dissolved solids (TDS) or conductivity
- Oxidation-reduction (redox) potential

3.2 pH

BACKGROUND

Pure water always contains a small number of molecules that have dissociated into hydrogen ions (H^+) and hydroxyl ions (OH^-), as illustrated by Equation 3.1.

$$H_2O \leftrightarrow H^+ + OH^-. \tag{3.1}$$

The water dissociation constant, K_w, is defined as the product of the concentrations of H^+ and OH^- ions, expressed in moles per liter:

$$K_w = [H^+][OH^-], \tag{3.2}$$

where enclosing a species in square brackets is chemical symbolism that represents the species concentration in moles per liter.

Because the degree of dissociation increases with temperature, K_w is temperature dependent. At 25°C,

$$K_{w,25C} = [H^+][OH^-] = 1.0 \times 10^{-14} \ (mol/L)^2, \tag{3.3}$$

while at 50°C,

$$K_{w,50C} = [H^+][OH^-] = 1.83 \times 10^{-13} \ (mol/L)^2. \tag{3.4}$$

If, for example, an acid is added to water at 25°C, the H^+ concentration increases but the product expressed by Equation 3.3 will always be equal to 1.0×10^{-14} $(mol/L)^2$. This means that if $[H^+]$ increases, $[OH^-]$ must decrease. Adding a base causes $[OH^-]$ to increase and $[H^+]$ to decrease correspondingly.

In pure water or in water with no other sources or sinks of H^+ or OH^-, Equation 3.1 leads to equal numbers of H^+ and OH^- species. Thus, at 25°C, the values of $[H^+]$ and $[OH^-]$ must each be equal to 1.0×10^{-7} mol/L, since:

$$K_{w,25C} = (1.0 \times 10^{-7} \ mol/L)(1.0 \times 10^{-7} \ mol/L) = 1.0 \times 10^{-14} \ (mol/L)^2.$$

Pure water is neither acidic nor basic. *Pure water defines the condition of acid-base neutrality.* Therefore, acid-base neutral water always has equal concentrations of H^+ and OH^-, or $[H^+] = [OH^-]$.

In neutral water at 25°C, $[H^+] = [OH^-] = 1 \times 10^{-7}$ mol/L.

In neutral water at 50°C, $[H^+] = [OH^-] = 4.3 \times 10^{-7}$ mol/L.

If $[H^+] > [OH^-]$, the water solution is acidic.

If $[H^+] < [OH^-]$, the water solution is basic.

Whatever their separate values, the product of hydrogen ion and hydroxyl ion concentrations must be equal to 1×10^{-14} at 25°C, as in Equation 3.3. If for example $[H^+] = 10^{-5}$ mol/L, then it is necessary that $[OH^-] = 10^{-9}$ mol/L, so that their product is 10^{-14} (mol/L)^2.

Many compounds dissociate in water to form ions. Those that form hydrogen ions, H^+, are called acids because when added to pure water they cause the condition $[H^+] > [OH^-]$. Compounds that cause the condition $[H^+] < [OH^-]$ when added to pure water are called bases. An acid water solution gets its acidic properties from the presence of H^+. Because H^+ is too reactive to exist alone, it is always attached to another molecular species. In water solutions, H^+ is often written as H_3O^+ because of the almost instantaneous reaction that attaches it to a water molecule

$$H^+ + H_2O \rightarrow H_3O^+. \tag{3.5}$$

H_3O^+ is called the *hydronium ion*. It does not make any difference to the meaning of a chemical equation whether the presence of an acid is indicated by H^+ or H_3O^+. For example, the addition of nitric acid, HNO_3, to water produces the ionic dissociation reaction

$$HNO_3 + H_2O \rightarrow H_3O^+ + NO_3^-,$$

or equivalently

$$HNO_3 \xrightarrow{H_2O} H^+ + NO_3^-.$$

Both equations are read "HNO_3 added to water forms H^+ (or H_3O^+) and NO_3^- ions."

DEFINING pH

The concentration of H^+ in water solutions commonly ranges from about 1 mol/L (equivalent to 1 g/L or 1000 ppm) for very acidic water, to about 10^{-14} mol/L (10^{-14} g/L or 10^{-11} ppm) for very basic water. Under special circumstances, the range can be even wider.

Rather than work with such a wide numerical range for a measurement that is so common, chemists have developed a way to use logarithmic units for expressing $[H^+]$ as a positive decimal number whose value normally lies between 0 and 14. This number is called the *pH*, and is defined in Equation 3.6 as the negative of the base$_{10}$ logarithm of the hydrogen ion concentration in moles per liter:

$$pH = -\log_{10}[H^+]. \tag{3.6}$$

Note that if $[H^+] = 10^{-7}$, then pH $= -\log_{10}(10^{-7}) = -(-7) = 7$. A higher concentration of H^+ such as $[H^+] = 10^{-5}$ yields a lower value for pH, i.e., pH $= -\log_{10}(10^{-5}) = 5$. Thus, if pH is less than 7, the solution contains more H^+ than OH^- and is acidic; if pH is greater than 7, the solution is basic.

ACID-BASE REACTIONS

In acid-base reactions, protons (H^+ ions) are transferred between chemical species, one of which is an acid and the other is a base. The proton donor is the acid and the proton acceptor is the base. For example, if an acid, such as hydrochloric acid (HCl), is dissolved in water, water acts as a base by accepting the proton donated by HCl. The acid-base reaction is written: $HCl + H_2O \rightarrow Cl^- + H_3O^+$. A water molecule that behaved as a base by accepting a proton is turned into an acid, H_3O^+, a species that has a proton available to donate. The species H_3O^+, as noted above, is called a *hydronium ion* and is the chemical species that gives acid water solutions their acidic characteristics. An HCl/water solution contains water molecules, hydronium ions, hydroxyl ions (in smaller concentration than H_3O^+), and chloride ions. The solution is termed acidic, with pH (at 25°C) < 7. The measurable parameter *pH* indicates the concentration of protons available for acid-base reactions.

Rules of Thumb

1. In an acid-base reaction, H^+ ions are exchanged between chemical species. The species that donates the H^+ is the acid. The species that accepts the H^+ is the base.
2. The concentration of H^+ in water solutions is an indication of how many hydrogen ions are available, at the time of measurement, for exchange between chemical species. The exchange of hydrogen ions changes the chemical properties of the species between which the exchange occurs.
3. pH is a measure of $[H^+]$, the hydrogen ion concentration, which determines the acidic or basic quality of water solutions. At 25°C:
   - When pH < 7, a water solution is acidic.
   - When pH = 7, a water solution is neutral.
   - When pH > 7, a water solution is basic.

Example 3.1

The $[H^+]$ of water in a stream = 3.5×10^{-6} mol/L. What is the pH?

Answer:

$$pH = -\log_{10}[H^+] = -\log_{10}(3.5 \times 10^{-6}) = -(-5.46) = 5.46.$$

Notice that since logarithms are dimensionless, the pH unit has no dimensions or units. Frequently, pH is unnecessarily assigned units called *SU*, or standard units, even though pH is unitless. This mainly serves to avoid blank spaces in a table that contains a column for units, or to satisfy a database that requires an entry in a units field. An alternate and useful form of Equation 3.6 is:

$$[H^+] = 10^{-pH}. \tag{3.7}$$

Example 3.2

The pH of water in a stream is 6.65. What is the hydrogen ion concentration?

Answer:

$$[H^+] = 10^{-pH} = 10^{-6.65} = 2.24 \times 10^{-7} \text{ mol/L.}$$

IMPORTANCE OF pH

Measurement of pH is one of the most important and frequently used tests in water chemistry. pH is an important factor in determining the chemical and biological properties of water. It affects the chemical forms and environmental impact of many chemical substances in water. For example, many metals dissolve as ions at lower pH values precipitate as hydroxides and oxides at higher pH and redissolve again at very high pH. Figure 3.1 shows the pH scale and typical pH values of some common substances.

pH also influences the degree of ionization, volatility, and toxicity to aquatic life of certain dissolved substances, such as ammonia, hydrogen sulfide, and hydrogen cyanide. The ionized form of ammonia, which predominates at low pH, is the less toxic ammonium ion NH_4^+. NH_4^+ transforms to the more toxic form of unionized ammonia NH_3, at higher pH. Both hydrogen sulfide (H_2S) and hydrogen cyanide (HCN) behave oppositely to ammonia; the less toxic ionized forms, S^{2-} and CN^-, are predominant at high pH, and the more toxic unionized forms, H_2S and HCN, are predominant at low pH. The pH value is an indicator of the chemical state in which these compounds will be found and must be considered when establishing water quality standards.

MEASURING pH

The pH of environmental waters is most commonly measured with electronic pH meters or by wetting with sample, special papers impregnated with color-changing dyes. Battery-operated field meters are common. A pH measurement of surface or groundwater is valid only when made in the field or very shortly after sampling. The pH is altered by many processes that occur after the sample is collected, such as loss or gain of dissolved carbon dioxide or the oxidation of dissolved iron. A laboratory determination of pH made hours or days after sampling may be more than a full pH unit (a factor of 10 in H^+ concentration) different from the value at the time of sampling.

Loss or gain of dissolved carbon dioxide (CO_2) is one of the most common causes for pH changes. When CO_2 dissolves into water, by diffusion from the atmosphere or from microbial activity in water or soil, the pH is lowered. Conversely, when CO_2 is lost, by diffusion to the atmosphere or consumption during photosynthesis of algae or water plants, the pH is raised.

Rules of Thumb

1. Under low pH conditions (acidic)
   a. Metals tend to dissolve.
   b. Cyanide and sulfide are *more* toxic to fish.
   c. Ammonia is *less* toxic to fish.
2. Under high pH conditions (basic)
   a. Metals tend to precipitate as hydroxides and oxides. *However, if the pH gets too high, some precipitates begin to dissolve again because soluble hydroxide complexes are formed (see Metals).*
   b. Cyanide and sulfide are *less* toxic to fish.
   c. Ammonia is *more* toxic to fish.

CRITERIA AND STANDARDS

The pH of pure water at 25°C is 7.0, but the pH of environmental waters is affected by dissolved carbon dioxide and exposure to minerals. Most unpolluted groundwaters and surface waters in the U.S. have pH values between about 6.0 and 8.5, although higher and lower values can occur because of special conditions such as sulfide oxidation which lowers the pH, or low carbon dioxide concentrations which raises the pH. During daylight, photosynthesis in surface waters by aquatic organisms may consume more carbon dioxide than is dissolved from the atmosphere, causing pH to rise. At night, after photosynthesis has ceased, carbon dioxide from the atmosphere continues to dissolve and lowers the pH again. In this manner, photosynthesis can cause diurnal pH fluctuations, the magnitude of which depends on the alkalinity buffering capacity of the water. In poorly buffered lakes or rivers, the daytime pH may reach 9.0 to 12.0.

The permissible pH range for fish depends on factors such as dissolved oxygen, temperature, and concentrations of dissolved anions and cations. A pH range of 6.5 to 9.0, with no short-term change greater than 0.5 units beyond the normal seasonal maximum or minimum, is deemed protective of freshwater aquatic life and considered harmless to fish. In irrigation waters, the pH should not fall outside a range of 4.5 to 9.0 to protect plants.

EPA Criteria

Domestic water supplies: 5.0–9.0.
Freshwater aquatic life: 6.5–9.0.

Rules of Thumb

1. The pH of natural unpolluted river water is generally between 6.5 and 8.5.
2. The pH of natural unpolluted groundwater is generally between 6.0 and 8.5.
3. Clean rainwater has a pH of about 5.7 because of dissolved CO_2.
4. After reaching the surface of the earth, rainwater usually acquires alkalinity while moving over and through the earth, which may raise the pH and buffer the water against severe pH changes.
5. The pH of drinking water supplies should be between 5.0 to 9.0.
6. Fish acclimate to ambient pH conditions. For aquatic life, pH should be between 6.5 to 9.0 and should not vary more than 0.5 units beyond the normal seasonal maximum or minimum.

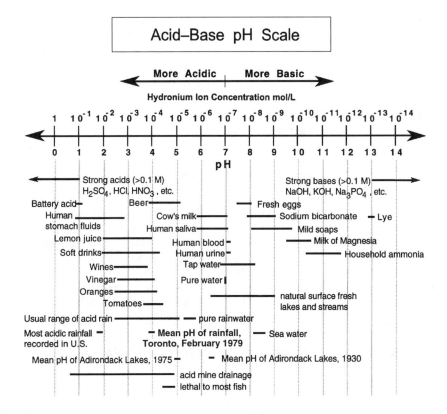

FIGURE 3.1 pH scale and typical pH values of some common substances.

3.3 OXIDATION-REDUCTION (REDOX) POTENTIAL

BACKGROUND

The redox potential measures the availability of electrons for exchange between chemical species. This may be viewed as analogous to pH, which measures the availability of protons (H^+ ions) for exchange between chemical species. When H^+ ions are exchanged, the acid or base properties of the species are changed. When electrons are exchanged, the oxidation states of the species and their chemical properties are changed, resulting in oxidation and reduction reactions. The electron donor is said to be *oxidized*. The electron acceptor is said to be *reduced*. For every electron donor, there must be an electron acceptor. For example, whenever one substance is oxidized, another must be reduced. Strong oxidizing agents, such as ozone, chlorine, or permanganate, are those that readily take electrons from many substances, causing the electron donor to be oxidized. By accepting electrons, the oxidizing agents are themselves reduced. In a similar manner, strong reducing agents are those that are easily oxidized, in other words, they readily give up electrons to other substances that in turn become reduced.

For example, chlorine is widely used to treat water and sewage. Chlorine oxidizes many pollutants to less objectionable forms. When chlorine reacts with hydrogen sulfide (H_2S) — a common sewage pollutant that smells like rotten eggs — it oxidizes the sulfur in H_2S to insoluble elemental sulfur, which is easily removed by settling or filtering. The reaction is

$$8\ Cl_2(g) + 8\ H_2S(aq) \rightarrow S_8(s) + 16\ HCl(aq). \tag{3.8}$$

The sulfur in H_2S donates two electrons that are accepted by the chlorine atoms in Cl_2. Chlorine is reduced (it accepts electrons), and sulfur is oxidized (it donates electrons). Because chlorine is the agent that causes the oxidation of H_2S, chlorine is called an *oxidizing agent*. Because H_2S is the agent that causes the reduction of chlorine, H_2S is called a *reducing agent*.

The class of oxidation-reduction reactions is very large. These include all combustion processes such as the burning of gasoline or wood, most microbial reactions such as those that occur in biodegradation, and all electrochemical reactions such as those that occur in batteries and metal corrosion. The use of subsurface groundwater treatment walls containing finely divided iron is based on the reducing properties of iron. Such treatment walls are placed in the path of groundwater contaminant plumes. The iron donates electrons to pollutants as they pass through the permeable barrier. Thus, the iron is oxidized and the pollutant reduced. This often causes the pollutant to decompose into less harmful or inert fragments.

Rules of Thumb

1. Oxygen gas (O_2) is always an oxidizing agent in its reactions with metals and most non-metals. If a compound has combined with O_2, it has been oxidized and the O_2 has been reduced. By accepting electrons, O_2 either is changed to the oxide ion (O^{2-}) or is combined in compounds such as CO_2 or H_2O.
2. Like O_2, the halogen gases (F_2, Cl_2, Br_2, and I_2) are always oxidizing agents in reactions with metals and most non-metals. They accept electrons to become halide ions (F^-, Cl^-, Br^-, and I^-) or are combined in compounds such as HCl or $CHBrCl_2$.
3. If an elemental metal (Fe, Al, Zn, etc.) reacts with a compound, the metal acts as a reducing agent by donating electrons, usually forming a soluble positive ion such as Fe^{2+}, Al^{3+}, or Zn^{2+}.

3.4 CARBON DIOXIDE, BICARBONATE, AND CARBONATE

BACKGROUND

The reactive inorganic forms of environmental carbon are carbon dioxide (CO_2), bicarbonate (HCO_3^-), and carbonate (CO_3^{2-}). Organic carbon, such as cellulose and starch, is made by plants from CO_2 and water during photosynthesis. Carbon dioxide is present in the atmosphere and in soil pore space as a gas, and in surface waters and groundwaters as a dissolved gas. The carbon cycle is based on the mobility of carbon dioxide, which is distributed readily through the environment as a gas in the atmosphere and dissolved in rain water, surface water, and groundwater. Most of the earth's carbon, however, is relatively immobile, being contained in ocean sediments and on continents as minerals. The atmosphere, with about 360 ppmv (parts per million by volume) of mobile CO_2, is the second smallest of the earth's global carbon reservoirs, after life forms which are the smallest.

On land, solid forms of carbon are mobilized as particulates mainly by weathering of carbonate minerals, biodegradation and burning of organic carbon, and burning of fossil fuels.

SOLUBILITY OF CO_2 IN WATER

Carbon dioxide plays a fundamental role in determining the pH of natural waters. Although CO_2 itself is not acidic, it reacts in water (reversibly) to make an acidic solution by forming carbonic acid (H_2CO_3), as shown in Equation 3.9. Carbonic acid can subsequently dissociate in two steps to release hydrogen ions, as shown in Equations 3.10 and 3.11:

$$CO_2 + H_2O \leftrightarrow H_2CO_3. \tag{3.9}$$

$$H_2CO_3 \leftrightarrow H^+ + HCO_3^-. \tag{3.10}$$

$$HCO_3^- \leftrightarrow H^+ + CO_3^{2-}. \tag{3.11}$$

As a result, pure water exposed to air is not acid-base neutral with a pH near 7.0 because dissolved CO_2 makes it acidic, with a pH around 5.7. The pH dependence of Equations 3.9–3.11 is shown in Figure 3.2 and Table 3.1.

Observations From Figure 3.2 and Table 3.1

- As pH increases, all equilibria in Equations 3.9–3.11 shift to the right.
- As pH decreases, all equilibria shift to the left.
- Above pH = 10.3, carbonate ion (CO_3^{2-}) is the dominant species.
- Below pH = 6.3, dissolved CO_2 is the dominant species.
- Between pH = 6.3 and 10.3, a range common to most environmental waters, bicarbonate ion (HCO_3^-) is the dominant species.

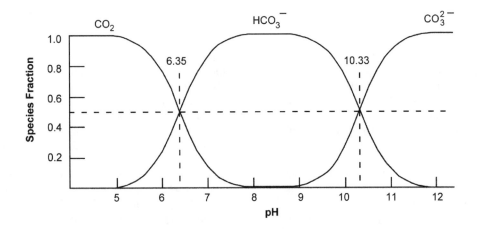

FIGURE 3.2   Distribution diagram showing pH dependence of carbonate species in water.

TABLE 3.1
pH Dependence of Carbonate Fractions (From Figure 3.2)

pH	fraction as CO_2	fraction as HCO_3^-	fraction as CO_3^{2-}
<< 6.35	essentially 1.00	essentially 0	essentially 0
6.35	0.50	0.50	essentially 0
$\frac{1}{2}(6.35 + 10.33) = 8.34$	0.01	0.98	0.01
10.33	essentially 0	0.50	0.50
>> 10.33	essentially 0	essentially 0	essentially 1.00

The equilibria among only the carbon species (omitting the display of H^+) are

$$CO_2(gas, atm) \leftrightarrow CO_2(aq) \leftrightarrow H_2CO_3(aq) \leftrightarrow HCO_3^-(aq) \leftrightarrow CO_3^{2-}(aq). \qquad (3.12)$$

These dissolved carbon species are sometimes referred to as *dissolved inorganic carbon (DIC)*.

SOIL CO_2

Processes such as biodegradation of organic matter and respiration of plants and organisms which commonly occur in the subsurface consume O_2 and produce CO_2. In the soil subsurface, air in the pore spaces cannot readily equilibrate with the atmosphere, and therefore pore space air becomes lower in O_2 and higher in CO_2 concentrations.

- Oxygen may decrease from about 21% (210,000 ppmv) in the atmosphere to between 15% and 0% (150,000 to 0 ppmv) in the soil.
- Carbon dioxide may increase from about 0.04% (~360 ppmv) in the atmosphere to between 0.1% and 10% (1000 to 100,000 ppmv) in the soil.

When water moves through the subsurface, it equilibrates with soil gases and may become more acidic because of a higher concentration of dissolved CO_2. Acidic groundwater has an increased capacity for dissolving minerals. The higher the CO_2 concentration in soil air, the lower is the pH of groundwater. Acidic groundwater may become buffered, minimizing pH changes, by dissolution of soil minerals, particularly calcium carbonate. Limestone (calcium carbonate, $CaCO_3$) is particularly susceptible to dissolution by low pH waters. Limestone caves are formed when low pH groundwaters move through limestone deposits and dissolve the limestone minerals.

Rules of Thumb

1. Unpolluted rainwater is acidic, about pH = 5.7, because of dissolved CO_2 from the atmosphere.
2. Acid rain has lower pH values, reaching pH = 2.0 or lower, because of dissolved sulfuric, nitric, and hydrochloric acids which result mainly from industrial air emissions.
3. The dissolved carbonate species, $CO_2(aq)$ (equivalent to H_2CO_3), HCO_3^-, and CO_3^{2-}, are present in any natural water system near the surface of the earth. The relative proportions depend on pH.
4. At pH values between 7.0 and 10.0, bicarbonate is the dominant dissolved inorganic carbon species in water. Between pH 7.8 and 9.2, bicarbonate is close to 100%; carbonate and dissolved CO_2 concentrations are essentially zero.
5. In subsurface soil pore space, oxygen is depleted and carbon dioxide increased, compared to the atmosphere. Oxygen typically decreases from 21% in atmospheric air to 15% or less in soil pore space air, and carbon dioxide typically increases from ~360 ppmv in atmospheric air to between 1000 and 100,000 ppmv in soil pore space air. Thus, unpolluted groundwaters tend to be more acidic than unpolluted surface waters because of higher dissolved concentrations of CO_2.

3.5 ACIDITY AND ALKALINITY

BACKGROUND

The *alkalinity* of water is its acid-neutralizing capacity. The *acidity* of water is its base-neutralizing capacity. Both parameters are related to the *buffering capacity* of water (the ability to resist changes in pH when an acid or base is added). Water with high alkalinity can neutralize a large quantity of acid without large changes in pH; on the other hand, water with high acidity can neutralize a large quantity of base without large changes in pH.

ACIDITY

Acidity is determined by measuring how much standard base must be added to raise the pH to a specified value. Acidity is a net effect of the presence of several constituents, including dissolved carbon dioxide, dissolved multivalent metal ions, strong mineral acids such as sulfuric, nitric, and hydrochloric acids, and weak organic acids such as acetic acid. Dissolved carbon dioxide (CO_2) is the main source of acidity in unpolluted waters. Acidity from sources other than dissolved CO_2 is not commonly encountered in unpolluted natural waters and is often an indicator of pollution.

Titrating an acidic water sample with base to pH 8.3 measures *phenolphthalein* acidity* or *total acidity*. Total acidity measures the neutralizing effects of essentially all the acid species present, both strong and weak.

Titrating with base to pH 3.7 measures *methyl orange* acidity*. Methyl orange acidity primarily measures acidity due to dissolved carbon dioxide and other weak acids that are present.

ALKALINITY

In natural waters that are not highly polluted, alkalinity is more commonly found than acidity. Alkalinity is often a good indicator of the total dissolved inorganic carbon (bicarbonate and carbonate anions) present. All unpolluted natural waters are expected to have some degree of alkalinity. Since all natural waters contain dissolved carbon dioxide, they all will have some degree of alkalinity contributed by carbonate species — unless acidic pollutants would have consumed the alkalinity. It is not unusual for alkalinity to range from 0 to 750 mg/L as $CaCO_3$. For surface waters, alkalinity levels less than 30 mg/L are considered low, and levels greater than 250 mg/L are considered high. Average values for rivers are around 100–150 mg/L. Alkalinity in environmental waters is beneficial because it minimizes pH changes, reduces the toxicity of many metals by forming complexes with them, and provides nutrient carbon for aquatic plants.

Alkalinity is determined by measuring how much standard acid must be added to a given amount of water in order to lower the pH to a specified value. Like acidity, alkalinity is a net effect of the presence of several constituents, but the most important are the bicarbonate (HCO_3^-), carbonate (CO_3^{2-}), and hydroxyl (OH^-) anions. Alkalinity is often taken as an indicator for the concentration of these constituents. There are other, usually minor, contributors to alkalinity, such as ammonia, phosphates, borates, silicates, and other basic substances.

Titrating a basic water sample with acid to pH 8.3 measures *phenolphthalein alkalinity*. Phenolphthalein alkalinity primarily measures the amount of carbonate ion (CO_3^{2-}) present. Titrating with acid to pH 3.7 measures *methyl orange alkalinity* or *total alkalinity*. Total alkalinity measures the neutralizing effects of essentially all the bases present.

Because alkalinity is a property caused by several constituents, some convention must be used for reporting it quantitatively as a concentration. The usual convention is to express alkalinity as ppm or mg/L of calcium carbonate ($CaCO_3$). This is done by calculating how much $CaCO_3$ would be neutralized by the same amount of acid as was used in titrating the water sample when measuring

---

* *Phenolphthalein* and *methyl orange* are pH-indicator dyes that change color at pH 8.3 and 3.7, respectively.

either phenolphthalein or methyl orange alkalinity. Whether it is present or not, $CaCO_3$ is used as a proxy for all the base species that are actually present in the water. *The alkalinity value is equivalent to the mg/L of $CaCO_3$ that would neutralize the same amount of acid as does the actual water sample.*

IMPORTANCE OF ALKALINITY

Alkalinity is important to fish and other aquatic life because it buffers both natural and human-induced pH changes. The chemical species that cause alkalinity, such as carbonate, bicarbonate, hydroxyl, and phosphate ions, can form chemical complexes with many toxic heavy metal ions, often reducing their toxicity. Water with high alkalinity generally has a high concentration of dissolved inorganic carbon (in the form of HCO_3^- and CO_3^{2-}) which can be converted to biomass by photosynthesis. A minimum alkalinity of 20 mg/L as $CaCO_3$ is recommended for environmental waters and levels between 25 and 400 mg/L are generally beneficial for aquatic life. More productive waterfowl habitats correlate with increased alkalinity above 25 mg/L as $CaCO_3$.

CRITERIA AND STANDARDS FOR ALKALINITY

Naturally occurring levels of alkalinity reaching at least 400 mg/L as $CaCO_3$ are not considered a health hazard. EPA guidelines recommend a minimum alkalinity level of 20 mg/L as $CaCO_3$, and that natural background alkalinity is not reduced by more than 25% by any discharge. For waters where the natural level is less than 20 mg/L, alkalinity should not be further reduced. Changes from natural alkalinity levels should be kept to a minimum. The volume of sample required for alkalinity analysis is 100 mL.

Rules of Thumb

1. Alkalinity is the mg/L of $CaCO_3$ that would neutralize the same amount of acid as does the actual water sample.
2. *Phenolphthalein alkalinity* (titration with acid to pH 8.3) measures the amount of carbonate ion (CO_3^{2-}) present.
3. *Total* or *methyl orange alkalinity* (titration with acid to pH 3.7) measures the neutralizing effects of essentially all the bases present.
4. Surface and groundwaters draining carbonate mineral formations become more alkaline due to dissolved minerals.
5. High alkalinity can partially mitigate the toxic effects of heavy metals to aquatic life.
6. Alkalinity greater than 25 mg/L $CaCO_3$ is beneficial to water quality.
7. Surface waters without carbonate buffering may be more acidic than pH 5.7 (the value established by equilibration of dissolved CO_2 with CO_2 in the atmosphere) because of water reactions with metals and organic substances, biochemical reactions, and acid rain.

CALCULATING ALKALINITY

Although alkalinity is usually determined by titration, the part due to carbonate species (carbonate alkalinity) is readily calculated from a measurement of pH, bicarbonate and/or carbonate. Carbonate alkalinity is equal to the sum of the concentrations of bicarbonate and carbonate ions, expressed as the equivalent concentration of $CaCO_3$.

Example 3.3

A groundwater sample contains 300 mg/L of bicarbonate at pH = 10.0. Calculate the carbonate alkalinity as $CaCO_3$.

Answer:

1. Use the measured values of bicarbonate and pH, with Figure 3.2, to determine the value of CO_3^{2-}. At pH = 10.0, total carbonate is about 73% bicarbonate ion and 27% carbonate ion. Although these percentages are related to moles/L rather than mg/L, the molecular weights of bicarbonate and carbonate ions differ by only about 1.7%; therefore, mg/L can be used in the calculation without significant error.

$$\text{Total carbonate} = \frac{300 \text{ mg/L}}{0.73} = 411 \text{ mg/L.}$$

$CO_3^{2-} = 0.27 \times 411 = 111$ mg/L, or alternatively, $411 - 300 = 111$ mg/L.

2. Determine the equivalent weights of HCO_3^-, CO_3^{2-}, and $CaCO_3$.

$$\text{eq. wt.} = \frac{\text{molecular or atomic weight}}{\text{magnitude of ionic charge or oxidation number}}.$$

$$\text{eq. wt. of } HCO_3^- = \frac{61.02}{1} = 61.0.$$

$$\text{eq. wt. of } CO_3^{2-} = \frac{61.01}{2} = 30.0.$$

$$\text{eq. wt. of } CaCO_3 = \frac{100.09}{2} = 50.0.$$

3. Determine the multiplying factors to obtain the equivalent concentration of $CaCO_3$.

$$\text{Multiplying factor of } HCO_3^- \text{ as } CaCO_3 = \frac{\text{eq. wt. of } CaCO_3}{\text{eq. wt. of } HCO_3^-} = \frac{50.0}{61.0} = 0.820.$$

$$\text{Multiplying factor of } CO_3^{2-} \text{ as } CaCO_3 = \frac{\text{eq. wt. of } CaCO_3}{\text{eq. wt. of } CO_3^{2-}} = \frac{50.0}{30.0} = 1.667.$$

4. Use the multiplying factors and concentrations to calculate the carbonate alkalinity, expressed as mg/L of $CaCO_3$.

$$\text{Carbonate alk. (as } CaCO_3) = 0.820 \, [HCO_3^-, \text{mg/L}] + 1.667 \, [CO_3^{2-}, \text{mg/L}]. \qquad (3.13)$$

$$\text{Carbonate alk.} = 0.820 \, [300 \text{ mg/L}] + 1.667 \, [111 \text{ mg/L}] = \underline{431 \text{ mg/L } CaCO_3}.$$

Equation 3.13 may be used to calculate carbonate alkalinity whenever pH and either bicarbonate or carbonate concentrations are known.

CALCULATING CHANGES IN ALKALINITY, CARBONATE, AND pH

A detailed calculation of how pH, total carbonate, and total alkalinity are related to one another is moderately complicated because of the three simultaneous carbonate equilibria reactions, Equations 3.9–3.11. However, the relations can be conveniently plotted on a total alkalinity/pH/total carbonate graph, also called a *Deffeyes diagram*, or *capacity diagram* (see Figures 3.3 and 3.4). Details of the construction of the diagrams may be found in Stumm and Morgan (1996) and Deffeyes (1965).

In a total alkalinity/pH/total carbonate graph shown in Figure 3.3, a vertical line represents adding strong base or acid without changing the total carbonate (C_T). The added base or acid changes the pH and, therefore, shifts the carbonate equilibrium, but does not add or remove any carbonate. The amount of strong base or acid in meq/L equals the vertical distance on the graph. You can see from Figure 3.3 that if the total carbonate is small, the system is poorly buffered, so a little base or acid makes large changes in pH. If total carbonate is large, the system buffering capacity is similarly large and it takes much more base or acid for the same pH change.

A horizontal line represents changing total carbonate, generally by adding or losing CO_2, without changing alkalinity. For alkalinity to remain constant when total carbonate changes, the pH must also change. Changes caused by adding bicarbonate or from simple dilution are indicated in the figure.

Figure 3.4 is a total acidity/pH/total carbonate graph. Note that changes in composition, caused by adding or removing carbon dioxide and carbonate, are indicated by different movement vectors in the acidity and alkalinity graphs. The examples below illustrate the uses of the diagrams.

Example 3.4

Designers of a wastewater treatment facility for a meat rendering plant planned to control ammonia concentrations in the wastewater by raising its pH to 11, in order to convert about 90% of the ammonia to the volatile form. The wastewater would then be passed through an air-stripping tower to transfer the ammonia to the atmosphere. Average initial conditions for alkalinity and pH in the wastewater were expected to be about 0.5 meq/L and 6.0, respectively.

In the preliminary design plan, four options for increasing the pH were considered:

1. Raise the pH by adding NaOH, a strong base.
2. Raise the pH by adding calcium carbonate, $CaCO_3$, in the form of limestone.
3. Raise the pH by adding sodium bicarbonate, $NaHCO_3$.
4. Raise the pH by removing CO_2, perhaps by aeration.

Addition of NaOH

In Figure 3.3, we find that the intersection of pH = 6.0 and alkalinity = 0.5 meq/L occurs at total carbonate = 0.0015 mol/L, point A. Assuming that no CO_2 is lost to the atmosphere, addition of the strong base NaOH represents a vertical displacement upward from point A. Enough NaOH must be added to intersect with the pH = 11.0 contour at point B. In Figure 3.3, the vertical line between points A and B has a length of about 3.3 meq/L. Thus, The quantity of NaOH needed to change the pH from 6.0 to 11.0 is 3.3 meq/L (132 mg/L).

Addition of CaCO₃

Addition of $CaCO_3$ is represented by a line of slope +2 from point A. The carbonate addition line rises by 2 meq/L of alkalinity for each increase of 1 mol/L of total carbonate (because one mole of carbonate = 2 equivalents). Notice in Figure 3.3 that the slope of the pH = 11.0 contour is very nearly 2. The $CaCO_3$ addition vector and the pH = 11.0 contour are nearly parallel. Therefore, a very large quantity of $CaCO_3$ would be needed, making this method impractical.

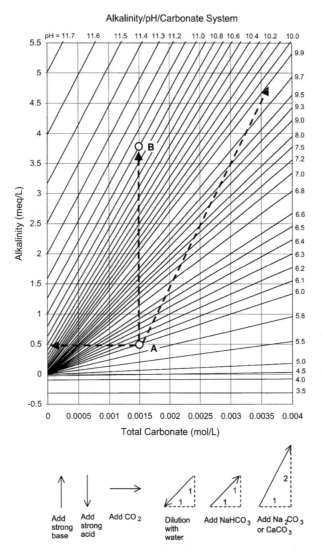

FIGURE 3.3   Total alkalinity-pH-total carbonate diagram (Deffeyes diagram): In this figure, the relationships among total alkalinity, pH, and total carbonate are shown. If any two of these quantities are known, the third may be determined from the plot. The composition changes indicated in the figure refer to Example 3.4.

Addition of NaHCO₃

Addition of $NaHCO_3$ is represented by a line of slope +1 from point A. Although this vector is not shown in Figure 3.3, it is evident it cannot cross the pH = 11 contour. Therefore, this method will not work.

Removing CO₂

Removal of CO_2 is represented by a horizontal displacement to the left. Loss or gain of CO_2 does not affect the alkalinity. Note that if CO_2 is removed, total carbonate is decreased correspondingly. However, pH and [OH⁻] also increase correspondingly, resulting in no net change in alkalinity. We see from Figure 3.3 that removal of CO_2 to the point of zero total carbonate cannot achieve pH = 11.0. Therefore, this method also will not work.

Of the four potential methods considered for raising the wastewater pH to 11.0, only addition of NaOH is useful.

Example 3.5

A large excavation at an abandoned mine site has filled with water. Because pyrite minerals were exposed in the pit, the water is acidic with pH = 3.2. The acidity was measured at 3.5 meq/L. Because the pit overflows into a stream during heavy rains, managers of the site must meet the conditions of a discharge permit, which include a requirement that pH of the overflow water be between 6.0 and 9.0. The site managers decide to treat the water to pH = 7.0 to provide a safety margin. Use Figure 3.4 to evaluate the same options for raising the pH as were considered in Example 3.4.

Addition of NaOH

In Figure 3.4, we find that the intersection of pH = 3.2 and acidity = 3.5 meq/L occurs at about total carbonate = 0.0014 mol/L, point A. Assuming that no CO_2 is lost to the atmosphere, addition of the strong base NaOH represents a vertical displacement downward from point A to point C. Enough NaOH must be added to intersect with the pH = 7.0 contour. The vertical line between points A and C has a length of about 1.8 meq/L. Thus, The quantity of NaOH needed to change the pH from 3.0 to 7.0 is 1.8 meq/L (72 mg/L).

Addition of $CaCO_3$

In the acidity diagram, addition of $CaCO_3$ is represented by a horizontal line to the right. In Figure 3.4, the $CaCO_3$ addition line intersects the pH = 7.0 contour at point B, where total carbonate = 0.0030 mol/L. Therefore, the quantity of $CaCO_3$ required to reach pH = 7.0 is 0.0030 – 0.0014 = 0.0016 mol/L (160 mg/L).

Addition of $NaHCO_3$

The addition of $NaHCO_3$ is represented by a line of slope +1 (the vector upward to the right from point A in Figure 3.4). Notice that the slope of the pH = 7.0 contour is just a little greater than +1. The $NaHCO_3$ addition vector and the pH = 7.0 contour are nearly parallel. Therefore, a very large quantity of $NaHCO_3$ would be needed, making this method impractical.

Removing CO_2

In the acidity diagram, the removal of CO_2 is represented by a line downward to the left with slope 2. We see from Figure 3.4 that removal of CO_2 to the point of zero total carbonate cannot achieve pH = 7.0. Therefore, this method will not work.

Of the four potential methods considered for raising the wastewater pH to 7.0, addition of either NaOH or $CaCO_3$ will work. The choice will be based on other considerations, such as costs or availability.

3.6   HARDNESS

BACKGROUND

Originally, water hardness was a measure of the ability of water to precipitate soap. It was measured by the amount of soap needed for adequate lathering and served also as an indicator of the rate of scale formation in hot water heaters and boilers. Soap is precipitated as a gray "bathtub ring" deposit mainly by reacting with the calcium and magnesium cations (Ca^{2+} and Mg^{2+}) present, although other polyvalent cations may play a minor role.

Hardness has some similarities to alkalinity. Like alkalinity, it is a water property that is not attributable to a single constituent and, therefore, some convention must be adopted to express hardness quantitatively as a concentration. As with alkalinity, hardness is usually expressed as an equivalent concentration of $CaCO_3$. However, hardness is a property of cations (Ca^{2+} and Mg^{2+}), while alkalinity is a property of anions (HCO_3^- and CO_3^{2-}).

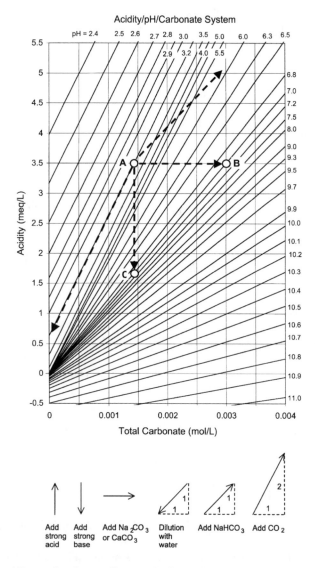

FIGURE 3.4 Total acidity-total carbonate diagram (Deffeyes diagram): In this figure, the relationships among total acidity, pH, and total carbonate are shown. If any two of these quantities are known, the third may be determined from the plot. The composition changes indicated in the figure refer to Example 3.5.

CALCULATING HARDNESS

Current practice is to define total hardness as the sum of the calcium and magnesium ion concentrations in mg/L, both expressed as calcium carbonate. Hardness usually is calculated from separate measurements of calcium and magnesium, rather than measured directly by colorimetric titration.

Calcium and magnesium ion concentrations are converted to equivalent concentrations of $CaCO_3$ as follows:

1. Find the equivalent weights of Ca^{2+}, Mg^{2+}, and $CaCO_3$.

$$\text{eq. wt.} = \frac{\text{molecular or atomic weight}}{\text{magnitude of ionic charge or oxidation number}}.$$

$$\text{eq. wt. of } Ca^{2+} = \frac{40.08}{2} = 20.04.$$

$$\text{eq. wt. of } Mg^{2+} = \frac{24.31}{2} = 12.15.$$

$$\text{eq. wt. of } CaCO_3 = \frac{100.09}{2} = 50.04.$$

2. Determine the multiplying factors to obtain the equivalent concentration of $CaCO_3$.

$$\text{Multiplying factor of } Ca^{2+} \text{ as } CaCO_3 = \frac{\text{eq. wt. of } CaCO_3}{\text{eq. wt. of } Ca^{2+}} = \frac{50.04}{20.04} = 2.497.$$

$$\text{Multiplying factor of } Mg^{2+} \text{ as } CaCO_3 = \frac{\text{eq. wt. of } CaCO_3}{\text{eq. wt. of } Mg^{2+}} = \frac{50.04}{12.15} = 4.118.$$

3. Calculate the total hardness.

$$\text{Total hardness (as } CaCO_3) = 2.497 \,[Ca^{2+}, \text{mg/L}] + 4.118 \,[Mg^{2+}, \text{mg/L}]. \tag{3.14}$$

Equation 3.14 may be used to calculate hardness whenever Ca^{2+} and Mg^{2+} concentrations are known.

Example 3.6

Calculate the total hardness as $CaCO_3$ of a water sample in which:

$$Ca^{2+} = 98 \text{ mg/L and } Mg^{2+} = 22 \text{ mg/L}.$$

Answer: From Equation 3.14,

$$\text{Total hardness} = 2.497 \,[98 \text{ mg/L}] + 4.118 \,[22 \text{ mg/L}] = 335 \text{ mg/L } CaCO_3.$$

Both alkalinity and hardness are expressed in terms of an equivalent concentration of calcium carbonate. As noted before, alkalinity results from reactions of the anions, CO_3^{2-} and HCO_3^{-}, whereas hardness results from reactions of the cations, Ca^{2+} and Mg^{2+}. It is possible for hardness as $CaCO_3$ to exceed the total alkalinity as $CaCO_3$. When this occurs, the portion of the hardness that is equal to the alkalinity is referred to as *carbonate hardness* or *temporary hardness*, and the amount in excess of alkalinity is referred to as *noncarbonate hardness* or *permanent hardness*.

IMPORTANCE OF HARDNESS

Hardness is sometimes useful as an indicator proportionate to the total dissolved solids present, since Ca^{2+}, Mg^{2+}, and HCO_3^{-} often represent the largest part of the total dissolved solids. No human health effects due to hardness have been proven; however, an inverse relation with cardiovascular disease has been reported. Higher levels of drinking water hardness correlates with lower incidence of cardiovascular disease. High levels of water hardness may limit the growth of fish; on the other hand, low hardness (soft water) may increase fish sensitivity to toxic metals. In general, higher hardness is beneficial by reducing metal toxicity to fish. Aquatic life water quality standards for many metals are calculated by using an equation that includes water hardness as a variable.

The main advantages in limiting hardness levels (by softening water) are economical: less soap requirements in domestic and industrial cleaning, and less scale formation in pipes and boilers. Water treatment by reverse osmosis (RO) often requires a water softening pretreatment to prevent scale formation on RO membranes. Increased use of detergents, which do not form precipitates with Ca^{2+} and Mg^{2+}, has lessened the importance of hardness for soap consumption. On the other hand, a drawback to soft water is that it is more "corrosive" or "aggressive" than hard water. In this context, "corrosive" means that soft water more readily dissolves metal ions from a plumbing system than does hard water. Thus, in plumbing systems where brass, copper, galvanized iron, or lead solders are present, a soft water system will carry higher levels of dissolved copper, zinc, lead, and iron, than will a hard water system.

Rules of Thumb

1. The higher the hardness, the more tolerant are many stream metal standards for aquatic life.
2. Hardness above 100 mg/L can cause significant scale deposits to form in boilers.
3. The softer the water, the greater the tendency to dissolve metals from the pipes of water distribution systems.
4. An ideal quality goal for total hardness is about 70–90 mg/L. Municipal treatment sometimes allows up to 150 mg/L of total hardness in order to reduce chemical costs and sludge production from precipitation of Ca^{2+} and Mg^{2+}.

Water will be "hard" wherever groundwater passes through calcium and magnesium carbonate mineral deposits. Such deposits are very widespread and hard to moderately hard groundwater is more common than soft groundwater. Very hard groundwater occurs frequently. Calcium and magnesium carbonates are the most common carbonate minerals and are the main sources of hard water. A geologic map showing the distribution of carbonate minerals serves also as an approximate map of the distribution of hard groundwater. The most common sources of soft water are where rain water is used directly, or where surface waters are fed more by precipitation than by groundwater.

Rules of Thumb

Degree of Hardness	mg $CaCO_3$/L	Effects
Soft	<75	May increase toxicity of dissolved metals. No scale deposits. Efficient use of soap.
Moderately Hard	75 – 120	Not objectionable for most purposes Requires somewhat more soap for cleaning. Above 100 mg/L will deposit significant scale in boilers.
Hard	120 – 200	Considerable scale buildup and staining. Generally softened if >200 mg/L.
Very Hard	>200	Requires softening for household or commercial use.

In industrial usage, hardness is sometimes expressed as *grains/gallon* or *gpg*. The conversion between gpg and mg/L is shown in Figure 3.5.

3.7 DISSOLVED OXYGEN (DO)

BACKGROUND

Sufficient *dissolved oxygen (DO)* is crucial for fish and many other aquatic life forms. DO is important for high quality water. It oxidizes many sources of objectionable tastes and odors. Oxygen

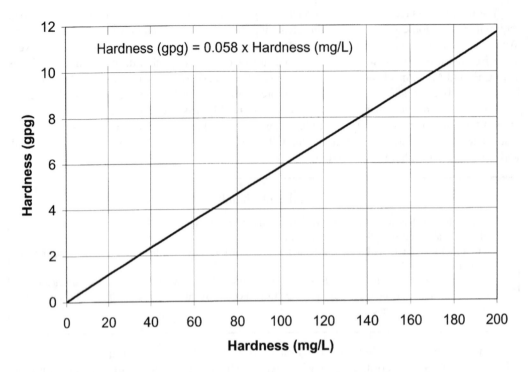

FIGURE 3.5   Relation between hardness expressed as mg/L and grains per gallon (gpg).

becomes dissolved in surface waters by diffusion from the atmosphere and from aquatic-plant photosynthesis.

On average, most oxygen dissolves into water from the atmosphere; only a little net DO is produced by aquatic-plant photosynthesis. Although water plants produce oxygen during the day, they consume oxygen at night as an energy source. When they die and decay, plants serve as energy sources for microbes which consume additional oxygen. The net change in DO is small during the life cycle of aquatic plants.

Rules of Thumb

1. The solubility of oxygen in water *decreases* as the water temperature *increases*.
2. Saturation concentration of O_2 in water at sea level = 14.7 mg/L (ppm) at 0°C, 8.3 mg/L (ppm) at 25°C, 7.0 mg/L (ppm) at 35°C.

Dissolved oxygen is consumed by the degradation (oxidation) of organic matter in water. Because the concentration of dissolved oxygen is never very large, oxygen-depleting processes can rapidly reduce it to near zero in the absence of efficient aeration mechanisms. Fish need at least 5–6 ppm DO to grow and thrive. They stop feeding if the level drops to around 3–4 ppm and die if DO falls to 1 ppm. Many fish kills are not caused by the direct toxicity of contaminants but instead by a deficiency of oxygen caused by the biodegradation of contaminants.

Typical state aquatic life standards for DO are

- 7.0 ppm for cold water spawning periods,
- 6.0 ppm for class 1 cold water biota, and
- 5.0 ppm for class 1 warm water biota.

TABLE 3.2
Dissolved Oxygen and Water Quality

Water Quality	Dissolved Oxygen (mg/L)
Good	Above 8.0
Slightly polluted	6.5 – 8.0
Moderately polluted	4.5 –6.5
Heavily polluted	4.0 – 4.5
Severely polluted	Below 4.0

3.8 BIOLOGICAL OXYGEN DEMAND (BOD) AND CHEMICAL OXYGEN DEMAND (COD)

BACKGROUND

Biological oxygen demand (BOD) refers to the amount of oxygen consumed when the organic matter in a given volume of water is biodegraded. BOD is an indicator of the *potential* for a water body to become depleted in oxygen and possibly become anaerobic because of biodegradation. BOD measurements do not take into account re-oxygenation of water by naturally occurring diffusion from the atmosphere or mechanical aeration. Water with a high BOD and a microbial population can become depleted in oxygen and may not support aquatic life, unless there is a means for rapidly replenishing dissolved oxygen.

Chemical oxygen demand (COD) refers to the amount of oxygen consumed when the organic matter in a given volume of water is chemically oxidized to CO_2 and H_2O by a strong chemical oxidant, such as permanganate or dichromate. COD is sometimes used as a measure of general pollution. For example, in an industrial area built on fill dirt, COD in the groundwater might be used as an indicator of organic materials leached from the fill material. Leachate from landfills often has high levels of COD.

BOD is a subset of COD. The COD analysis oxidizes organic matter that is both chemically and biologically oxidizable. If a reliable correlation between COD and BOD can be established at a particular site, the simpler COD test may be used in place of the more complicated BOD analysis.

BOD$_5$

BOD$_5$ refers to a particular empirical test, accepted as a standard, in which a specified volume of sample water is seeded with bacteria and nutrients (nitrogen and phosphorus) and then incubated for 5 days at 20°C in the dark. BOD$_5$ is measured as the *decrease* in dissolved oxygen (in mg/L) after 5 days of incubation. The BOD$_5$ test originated in England, where any river contaminant not decomposed within 5 days will have reached the ocean.

Water surface turbulence helps to dissolve oxygen from the atmosphere by increasing the water surface area. A BOD$_5$ of 5 mg/L in a slow-moving stream might be enough to produce anaerobic conditions, while a turbulent mountain stream might be able to assimilate a BOD$_5$ of 50 mg/L without appreciable oxygen depletion.

BOD CALCULATION

Example 3.7

When a liter water sample is collected for analysis, an insect weighing 0.1g is accidentally trapped in the bottle. The initial DO is 10 mg/L. Assume that 10% of the insect's fresh weight is readily

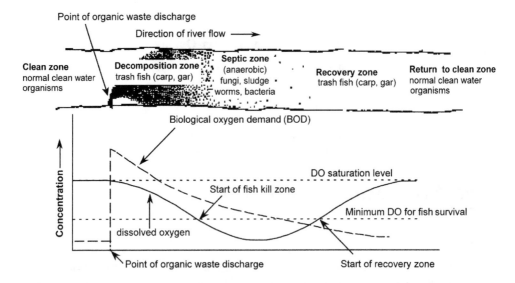

FIGURE 3.6   Dissolved oxygen sag curve caused by discharge of organic wastes into a river.

biodegradable and has the approximate unit formula CH_2O.* Also, assume that microbes that will metabolize the insect are present. If the laboratory does not analyze the sample until biodegradation is complete, what DO will they measure?

Answer: The chemical reaction for oxidation of organic matter is

$$CH_2O + O_2 \rightarrow CO_2 + H_2O.$$

This equation shows that one mole of O_2 oxidizes one mole of CH_2O. Therefore, the moles of CH_2O in the insect will equal the moles of O_2 consumed during biodegradation. Find the moles of O_2 initially present and the moles of CH_2O in the insect.

Molecular weight of $CH_2O = 12 + 2 + 16 = 30$.

$$\text{Moles } O_2 \text{ initially present} = \frac{10 \times 10^{-3} \text{ g/L}}{32 \text{ g/mol}} = 3.1 \times 10^{-4} \text{ mol/L}.$$

$$\text{Moles } CH_2O \text{ in roach} = \text{moles of } O_2 \text{ consumed} = \frac{0.01 \text{ g}}{30 \text{ g/mol}} = 3.3 \times 10^{-4} \text{ mol}.$$

Biodegradation of the insect will consume all of the DO present and there will be about 0.2×10^{-4} mol of insect tissue left undegraded or $(0.2 \times 10^{-4} \text{ mol})(30 \text{ g/mol}) = 0.6$ mg insect tissue left over. The laboratory will find the water anaerobic.

---

* Organic biomass contains carbon, hydrogen, and oxygen atoms in approximately the ratio of 1:2:1, so that CH_2O serves as a convenient unit molecule of organic matter.

COD CALCULATION

Example 3.8

COD levels of 60 mg/L were measured in groundwater. It is suspected that fuel contamination is the main cause. What concentration of hydrocarbons from fuel is necessary to account for all of the COD observed?

Answer: For simplicity, assume fuel hydrocarbons to have an average unit formula of CH_2. The oxidation reaction is

$$CH_2 + 1.5\ O_2 \rightarrow CO_2 + H_2O.$$

For each carbon atom in the fuel, 1.5 oxygen molecules are consumed.

Weight of 1 mole of unit fuel = 12 + 2 = 14 g.

Weight of 1.5 mole of O_2 = 1.5 × 32 = 48 g.

Weight ratio of oxygen to fuel is: $\dfrac{48}{14}$ = 3.4.

A COD of 60 mg/L requires: $\dfrac{60}{3.4}$ = *18 mg/L* fuel hydrocarbons.

If dissolved fuel hydrocarbons in the groundwater are 18 mg/L or greater, the fuel alone could account for all the measured COD. If dissolved fuel hydrocarbons in the groundwater are less than 18 mg/L, then fuels could account for only a part of the COD; other organic substances, such as pesticides, fertilizers, solvents, PCBs, etc., must account for the rest.

3.9 NITROGEN: AMMONIA (NH_3), NITRITE (NO_2^-), AND NITRATE (NO_3^-)

BACKGROUND

Nitrogen compounds of greatest interest to water quality are those that are biologically available as nutrients to plants or exhibit toxicity to humans or aquatic life. Atmospheric nitrogen (N_2) is the primary source of all nitrogen species, but it is not directly available to plants because the N≡N triple bond is too strong to be broken by photosynthesis. Atmospheric nitrogen must be converted to other nitrogen compounds before it can become available as a plant nutrient.

The conversion of atmospheric nitrogen to other chemical forms is called *nitrogen fixation* and is accomplished by a few types of bacteria that are present in water, soil, and root nodules of alfalfa, clover, peas, beans, and other legumes. Atmospheric lightning is another significant source of fixed nitrogen because the high temperatures generated in lightning strikes are sufficient to break N_2 and O_2 bonds making possible the formation of nitrogen oxides. Nitrogen oxides created within lightning bolts are dissolved in rainwater and absorbed by plant roots, thus entering the nitrogen nutrient sub-cycles, (see Figure 3.5). The rate at which atmospheric nitrogen can enter the nitrogen cycle by natural processes is too low to support today's intensive agricultural production. The shortage of fixed nitrogen must be made up with fertilizers containing nitrogen fixed by industrial processes, which are dependent on petroleum fuel. Modern large-scale farming has been called a method for converting petroleum into food.

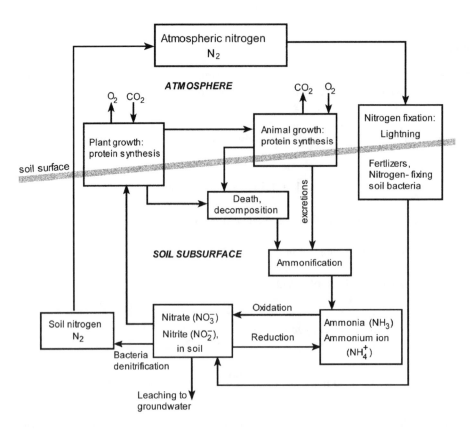

FIGURE 3.7   Nitrogen cycle.

THE NITROGEN CYCLE

As illustrated in Figure 3.7, in the nitrogen cycle, plants take up ammonia and nitrogen oxides dissolved in soil pore water and convert them into proteins, DNA, and other nitrogen compounds. Animals get their nitrogen by eating plants or other plant-eating animals. Once in terrestrial ecosystems, nitrogen is recycled through repeated biological birth, growth, death, and decay steps. There is a continual and relatively small loss of fixed nitrogen when specialized soil bacteria convert fixed nitrogen back into nitrogen gas (denitrification), which then is released to the atmosphere until it can reenter the nutrient sub-cycles again.

When nitrogen is circulating in the nutrient sub-cycles, it undergoes a series of reversible oxidation-reduction reactions that convert it from nitrogenous organic molecules, such as proteins, to ammonia (NH_3), nitrite (NO_2^-), and nitrate (NO_3^-). Ammonia is the first product in the oxidative decay of nitrogenous organic compounds. Further oxidation leads to nitrite and then to nitrate. Ammonia is naturally present in most surface and wastewaters. Its further degradation to nitrites and nitrates consumes dissolved oxygen.

$$\text{Organic N} \longrightarrow NH_3 \xrightarrow{\;O_2\;} NO_2^- \xrightarrow{\;O_2\;} NO_3^-. \tag{3.15}$$

AMMONIA/AMMONIUM ION (NH_3/NH_4^+)

In water, ammonia reacts as a base, raising the pH by generating OH^- ions, as in Equation 3.16.

$$NH_3 + H_2O \leftrightarrow NH_4^+ + OH^-. \tag{3.16}$$

The equilibrium of Equation 3.16 depends on pH and temperature, (see Figure 3.6). In a laboratory analysis, total ammonia ($NH_3 + NH_4^+$) is measured and the distribution between unionized ammonia (NH_3) and ionized ammonia (NH_4^+) is calculated from knowledge of the water pH and temperature at the sampling site. Since the unionized form is far more toxic to aquatic life than the ionized form, field measurements of water pH and temperature at the sampling site are very important. The two forms of ammonia have different mobilities in the environment. Ionized ammonia is strongly adsorbed on mineral surfaces where it is effectively immobilized. In contrast, unionized ammonia is only weakly adsorbed and is transported readily by water movement. If a suspended sediment carrying sorbed NH_4^+ is carried by a stream into a zone with a higher pH, a portion will be converted to unionized NH_3, which can then desorb and become available to aquatic life forms as a toxic pollutant. Unionized ammonia is also volatile and a fraction of it is transported as a gas.

As discussed above, nitrogen passes through several different chemical forms in the nutrient sub-cycle. In order to allow quantities of these different forms to be directly compared with one another, analytical results often report their concentrations in terms of their nitrogen content. For example, 10.0 mg/L of unionized ammonia may be reported as 8.23 mg/L NH_3–N (ammonia nitrogen),* or 10.0 mg/L of nitrate reported as 2.26 mg/L NO_3–N (nitrate nitrogen).**

Rules of Thumb

1. Ammonia toxicity increases with pH and temperature.
2. At 20°C and pH > 9.4, the equilibrium of Equation 3.16 is to the left, favoring NH_3, the toxic form.
3. At 20°C and pH < 9.4, the equilibrium of Equation 3.16 is to the right, favoring NH_4^+, the nontoxic form.
4. A temperature increase shifts the equilibrium to the left, favoring the NH_3 form.
5. NH_3 concentrations >0.5 mg NH_3–N/L cause significant toxicity to fish.
6. The unionized form is volatile, or air-strippable. The ionized form is nonvolatile.

Changes in environmental conditions can cause an initially acceptable concentration of total ammonia to become unacceptable and in violation of a stream standard. For example, consider the case of a wastewater treatment plant that discharges its effluent into a detention pond that, in turn, periodically releases its water into a stream. The treatment plant is meeting its discharge limit for unionized ammonia when its effluent is measured at the end of its discharge pipe. However, the detention pond is prone to support algal growth. In such a situation, it is not unusual for algae to grow to a level that influences the pond's pH. During daytime photosynthesis, algae may remove enough dissolved carbon dioxide from the pond to raise the pH and shift the equilibrium of Equation 3.16 to the left, far enough that the pond concentration of NH_3 becomes higher than the discharge permit limit. In this case, discharges from the pond could exceed the stream standard for unionized ammonia even though the total ammonia concentration is unchanged.

Example 3.9

Ammonia is removed from an industrial wastewater stream by an air-stripping tower. To meet the effluent discharge limit of 5-ppm ammonia, the influent must be adjusted so that 60% of the total ammonia is in the volatile form. To what pH must the influent be adjusted if the wastewater in the stripping tower is at 10°C? Use Figure 3.8.

---

* Calculated as follows: $\dfrac{\text{atomic wt. of N}}{\text{molecular wt. of } NH_3} \times \text{conc. of } NH_3 = \dfrac{14}{17} \times 10 \text{ mg/L} = 8.23 \text{ mg/L } NH_3\text{–N}.$

** Calculated as follows: $\dfrac{\text{atomic wt. of N}}{\text{molecular wt. of } NO_3} \times \text{conc. of } NH_3 = \dfrac{14}{62} \times 10 \text{ mg/L} = 2.26 \text{ mg/L } NO_3\text{–N}.$

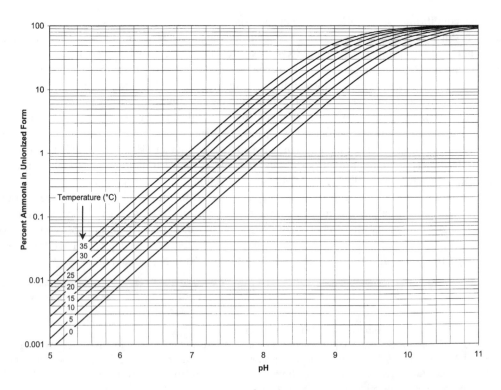

FIGURE 3.8 Percent unionized ammonia (NH_3) as a function of pH and temperature.

Answer: In Figure 3.8, the 60% unionized ammonia gridline crosses the 10°C curve between pH = 9.7 and 9.8. Thus, the influent must be adjusted to pH = 9.8 or higher to meet the discharge limit.

CRITERIA AND STANDARDS FOR AMMONIA

Typical state standards for unionized ammonia (NH_3) are

- Aquatic life: Cold water biota = 0.02 mg/L NH_3–N, chronic; warm water biota = 0.06 mg/L NH_3–N, chronic; acute standard calculated from temperature and pH.
- Domestic water supply: 0.05 mg N/L total ($NH_3 + NH_4^+$), for a 30-day average.

NITRITE (NO_2^-) AND NITRATE (NO_3^-)

Ammonia and other nitrogenous materials in natural waters tend to be oxidized by aerobic bacteria, first to nitrite and then to nitrate. Therefore, all organic compounds containing nitrogen should be considered as potential nitrate sources. Organic nitrogen compounds enter the environment from wild animal and fish excretions, human sewage, and livestock manure. Inorganic nitrates come primarily from manufactured fertilizers containing ammonium nitrate and potassium nitrate.

In oxygenated waters, nitrite is rapidly oxidized to nitrate, so normally there is little nitrite present in surface waters. Manufactured fertilizers are another source of nitrate and ammonia. Both nitrite and nitrate are important nutrients for plants, but they are toxic to fish and humans at high concentrations. Nitrates and nitrites are very soluble, do not adsorb readily to mineral and soil surfaces, and are very mobile in the environment. Consequently, where soil nitrate levels are high, contamination of groundwater by nitrate leaching is a serious problem. Unlike ammonia, nitrites and nitrates do not evaporate and remain in water until they are consumed by plants and microorganisms.

Rules of Thumb

1. Unpolluted surface waters normally contain only trace amounts of nitrite.
2. Measurable groundwater nitrite contamination is more common because of low oxygen concentrations in the soil's subsurface.
3. Nitrate and nitrite leach readily from soils to surface and groundwaters.
4. High concentrations (>1–2 mg/L) of nitrate or nitrite in surface or groundwater generally indicate agricultural contamination from fertilizers and manure seepage.
5. Greater than 10 mg/L of nitrite and nitrate in drinking water is a human health hazard.

Drinking water standards for nitrate are strict because the nitrate ion is reduced to nitrite ion in the saliva of all humans and in the intestinal tracts of infants during the first six months of life. Nitrite oxidizes iron in blood hemoglobin from Fe^{2+} to Fe^{3+}. The resulting compound, called methemoglobin, cannot carry oxygen. The resulting oxygen deficiency is called methemoglobinemia. It is especially dangerous in infants (blue baby syndrome) because of their small total blood volume.

CRITERIA AND STANDARDS FOR NITRATE

Typical state standards for nitrate (NO_3^-) are

- Agriculture MCLs: Nitrate = 100 mg/L NO_3–N; Nitrite = 10 mg/L NO_2–N (1-day average).
- Domestic water supply MCLs: Nitrate = 10 mg/L NO_3–N; Nitrite = 1.0 mg/L NO_2–N (1-day average).

Example 3.10

COD caused by sodium nitrite disposal
A chemical company wished to dispose of 250,000 gallons of water containing 500 mg/L of nitrite into a municipal sewer system. The manager of the municipal wastewater treatment plant had to determine whether this waste might be detrimental to the operation of his plant.

Under oxidizing conditions that exist in the treatment plant, nitrite is oxidized to nitrate as follows:

$$2\ NO_2^- + O_2 \rightarrow 2\ NO_3^-. \tag{3.17}$$

The consumption of oxygen shown in Equation 3.17 makes nitrite useful as a rust–inhibiting additive in boilers, heat exchangers, and storage tanks by deoxygenating the water. When nitrite is added to a wastewater stream, it is the same as adding chemical oxygen demand. More oxygen will be needed to maintain aerobic treatment steps at their optimum performance level. In addition, it will produce additional nitrate that may have to be denitrified before it can be discharged.

Calculation
For a wastewater stream of a total volume of 250,000 gallons that contains 500 mg/L of nitrite, the net weight of nitrite is

500 mg/L × 250,000 gal. × 3.79 L/gal. = 474 × 10^6 mg = 474,000 g of nitrite.

From Equation 3.17, stoichiometric consumption of oxygen is 1 mole (32 g) for each 2 moles (92 g) of nitrite oxidized, resulting in a 1 to 2.9 ratio of O_2 to NO_2^- by weight. Therefore, 474,000 g of nitrite will potentially consume 163,000 g of dissolved oxygen. Whether or not this represents a significant additional COD depends on the operating specifications of the treatment plant.

Also from Equation 3.17, stoichiometric production of nitrate is 2 moles (124 g) for each 2 moles (92 g) of nitrite oxidized, resulting in a 1.3 to 1 ratio of NO_3^- to NO_2^- by weight. Therefore, 474,000 g of nitrite will produce 616,000 g (1,362 lbs) of nitrate that might have to be denitrified before release.

METHODS FOR REMOVING NITROGEN FROM WASTEWATER

After the activated-sludge treatment stage, municipal wastewater generally still contains some nitrogen in the forms of organic nitrogen and ammonia. Additional treatment is required to remove nitrogen from the waste stream.

Air-stripping Ammonia

See Example 3.9. Air-stripping can follow the activated-sludge process. pH must be raised with lime to about 10 or higher to convert all ammoniacal nitrogen to the volatile NH_3 form. Scaling, icing, and air pollution are some of the disadvantages of air-stripping; whereas an advantage is that raising the pH precipitates phosphorus in the form of calcium phosphate compounds.

Nitrification-denitrification

This is a two-step process:

1. Ammonia and organic nitrogen are first biologically oxidized completely to nitrate under strongly *aerobic* conditions *(nitrification)*. This is achieved by more than normal and extensive aeration of the sewage:

$$2\ NH_4^+ + 3\ O_2 \xrightarrow{\ Nitrosomas\ } 4\ H^+ + 2\ NO_2^- + 2\ H_2O.$$

$$2\ NO_2^- + O_2 \xrightarrow{\ Nitrobacter\ } 2\ NO_3^-.$$

2. Nitrate is then biologically converted to gaseous nitrogen under *anaerobic* conditions *(denitrification)*. This requires a carbon nutrient source. Water that is low in total organic carbon (TOC) may require the addition of methanol or other carbon source.

$$4\ NO_3^- + 5\ \{CH_2O\} + 4\ H^+ \xrightarrow{\ denitrifying\ bacteria\ } 2\ N_2(g) + 5\ CO_2(g) + 7\ H_2O.$$

Break-point Chlorination

The chemical reaction of ammonia with dissolved chlorine results in denitrification by converting ammonia to chloramines and nitrogen gas (see Chapter 6). With continued addition of Cl_2, nitrogen gas and a small amount of nitrate are formed. Any chloramine remaining serves as a weak disinfectant and is relatively nontoxic to aquatic life.

Ammonium Ion-exchange

This is a good alternative to air-stripping because an exchange resin, the natural zeolite *clinoptilolite*, has been developed and is selective for ammonia. NH_4^+ is exchanged for Na^+ or Ca^{2+} on the resin. The zeolite can be regenerated with sodium or calcium salts.

Biosynthesis

The removal of biomass, produced in the sewage treatment system by filtering to reduce suspended solids, results in a net loss of nitrogen that has been incorporated in the biomass cell structure.

3.10   SULFIDE (S²⁻)

BACKGROUND

Sulfide is often present naturally in groundwater as the dissolved anion S^{2-}, especially in natural hot springs. There, it arises from soluble sulfide minerals and anaerobic bioreduction of dissolved sulfates. Sulfide is also formed in surface waters from anaerobic decomposition of organic matter containing sulfur. It is a common product of wetlands and eutrophic lakes and ponds. Sulfide reacts with water to form hydrogen sulfide, H_2S, a colorless, highly toxic gas that smells like rotten eggs. The human nose is very sensitive to the odor of low levels of H_2S. The odor threshold for H_2S dissolved in water is 0.03 to 0.3 µg/L.

There are two important sources of H_2S in the environment: the anaerobic decomposition of organic matter containing sulfur, and the reduction of mineral sulfates and sulfites to sulfide. Both mechanisms require reducing, or anaerobic conditions, and are strongly accelerated by the presence of sulfur-reducing bacteria. H_2S is not formed in the presence of an abundant supply of oxygen.

Blackening of soils, wastewater, sludge, and sediments in locations with standing water, in addition to the odor of rotten eggs, is an indication that sulfide is present. The black material results from a reaction of H_2S with dissolved iron and other metals to form precipitated ferrous sulfide (FeS), along with other metal sulfides.

H_2S can have two stages of dissociation under reducing conditions in water, depending on the pH:

$$H_2S \leftrightarrow H^+ + HS^- \leftrightarrow 2\,H^+ + S^{2-}. \tag{3.18}$$

- At pH = 5, about 99% of dissolved sulfide is in the form of H_2S, the unionized form.
- At pH = 7, it is 50% HS^- and 50% H_2S.
- At pH = 9, about 99% is in the form of HS^-.
- S^{2-} becomes measurable only above pH = 12.

H_2S is the most toxic and volatile form; HS^- and S^{2-} are nonvolatile and much less toxic. $H_2S > 2.0$ µg/L constitutes a long-term hazard to fish.

Rules of Thumb

1. Well water smelling of H_2S is usually a sign of sulfate-reducing bacteria. Look for a water redox potential <–200 mV and a sulfate (SO_4^{2-}) concentration in groundwater >100 mg/L.
2. A typical concentration of H_2S in unpolluted surface water is <0.25 µg/L.
3. $H_2S > 2.0$ µg/L constitutes a chronic hazard to aquatic life.
4. In aerated water, H_2S is bio-oxidized to sulfates and elemental sulfur.
5. Unionized H_2S is volatile and air-strippable. The ionized forms, HS^- and S^{2-}, are nonvolatile.

Typical state standards for unionized H_2S are

- Aquatic life (cold and warm water biota): 2.0 µg/L (30–day).
- Domestic water supply: 0.05 µg/L (30–day).

3.11   PHOSPHORUS (P)

BACKGROUND

Phosphorus is a common element in igneous and sedimentary rocks and in sediments but it tends to be a minor element in natural waters because most inorganic phosphorus compounds have low solubility. Dissolved concentrations are generally in the range of 0.01–0.1 mg/L and seldom exceed 0.2 mg/L. The environmental behavior of phosphorus is largely governed by the low solubility of most of its inorganic compounds, its strong adsorption to soil particles, and its importance as a nutrient for biota.

Because of its low dissolved concentrations, phosphorus is usually the limiting nutrient in natural waters. The dissolved phosphorus concentration is often low enough to limit algal growth. Because phosphorus is essential to metabolism, it is always present in animal wastes and sewage. Too much phosphorus in wastewater effluent is frequently the main cause of algal blooms and other precursors of eutrophication.

IMPORTANT USES FOR PHOSPHORUS

Phosphorus compounds are used for corrosion control in water supply and industrial cooling water systems. Certain organic phosphorus compounds are used in insecticides. Perhaps the major commercial uses of phosphorus compounds are in fertilizers and in the production of synthetic detergents. Detergent formulations may contain large amounts of polyphosphates as "builders," to sequester metal ions and maintain alkaline conditions. The widespread use of detergents instead of soap has caused a sharp increase in available phosphorus in domestic wastewater.

Prior to the use of phosphate detergents, most wastewater inorganic phosphorus was contributed from human wastes; about 1.5 g/day per person is released in urine. As a consequence of detergent use, the concentration of phosphorus in treated municipal wastewaters has increased from 3–4 mg/L in pre-detergent days, to the present values of 10–20 mg/L. Since phosphorus is an essential element for the growth of algae and other aquatic organisms, rapid growth of aquatic plants can be a serious problem when effluents containing excessive phosphorus are discharged to the environment.

THE PHOSPHORUS CYCLE

In a manner similar to nitrogen, phosphorus in the environment is cycled between organic and inorganic forms. An important difference is that under certain soil conditions, some nitrogen is lost to the atmosphere by ammonia volatilization and microbial denitrification. There are no analogous gaseous loss mechanisms for phosphorus. Also important are the differences in mobility of the two nutrients. Both exist in anionic forms (NO_2^-/NO_3^- and $H_2PO_4^-/HPO_4^{2-}$) which are not subject to retention by cation exchange reactions. However, nitrate anions do not form insoluble compounds with metals and, therefore, readily leach from soil into surface and groundwaters. Phosphate anions are largely immobilized in the soil by the formation of insoluble compounds — chiefly iron, calcium, and aluminum phosphates — and by adsorption to soil particles. Nitrogen compounds leach more readily than phosphorus compounds from soils into ground and surface waters which contribute to a phosphorus-limited algal growth in most surface waters. The critical level of inorganic phosphorus for forming algal blooms can be as low as 0.01 to 0.005 mg/L under summer growing conditions but more frequently is around 0.05 mg/L.

Organic compounds containing phosphorus are found in all living matter. Orthophosphate (PO_4^{3-}) is the only form readily used as a nutrient by most plants and organisms. The two major steps of the phosphorus cycle, conversion of organic phosphorus to inorganic phosphorus and back to organic phosphorus, are both bacterially mediated. Conversion of insoluble forms of phosphorus, such as calcium phosphate, $Ca(HPO_4)_2$, into soluble forms, principally PO_4^{3-}, is also carried out by microorganisms. Organic phosphorus in tissues of dead plants and animals, and in animal waste products is converted bacterially to PO_4^{3-}. The PO_4^{3-} thus released to the environment is taken up again into plant and animal tissue.

MOBILITY IN THE ENVIRONMENT

Phosphorus is an important plant nutrient and is often present in fertilizers to augment the natural concentration in soils. Phosphorus is also a constituent of animal wastes. Runoff from agricultural

---

Rules of Thumb

1. In surface waters, phosphorus concentrations are influenced by the sediments, which serve as a reservoir for adsorbed and precipitated phosphorus. Sediments are an important part of the phosphorus cycle in streams. Bacterially mediated exchange between dissolved and sediment-adsorbed forms plays a role in making phosphorus available for algae and therefore contributes to eutrophication.
2. In streams, dissolved phosphorus from all sources, natural and anthropogenic, is generally present in low concentrations, around 0.1 mg/L or less.
3. The natural background of total dissolved phosphorus has been estimated to be about 0.025 mg–P/L; that of dissolved phosphates about 0.01 mg–P/L.
4. The solubility of phosphates increases at low pH and decreases at high pH.
5. Particulate phosphorus (sediment-adsorbed and insoluble compounds) is about 95% of the total phosphorus in most cases.
6. In carbonate soils, dissolved phosphorus can react with carbonate to form the mineral precipitate hydroxyapatite (calcium phosphate hydroxide), $Ca_{10}(PO_4)_6(OH)_2$.

---

areas is a major contributor to total phosphorus in surface waters, where it occurs mainly in sediments because of the low solubility of its inorganic compounds and its tendency to adsorb strongly to soil particles.

Dissolved phosphorus is removed from solution by

- Precipitation
- Strong adsorption to clay minerals and oxides of aluminum and iron
- Adsorption to organic components of soil

Reducing and anaerobic conditions, as in water-saturated soil, increase phosphorus mobility because insoluble ferric iron, to which phosphorus is strongly adsorbed, is reduced to soluble ferrous iron, thereby releasing adsorbed phosphorus. In acid soils, aluminum and iron phosphates precipitate, while in basic soils, calcium phosphates precipitate. The immobilization of phosphorus is therefore dependent on soil properties, such as pH, aeration, texture, cation-exchange capacity, the amount of calcium, aluminum and iron oxides present, and the uptake of phosphorus by plants.

Because of these removal mechanisms for dissolved phosphorus, phosphorus compounds resist leaching, and there is little movement of phosphorus with water drainage through most soils. It is mobilized mainly with erosion sediments. Phosphorus transport into surface waters is controlled chiefly by preventing soil erosion and controlling sediment transport. In most soils, except for those that are nearly all sand, almost all the phosphorus applied to the surface is retained in the top 1 to 2 ft. The adsorption capacity for phosphorus has been estimated for several soils to be in the range of 77 to over 900 lbs/acre-ft of soil profile. Often, the total phosphorus removal capacity for a soil will exceed the planning life of a typical land application project. If the phosphorus-removing capacity of a soil becomes saturated, it usually can be restored in a few months, during which adsorbed phosphorus is precipitated with metals or removed by crops.

Dissolved phosphate species exhibit the following pH-dependent equilibria (see Figure 3.9):

$$H_3PO_4 \leftrightarrow H_2PO_4^- + H^+ \leftrightarrow HPO_4^{2-} + 2\ H^+ \leftrightarrow PO_4^{3-} + 3\ H^+ \qquad (3.19)$$

- Below pH 2, H_3PO_4 is the dominant species.
- Between pH 2 and pH 7, $H_2PO_4^-$ is the dominant species.
- Between pH 7 and pH 12, HPO_4^{2-} is the dominant species.
- Above pH 12, PO_4^{3-} is the dominant species.

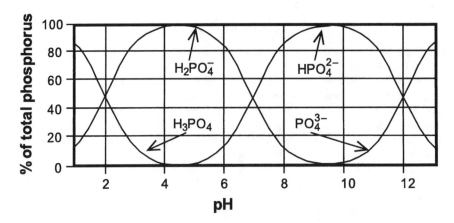

FIGURE 3.9   pH dependence of phosphate species.

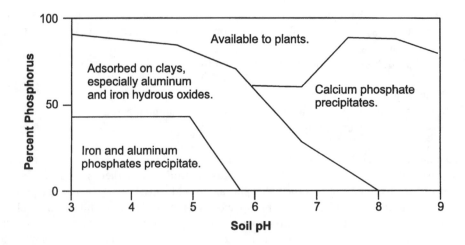

FIGURE 3.10   Forms of immobile phosphorus. General relationships between soil pH and phosphorus reactions are
- In the acid range, dissolved phosphorus is predominantly $H_2PO_4^-$, and immobile phosphorus is bound with iron and aluminum compounds.
- In the basic range, dissolved phosphorus is predominantly HPO_4^{2-}, and immobile phosphorus is mainly in the form of calcium phosphate.
- Maximum availability of phosphorus for plant uptake (as well as leaching) occurs between pH 6 and 7.

    Whole-lake experiments [D.W. Schindler, *Science*, 184, 897 (1974); 195, 260 (1977)] have demonstrated that, even when algal growth in lakes is temporarily limited by carbon or nitrogen instead of phosphorus, natural long-term mechanisms act to compensate for these deficiencies. Carbon deficiencies are corrected by CO_2 diffusion from the atmosphere, and nitrogen deficiencies are corrected by changes in biological growth mechanisms. Therefore, even if a sudden increase in phosphorus occurs temporarily causing algal growth to be limited by carbon or nitrogen, eventually these deficiencies are corrected. Then, algal growth becomes proportional to the phosphorus concentration as the system becomes once more phosphorus-limited.

PHOSPHORUS COMPOUNDS

Compounds containing phosphorus that are of interest to water quality include:

- Orthophosphates (all contain PO_4^{3-})
    Trisodium phosphate — Na_3PO_4
    Disodium phosphate — Na_2HPO_4
    Monosodium phosphate — NaH_2PO_4
    Diammonium phosphate — $(NH_4)_2HPO_4$.

Orthophosphates are soluble and are considered the only biologically available form. In the environment, hydrolysis slowly converts polyphosphates to orthophosphates. Analytical methods measure orthophosphate. To measure total phosphate, all forms of phosphate are chemically converted to orthophosphates (hydrated forms).

- Polyphosphates (also called *condensed phosphates*, meaning dehydrated)
    Sodium hexametaphosphate — $Na_3(PO_4)_6$
    Sodium tripolyphosphate — $Na_5P_3O_{10}$
    Tetrasodium pyrophosphate — $Na_4P_2O_7$
    Organic phosphate (biodegradation or oxidation of organic phosphates releases orthophosphates).

Rules of Thumb

1. The critical level of inorganic phosphorus for algae bloom formation can be as low as 0.01 to 0.005 mg/L under summer growing conditions but more frequently is around 0.05 mg/L.
2. Lakes are *nitrogen-limited* if the ratio of total nitrogen to total phosphorus (N/P) is less than 13, *nutrient-balanced* if 13 < N/P < 21, and *phosphorus-limited* if N/P > 21. Exact ranges depend on the particular algae species. Most lakes are phosphorus limited; in other words, additional phosphorus is needed to sustain further algal growth.
3. Different N/P ratios and pH values favor the growth of different kinds of algae.
4. Low N/P ratios favor N-fixing blue-green algae.
5. High N/P ratios, often achieved by controlling phosphorus input by means of additional wastewater treatment, cause a shift from blue-green algae to less objectionable species.
6. Lower pH (or increased CO_2) gives green algae a competitive advantage over blue-green algae.

Sedimentary phosphorus occurs in the following forms:

- Phosphate minerals: Mainly hydroxyapatite, $Ca_5OH(PO_4)_3$.
- Nonoccluded phosphorus: Phosphate ions (usually orthophosphate) bound to the surface of SiO_2 or $CaCO_3$. Nonoccluded phosphorus is generally more soluble and more available than occluded phosphorus (below).
- Occluded phosphorus: Phosphate ions (usually orthophosphate) contained within the matrix structures of amorphous hydrated oxides of iron and aluminum and amorphous aluminosilicates. Occluded phosphorus is generally less available than nonoccluded phosphorus.
- Organic phosphorus: Phosphorus incorporated with aquatic biomass, usually algal or bacterial.

REMOVAL OF DISSOLVED PHOSPHATE

Current remedies for phosphate-caused foaming and eutrophication are

- Using lower phosphate formulas in detergents,
- Precipitating the phosphate with Fe^{3+}, Al^{3+}, or Ca^{2+}, and
- Diverting the discharge to a less sensitive location.

Phosphate removal is carried out in a manner similar to water softening by precipitation of Ca^{2+} and Mg^{2+}. The usual precipitants for removing phosphate are alum ($Al_2(SO_4)_3$), lime ($Ca(OH)_2$), and ferric chloride ($FeCl_3$). The choice of precipitant depends on the discharge requirements, wastewater pH, and chemical costs.

The pertinent reactions for the precipitation of phosphate with alum, ferric chloride, and lime are

$$\text{Alum: } Al_2(SO_4)_3 + 2\ HPO_4^{2-} \rightarrow 2\ AlPO_4(s) + 3\ SO_4^{2-} + 2\ H^+.$$

$$\text{Ferric chloride: } FeCl_3 + HPO_4^{2-} \rightarrow FePO_4(s) + 3\ Cl^- + H^+.$$

$$\text{Lime: } 3\ Ca^{2+} + 2\ OH^- + 2\ HPO_4^{2-} \rightarrow Ca_3(PO_4)_2(s) + 2\ H_2O.$$

$$4\ Ca^{2+} + 2\ OH^- + 3\ HPO_4^{2-} \rightarrow Ca_4H(PO_4)_3(s) + 2\ H_2O.$$

$$5\ Ca^{2+} + 4\ OH^- + 3\ HPO_4^{2-} \rightarrow Ca_5(OH)(PO_4)_3(s) + 3\ H_2O.$$

Where effluent concentrations of phosphorus up to 1.0 mg/L are acceptable, the use of iron or aluminum salts in a wastewater secondary treatment system is often the process of choice. If very low levels of effluent phosphorus are required, precipitation at high pH by lime in a tertiary unit is necessary. The lowest levels of phosphorus are achieved by adding NaF with lime to form $Ca_5(PO_4)_3F$ (fluorapatite). The operating pH for phosphate removal with lime is usually above 11 because flocculation is best in this range.

If alkalinity is present, aluminum and iron ions are consumed in the formation of metal-hydroxide flocs. This may increase required dosages by up to a factor of 3. Calcium ions react with alkalinity to form calcium carbonate. Thus, the amount of precipitant needed for phosphate precipitation is controlled more by the alkalinity than the stoichiometry of the reaction. In the case of aluminum and iron precipitants, the reaction with alkalinity is not totally wasted because the hydroxide flocs assist in the settling and removal of metal-phosphate precipitates, along with other suspended and colloidal solids in the wastewater.

Biological phosphorus removal can be accomplished by operating an activated sludge process in an anaerobic-aerobic sequence. A number of bacteria respond to this sequence by accumulating large excesses of polyphosphate within their cells in volutin granules. During the anaerobic phase, a release of phosphate occurs. In the aerobic phase, the released phosphate and an additional increment is taken up and stored as polyphosphate, giving a net removal, coincident with organic removal and metabolism. Phosphate can be removed from the waste stream as sludge or through use of a second anaerobic step. During the second anaerobic step, the stored phosphate is released in dissolved form. Then, the bacterial cells can be separated and recycled and the released soluble phosphate removed by precipitation.

3.12   METALS IN WATER

BACKGROUND

The commonly encountered elemental metals may be divided into three general classes:

1. Alkali metals: Li, Na, K (Periodic Table Group 1A).
2. Alkaline metals: Be, Mg, Ca, Sr, Ba (Periodic Table Group 2A).
3. Heavy metals: All metals to the right of the alkali and alkaline metals in the Periodic Table.

Metals in natural waters may be in dissolved or particulate forms.

Dissolved forms are

- Cations: Ca^{2+}, Fe^{2+}, K^+, Al^{3+}, Ag^+, etc.
- Complexes: $Zn(OH)_4^{2+}$, $Au(CN)_2^-$, $Ca(P_2O_7)^{2-}$, PuEDTA, etc.
- Organometallics: $Hg(CH_3)_2$, $B(C_2H_5)_3$, $Al(C_2H_5)_3$, etc.

Particulate forms are

- Mineral sediments
- Precipitated oxides, hydroxides, sulfides, carbonates, etc.
- Cations sorbed to sediments

A metal water quality standard may be written for the *dissolved, potentially dissolved, total recoverable,* or *total* form.

- *Dissolved:* Sample is filtered on site through a 0.45-micron filter, then acidified to pH 2 for preservation before analysis. Acidification prevents precipitation of any dissolved metal before analysis. This procedure omits from the analysis metals adsorbed on suspended sediments.
- *Potentially dissolved:* Sample is acidified to pH 2, held for 72 to 90 hours, then filtered through a 0.45-micron filter and analyzed. This procedure is intended to simulate the possibility that metals bound in suspended sediments might be transported into more acidic conditions and might partially dissolve. It measures the metals dissolved at the time of sampling, in addition to a portion of the metals bound to suspended sediments.
- *Total recoverable:* Sample is acidified to pH 2 and analyzed without filtering. This procedure measures all metals, dissolved and bound to suspended sediments.
- *Total:* Sample is "digested" in an acidic solution until essentially all the metals present are extracted into soluble forms for analysis.

GENERAL BEHAVIOR OF DISSOLVED METALS IN WATER

Because water molecules are polar, metal cations always attract a hydration shell of water molecules by electrostatic attraction to the positive charge of the cation, as illustrated in Figure 3.11.

$$M^{+n} \xrightarrow{\ H_2O\ } M(H_2O)_x^{+n}, \quad x = 6 \text{ for most cations.} \tag{3.20}$$

Hydrated metal ions behave as acids by donating protons (H^+) to H_2O molecules, forming the acidic H_3O^+ hydronium ion. The process can continue stepwise up to n times to make a neutral metal hydroxide:

$$M(H_2O)_6^{+n} + H_2O \leftrightarrow M(H_2O)_5OH^{+(n-1)} + H_3O^+. \tag{3.21}$$

$$M(H_2O)_5OH^{+(n-1)} + H_2O \leftrightarrow M(H_2O)_4(OH)_2^{+(n-2)} + H_3O^+, \quad \text{etc. up to } n \text{ times.} \tag{3.22}$$

For example, with Fe^{3+}, it takes 3 proton transfer steps to form neutral ferric hydroxide:

$$Fe(H_2O)_3^{3+} + H_2O \leftrightarrow Fe(H_2O)_2OH^{2+} + H_3O^+. \tag{3.23}$$

$$Fe(H_2O)_2OH^{2+} + H_2O \leftrightarrow Fe(H_2O)(OH)_2^+ + H_3O^+. \tag{3.24}$$

$$Fe(H_2O)(OH)_2^+ + H_2O \leftrightarrow Fe(OH)_3(s) + H_3O^+. \tag{3.25}$$

The overall reaction is

$$Fe(H_2O)_3^{3+} + 3\ H_2O \leftrightarrow Fe(OH)_3 + 3\ H_3O^+. \tag{3.26}$$

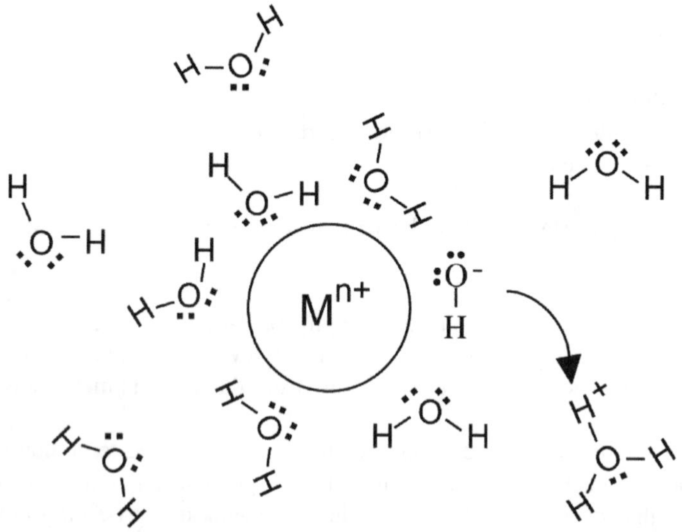

FIGURE 3.11 Water molecules form a hydration shell around dissolved metal cations. Molecules in the hydration shell can lose a proton to bulk water molecules, as indicated by the arrow, leaving a hydroxide group bonded to the metal. This way, the hydrated metal behaves as an acid. Eventually, the metal may precipitate as a hydroxide compound of low solubility.

With each step, the hydrated metal is progressively deprotonated, forming polyhydroxides and becoming increasingly insoluble. At the same time, the solution becomes increasingly acidic. Eventually, the metal precipitates as a low solubility hydroxide. The degree of acidity induced by metal hydration is greatest for cations of high charge and small size. All metal cations with a charge of +3 or more are moderately strong acids. This process is one source of acidic water draining from mines.

Rule of Thumb

Only polyvalent cations (e.g., Fe^{3+}, Zn^{2+}, Mn^{2+}, Cr^{3+}) attract water molecules strongly enough to act as acids by causing the release of H^+ from water molecules in the hydration sphere. Monovalent cations, such as Na^+, do not act as acids at all.

Lowering the pH (increasing H_3O^+) shifts the equilibrium of Equation 3.26 to the left, tending to dissolve any solid metal hydroxide that has precipitated. Raising the pH (adding OH^-) consumes H_3O^+ and shifts the equilibrium of Equation 3.26 to the right, precipitating an insoluble metal hydroxide. However, if the pH is raised too high, precipitated metal hydroxides can redissolve (see Figure 3.12). At high pH values, a metal hydroxide may form complexes with OH^- anions to become a negatively charged ion having increased solubility. For example, precipitated $Fe(OH)_3$ can react with OH^- anions as follows:

$$Fe(OH)_3 + OH^- \rightarrow Fe(OH)_4^-. \tag{3.27}$$

$$Fe(OH)_4^- + OH^- \rightarrow Fe(OH)_5^{2-}, \text{ etc.} \tag{3.28}$$

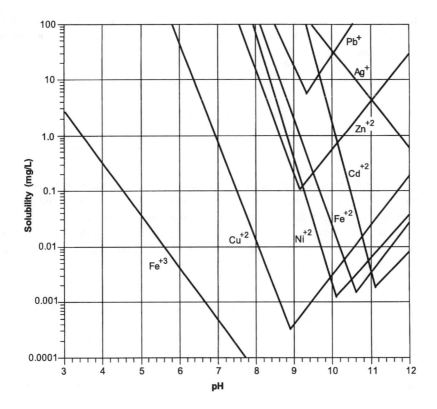

FIGURE 3.12 Solubilities of some metals precipitating as hydroxides vs. pH.

The negatively charged polyhydroxide anions are more soluble because their ionic charge attracts them strongly to polar water molecules. As shown in Figure 3.10, the high value of pH where solubility begins to increase again varies from metal to metal. Hard and alkaline water provides a buffer against pH changes. In hard or alkaline water, the tendency of metals to make water acidic is diminished.

Example 3.11: Effect of Dissolved Metal on Alkalinity

A sample of groundwater contains a high concentration of dissolved iron, about 20 mg/L. *At the laboratory,* alkalinity is measured to be 150 mg/L as $CaCO_3$. Does this laboratory measurement of alkalinity accurately represent the groundwater alkalinity?

Answer: Soluble inorganic iron is in the *ferrous* form, Fe^{2+}. When a groundwater sample is exposed to air, oxygen oxidizes Fe^{2+} to the *ferric* form, Fe^{3+}. This process is often enhanced by aerobic iron bacteria. Depending on the pH, hydrated Fe^{3+} can lose protons from its hydration sphere to any bases present, forming ferric hydroxide species and making the solution more acidic. Loss of protons from the hydration sphere is not a significant process for hydrated ferrous iron Fe^{2+}. The acidic behavior of hydrated Fe^{3+} occurs to a greater extent at a higher pH. Equation 3.29 represents the overall reaction converting dissolved ferrous iron to precipitated ferric hydroxide:

$$4\ Fe(H_2O)_6^{2+} + O_2 \leftrightarrow 4\ Fe(OH)_3(s) + 14\ H_2O + 8\ H^+. \tag{3.29}$$

$Fe(OH)_3$ is a yellow to red-brown precipitate often seen on rocks and sediments in surface waters with high iron concentrations.

The concentration of H^+ formed by Equation 3.29 can be up to 2 times the Fe^{2+} concentration, depending on the final pH. Each H^+ released will neutralize a molecule of base, consuming some alkalinity, by reactions such as

$$H^+ + OH^- \leftrightarrow H_2O. \tag{3.1}$$

$$H^+ + HCO_3^- \leftrightarrow H_2CO_3. \tag{3.30}$$

$$H^+ + CO_3^{2-} \leftrightarrow HCO_3^-. \tag{3.31}$$

We will assume the pH is high enough that the equilibrium of Equation 3.29 goes essentially to completion to the right side, a worst case scenario from the viewpoint of affecting the alkalinity. The atomic weights of hydrogen and iron are 1 g/mol and 56 g/mol, respectively. If the equilibrium of Equation 3.29 is completely to the right, one mole (56 g) of Fe^{3+} will make 2 moles of H^+ (2 g). At the time of sampling, the concentration of dissolved Fe^{2+} was about 20 mg/L and all is eventually oxidized to Fe^{3+}. The molar concentration of iron is

$$\frac{0.020 \text{ g/L}}{56 \text{ g/mol}} = 0.00036 \text{ mol/L, or } 0.36 \text{ mmol/L.}$$

The moles of H^+ produced are two times the moles of iron:

Moles of H^+ = 2×0.36 mmol/L = 0.72 mmol/L, or 1 mg/mmol $\times$ 0.72 mmol/L = 0.72 mg/L.

We must now determine what effect this quantity of H^+ will have on the alkalinity. Alkalinity is measured in terms of a comparable quantity of $CaCO_3$. The molecular weight of $CaCO_3$ is 100, and it dissolves to form the doubly charged ions Ca^{2+} and CO_3^{2-}. Alkalinity is a property of the CO_3^{2-} anion, which reacts to accept 2 H^+ cations:

$$2 H^+ + CO_3^{2-} \leftrightarrow H_2CO_3;$$

therefore, 0.72 mmol/L of H^+ will react with 0.072/2 = 0.36 mmol/L of CO_3^{2-}, and 0.36 mmol/L of $CaCO_3$ are required as a source of the CO_3^{2-}. From the definition of alkalinity, the change in alkalinity is equal to the change in concentration of $CaCO_3$, in mg/L.

$$0.36 \text{ mmol/L of } CaCO_3 = 0.36 \text{ mmol/L} \times \frac{100 \text{ mg}}{1 \text{ mmol}} = 36 \text{ mg/L} = \text{change in alkalinity}$$

Groundwater alkalinity at time of sampling =
Laboratory measured alkalinity + Alkalinity lost by Equation 3.29

The original alkalinity of the groundwater before exposure to air was

150 mg/L + 36 mg/L = *186 mg/L as CaCO₃.*

This example illustrates the pH buffering effect of alkalinity. The addition of H^+ to the solution by Equation 3.29 does not change the pH greatly as long as some alkalinity remains because the added H^+ is taken up by carbonate species in the water.

3.13 SOLIDS (TOTAL, SUSPENDED, AND DISSOLVED)

BACKGROUND

The general term *solids* refers to matter that is suspended (insoluble solids) or dissolved (soluble solids) in water. Solids can affect water quality in several ways. Drinking water with high dissolved solids may not taste good and may have a laxative effect. Boiler water with high dissolved solids requires pretreatment to prevent scale formation. Water high in suspended solids may harm aquatic life by causing abrasion damage, clogging fish gills, harming spawning beds, and reducing photosynthesis by blocking sunlight penetration, among other consequences. On the other hand, hard water (caused mainly by dissolved calcium and magnesium compounds) reduces the toxicity of metals to aquatic life.

Total solids (sometimes called *residue*) are the solids remaining after evaporating the water from an unfiltered sample. It includes two subclasses of solids that are separated by filtering (generally with a filter having a nominal 0.45-micron or smaller pore size):

1. *Total suspended solids (TSS,* sometimes called *filterable solids*) in water are organic and mineral particulate matter that do not pass through a filter. They may include silt, clay, metal oxides, sulfides, algae, bacteria, and fungi. TSS is generally removed by flocculation and filtering. TSS contributes to turbidity, which limits light penetration for photosynthesis and visibility in recreational waters.
2. *Total dissolved solids (TDS,* sometimes called *nonfilterable solids*) are substances that would pass through a 0.45 micron filter but will remain as residue when the water evaporates. They may include dissolved minerals and salts, humic acids, tannin, and pyrogens. TDS is removed by precipitation, ion-exchange, and reverse osmosis. In natural waters, the major contributors to TDS are carbonate, bicarbonate, chloride, sulfate, phosphate, and nitrate salts. Taste problems in water often arise from the presence of high TDS levels with certain metals present, particularly iron, copper, manganese, and zinc.

The difference between suspended and dissolved solids is a matter of definition based on the filtering procedure. Solids are always measured as the dry weight, and careful attention must be paid to the drying procedure to avoid errors caused by retained moisture or loss of material by volatilization or oxidation.

Rules of Thumb

1. TSS is detrimental to fish health by decreasing growth, disease resistance, and egg development.
2. Suspended solids should be restricted so they do not reduce the maximum depth of photosynthetic activity by more than 10% from the seasonally established norm.
3. Water with a TDS < 1200 mg/L generally has an acceptable taste. Higher TDS can adversely influence the taste of drinking water and may have a laxative effect.
4. In water to be treated for domestic potable supply, a TDS < 650 mg/L is a preferred goal.
5. For drinking water, recommended TDS is < 500 mg/L; the upper limit is 1000 mg/L.

TDS AND SALINITY

TDS and *salinity* both indicate dissolved salts. Table 3.1 offers a qualitative comparison between the terms.

TABLE 3.1
Comparison of TDS and Salinity

TDS	Degree of Salinity
1000 – 3000 mg/L	Slightly saline
3000 – 10,000 mg/L	Moderately saline
10,000 – 35,000 mg/L	Very saline
>35,000 mg/L	Briny

SPECIFIC CONDUCTIVITY AND TDS

Specific conductivity is directly related to TDS and serves as a check on TDS measurements.

Rules of Thumb

1. Conductivity units are μmhos/cm or μSiemens/cm:

$$1 \text{ μmhos/cm} = 1 \text{ μSiemens/cm (or μS/cm)}.$$

2. TDS in mg/L can be estimated from a measurement of specific conductivity.
3. For seawater (NaCl–based)

$$\text{TDS (mg/L)} \approx (0.5) \times (\text{Sp. Cond. in μS/cm}).$$

4. For groundwater (carbonate or sulfate-based)

$$\text{TDS (mg/L)} \approx (0.55 \text{ to } 0.7) \times (\text{Sp. Cond. in μS/cm}).$$

5. If the ratio $\dfrac{\text{TDS}_{meas}}{\text{Sp. Cond.}}$ is demonstrated to be consistent, the simpler specific conductivity measurement may sometimes be substituted for TDS analysis.

TDS TEST FOR ANALYTICAL RELIABILITY

A calculated value for TDS may be used for judging the reliability of a sample analysis if all the important ions have also been measured. The TDS concentration should be equal to the sum of the concentrations of all the ions present plus silica. You can use either of the following equations to calculate TDS from an analysis or to check on the validity of analytical results. All concentrations are in mg/L.

$$\text{TDS} = \text{sum of cations} + \text{sum of anions} + \text{silica}$$

or

$$\text{TDS} = 0.6(\text{alkalinity}) + Na^+ + K^+ + Ca^{2+} + Mg^{2+} + Cl^- + SO_4^{2-} + SiO_3.$$

In any given analysis, it is unlikely that all the ions have been measured. Frequently, only the major ions (Na^+, K^+, Ca^{2+}, Mg^{2+}, Cl^-, HCO_3^-, SO_4^{2-}) are necessary for the calculations, as other ion concentrations are likely to be insignificant by comparison.

Use the following guidelines for checking accuracy of a TDS analysis:

1. TDS_{meas} should always be equal to or somewhat larger than TDS_{calc} because a significant ion contributor might not have been included in the calculation.
2. An analysis is acceptable if the ratio of measured-to-calculated TDS is in the range

$$1.0 < \frac{\text{measured TDS}}{\text{calculated TDS}} < 1.2.$$

3. If $TDS_{meas} < TDS_{calc}$, the sample should be reanalyzed.
4. If $TDS_{meas} > 1.2 \times TDS_{calc}$, the sample should be reanalyzed, perhaps with a more complete set of ions.

3.14   TEMPERATURE

Temperature affects all water uses.

- The solubility of gases such as oxygen and carbon dioxide decreases as water temperature increases.
- Biodegradation of organic material in water and sediments is accelerated with increased temperatures, increasing the demand on dissolved oxygen.
- Fish and plant metabolism depends on temperature.

Most chemical equilibria are temperature dependent. Important environmental examples are the equilibria between ionized and unionized forms of ammonia, hydrogen cyanide, and hydrogen sulfide.

Temperature regulatory limits are set to maintain a normal pattern of diurnal and seasonal fluctuations, with no changes deleterious to aquatic life. Maximum induced change is limited to a 3°C increase over a 4-hour period, lasting for 12 hours maximum.

4 Soil, Groundwater, and Subsurface Contamination

CONTENTS

4.1   THE NATURE OF SOILS

Whereas water is always a potential conveyor of contaminants, soil can be either an obstacle to contaminant movement or a contaminant transporter. The stationary soil matrix slows the passage of groundwater and provides solid surfaces to which contaminants can sorb, delaying or stopping their movement. On the other hand, soil can also move, carried by wind, water flow, and construction equipment. Moving soil, like moving water, transports the contaminants it carries. Predicting and controlling pollutant behavior in the environment requires understanding how soil, water, and contaminants interact. That is the subject of this chapter.

SOIL FORMATION

Soil is the weathered and fragmented outer layer of the earth's solid surface, initially formed from the original rocks and then amended by growth and decay of plants and organisms. The initial step from rock to soil is destructive "weathering." Weathering is the disintegration and decomposition of rocks by physical and chemical processes.

Physical Weathering

Physical weathering causes fragmentation of rocks, increasing the exposed surface area. Common causes of physical weathering are

- Expansion and contraction caused by heating and cooling
- Stresses from mineral crystal growth and freezing and thawing of water in cracks and pores
- Penetration of tree and plant roots
- Scouring and grinding by abrasive particles carried by wind, water, and moving ice
- Unloading: When confining pressures are lessened by uplift, erosion, or changes in fluid pressures, unloading can cause cracks at thousands of feet deep

Chemical Weathering

Chemical weathering of rocks causes changes in their mineral composition. Common causes of chemical weathering are

- Hydrolysis and hydration reactions (water reacting with mineral structures)
- Oxidation (usually by oxygen in the atmosphere and in water) and reduction (usually by microbes)
- Dissolution and dissociation of minerals
- Immobilization by precipitation, e.g., the formation of solid oxides, hydroxides, carbonates, and sulfides
- Loss of mineral components by leaching and volatilization
- Chemical exchange processes, such as cation exchange

Physical and chemical weathering processes often produce loose materials that can be deposited elsewhere after being transported by wind (aeolian deposits), running water (alluvial deposits), and glaciers (glacial deposits).

The next steps, after the rocks have been fractured and broken down, are the formation of secondary minerals (e.g., clays, mineral precipitates, etc.) and changes caused by plants and bioorganisms.

Secondary Mineral Formation

Secondary minerals are formed within the soil by chemical reactions of the primary (original) minerals. The reactions forming secondary minerals are always in the direction of greater chemical stability under local environmental conditions. These reactions are facilitated by the presence of water which dissolves and mobilizes different components of the original rocks and allows them to react to form new compounds.

Roles of Plants and Bioorganisms

Organic materials play many complex roles. Roots form extensive networks permeating soil. They can exert pressures that compress aggregates in one location and separate them in another. Water uptake by roots causes differential dehydration, initiating the shrinkage and opening of many small cracks.

The plant root zone in the soil is called the *rhizosphere*. It is the soil region where plants, microbes, and other soil organisms support one another. Soil organisms include thousands of species of bacteria, fungi, actinomycetes, worms, slugs, insects, mites, etc. The number of organisms in the rhizosphere can be 100 times larger than in non-rhizosphere soil zones. Root secretions and dead roots promote microbial activity that produces humic cements. Root secretions include *mucigel,* a gelatinous substance that lubricates root penetration, various sugars, as well as aliphatic, aromatic, and amino acids. These substances and dead root material are nutrients for rhizosphere microorganisms. The root structure itself provides surface area for microbial colonization.

4.2   SOIL PROFILES

A vertical profile through soil tells much about how the soil was formed. It usually consists of a succession of more-or-less distinct layers, or strata. The layers can form from aeolian or alluvial deposition of material, or from *in situ* weathering processes.

SOIL HORIZONS

When the layers develop *in situ* by the weathering processes described above, they form a sequence of *horizons*. The horizons are designated by the U.S. Department of Agriculture by the capital letters O, A, E, B, C, and R, in order of farthest distance from the surface (see Figure 4.1).

O-horizon: Organic

- The top horizon: starts at the soil's surface
- Formed from surface litter
- Dominated by fresh or partly decomposed organic matter

A-horizon: Topsoil

- The zone of greatest biological activity (rhizosphere)
- Contains an accumulation of finely divided, decomposed, organic matter, which imparts a dark color
- Clays, carbonates, and most metal cations are leached out by downward percolating water; less soluble minerals (such as quartz) of sand or silt size become concentrated in the A-horizon

E-horizon: Leaching zone (sometimes called the A-2 horizon)

- Light-colored region below the rhizosphere where clays and metal cations are leached out and organic matter is sparse

B-horizon: Subsoil

- Dark-colored zone where migrating materials from the A-horizon accumulate

C-horizon: Soil parent material

- Fragmented and weathered rock, either from bedrock or base material that has deposited from water or wind

R-layer: Bedrock

- Below all the horizons; consists of consolidated bedrock
- Impenetrable, except for fractures

STEPS IN THE TYPICAL DEVELOPMENT OF A SOIL AND ITS PROFILE (*PEDOGENESIS*)

- Physical disintegration (weathering) of exposed rock formations. This forms the soil parent material, the C-horizon.
- The gradual accumulation of organic residues near the surface begins to form the A-horizon, which might acquire a granular structure, stabilized to some degree by organic matter cementation. This process is retarded in desert regions where organic growth and decay are slow.
- Continued chemical weathering (oxidation, hydrolysis, etc.), dissolution, and precipitation begin to form clays.
- Clays, soluble salts, chelated metals, etc. migrate downward through the A-horizon, carried by permeating water, to accumulate in the B-horizon.
- The C-horizon, now below the O, A, and B horizons, continues to undergo physical and chemical weathering slowly transforming into B and A horizons, deepening the entire horizon structure.
- A quasi-stable condition is approached in which the opposing processes of soil formation and soil erosion are more or less balanced.

4.3   ORGANIC MATTER IN SOIL

Soil organic matter influences the weathering of minerals, provides food for microorganisms, and provides sites to which ions are attracted for ion exchange. Only two types of organisms can synthesize organic matter from non-organic materials. These are certain bacteria called *autotrophs* and chlorophyll-containing plants. Organic matter is developed in soil from the metabolism, wastes, and decay products of plants and soil organisms. For example, soil fungi metabolism produces excellent complexing agents, such as oxalate ion, and citric and other chelating organic acids. These promote the dissolution of minerals and increase nutrient availability. Some soil bacteria release the strong organic chelating agent 2-ketogluconic acid. This reacts with insoluble metal phosphates to solubilize the metal ions and release soluble phosphate.

The amount of organic matter has a strong influence on the properties of soil and on the behavior of soil contaminants. For example, plants have to compete with soil for water. In sandy soils, pore space is large and particle surface area is small. Water is not strongly adsorbed to sands and is easily available to plants. However in sandy soils water drains off quickly. On the other hand, water binds strongly to organic matter in soil. Soils with high organic content hold more water; but the water is less available to plants.

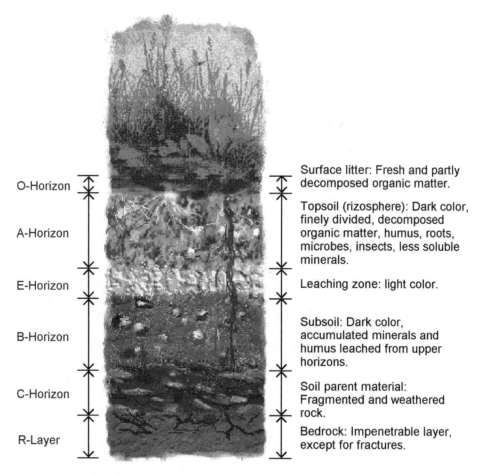

O-Horizon

A-Horizon

E-Horizon

B-Horizon

C-Horizon

R-Layer

Surface litter: Fresh and partly decomposed organic matter.

Topsoil (rizosphere): Dark color, finely divided, decomposed organic matter, humus, roots, microbes, insects, less soluble minerals.

Leaching zone: light color.

Subsoil: Dark color, accumulated minerals and humus leached from upper horizons.

Soil parent material: Fragmented and weathered rock.

Bedrock: Impenetrable layer, except for fractures.

FIGURE 4.1   Generalized soil profile showing the horizon sequence.

As another example, oxalate ion is a metabolite of certain soil organisms. In calcium soils, oxalate forms calcium oxalate, $Ca(C_2O_4)$, which then reacts with precipitated metals (particularly Fe or Al) to complex and mobilize them. The reaction with precipitated aluminum is

$$3\ H^+ + Al(OH)_3(s) + 2\ Ca(C_2O_4) \rightarrow Al(C_2O_4)_2^-(aq) + 2\ Ca^{2+}(aq) + 3\ H_2O. \qquad (4.1)$$

Because hydrogen ions are consumed, this reaction raises the pH of acidic soil. It also weathers minerals by dissolving some metals and provides Ca^{2+} as a plant and biota nutrient. Similar processes with silicate minerals release K^+ and other nutrient cations.

Rule of Thumb

Organic matter is typically less than 5% in most soils but is the main factor in plant productivity. Peat soils can be 95% organic. Mineral soils can be less than 1% organic.

HUMIC SUBSTANCES

The most important organic substance in soil is *humus*, a collection of variously sized polymeric molecules consisting of soluble fractions (humic and fulvic acids) and an insoluble fraction (humin).

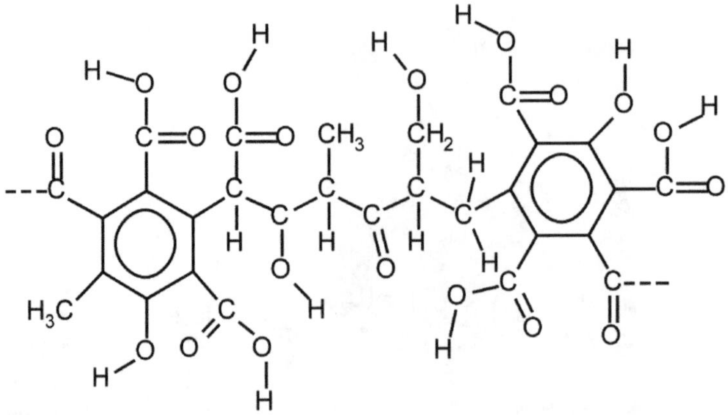

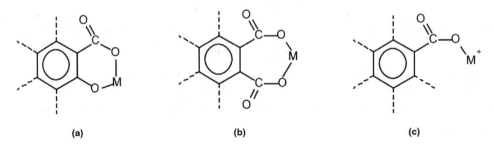

FIGURE 4.2 Characteristic structural portion of an unionized humic or fulvic acid.

Humus is the near-final residue of plant biodegradation and consists largely of protein and lignin. It is what remains after the more easily degradable components of plant biomass have degraded, leaving only parts that are most resistant to further degradation. Humic materials are not well defined chemically and have variable composition. Percent by weight for the most abundant elements are C: 45–55%, O: 30–45%, H: 3–6%, N: 1–5%, and S: 0–1%. The exact chemical structure depends on the source plant materials and the history of biodegradation.

Humic and fulvic acids are soluble organic acid macromolecules containing many –COOH and –OH functional groups that ionize in water, releasing H^+ ions and providing negative charge centers on the macromolecule to which cations are strongly attracted (see Figure 4.2). Humic materials are the most important class of soil complexing agents and are found where vegetation has decayed.

Some Properties of Humic Materials

Binding to Dissolved Species

Humic materials are effective at removing metals from water by sorption to negative charge sites, mainly at the structural oxygen atoms. They strongly adsorb heavy polyvalent metal cations. The cation exchange capacity of humic materials can be as high as 500 meq/100 mL. Humic materials may contain metals like uranium in concentrations 10,000 times greater than adjacent water. Humic materials also bind to organic pollutants, especially low solubility compounds like DDT and atrazine. Much of the utility of wetlands for water treatment arises from their high concentrations of humic materials. Figure 4.3 shows several ways by which metal cations bind to humic and fulvic acids.

FIGURE 4.3 Some types of binding metal ions (M^{2+}) to humic or fulvic acids.

Light Absorption

Humic materials absorb sunlight in the blue region (transmitting yellow) and can transfer the solar energy to sorbed molecules, initiating reactions. This energy transfer process can be effective in degrading pesticides and other organic compounds.

4.4 SOIL ZONES

The soil subsurface is commonly divided into three zones, based on their air and water content (see Figure 4.4). From the ground surface down to an aquifer water table, soils contain mostly air in the pore spaces, with some adsorbed and capillary-held water. This region is called the *water-unsaturated zone* or *vadose zone*. From the top of the water table to bedrock, soils contain mostly water in the pore spaces. This region is called the *saturated zone*.

Between the vadose and saturated zones, there is a transition region called the *capillary zone,* where water is drawn upward from the water table by capillary forces. The thickness of the capillary zone depends on the soil texture — the smaller the pore size, the greater the capillary rise. In fine gravel (2-5 mm grain size), the capillary zone will be of the order of 2.5 centimeters thick. In fine silt (0.02-0.05 mm grain size), the capillary zone can be 200 centimeters or greater.

The saturated zone lies above the solid bedrock, which is impermeable except for fractures and cracks. The region of the subsurface overlying the bedrock is generally unconsolidated porous, granular mineral material.

AIR IN SOIL

Air in soil has a different composition from atmospheric air because of biodegradation of organic matter by soil organisms. Biodegradation occurs in many small steps, but the net overall reaction is shown in Equation 4.2, where organic matter in soil is represented by the approximate generic unit formula $\{CH_2O\}$. An actual molecule of soil organic matter would have a formula that is approximately some whole number multiple of the $\{CH_2O\}$ unit.

$$\{CH_2O\} + O_2 \rightarrow CO_2 + H_2O. \tag{4.2}$$

Oxygen from soil pore space air is consumed and CO_2 released by microbial metabolism. Much of the soil air is semitrapped in pores and cannot readily equilibrate with the atmosphere. As a result, O_2 content is decreased in soil air from its atmospheric value of 21% to about 15%, and CO_2 content is increased from its atmospheric value of about 0.03% to about 3%. This, in turn increases the dissolved CO_2 concentration in groundwater, making it more acidic. Acidic groundwater contributes to the weathering of soils, especially calcium carbonate ($CaCO_3$) minerals.

When soil becomes waterlogged, as in the saturated zone, many changes occur:

- Oxygen becomes used up by respiration of microorganisms.
- Anaerobic processes lower the oxidation potential of water so that reducing conditions (electron gain) prevail, whereas oxidation conditions (electron loss) dominate in the unsaturated zone.
- Certain metals, particularly iron and manganese, become mobilized by chemical reduction reactions changing them from insoluble to soluble forms:

$$Fe(OH)_3(s) + 3\ H^+ + e^- \rightarrow Fe^{2+}(aq) + 3\ H_2O. \tag{4.3}$$

$$Fe_2O_3(s) + 6\ H^+ + 2\ e^- \rightarrow 2\ Fe^{2+}(aq) + 3\ H_2O. \tag{4.4}$$

$$MnO_2(s) + 4\ H^+ + 2\ e^- \rightarrow Mn^{2+}(aq) + 2\ H_2O. \tag{4.5}$$

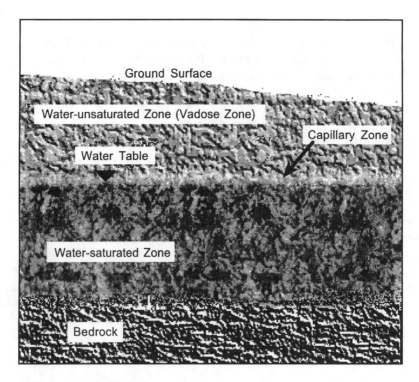

FIGURE 4.4   Soil zones in the subsurface region.

Groundwater, moving under gravity, can transport dissolved Fe^{2+} and Mn^{2+} into zones where oxidizing conditions prevail, e.g., by surfacing to a spring or lake. There, Equations 4.3–4.5 are reversed and the metals redeposit as solid precipitates, mainly $Fe(OH)_3$ and MnO_2. Precipitation of $Fe(OH)_3$ often causes "red water" and red or yellow deposits on rocks and soil. MnO_2 deposits are black. These deposits can clog underdrains in fields and water treatment filters.

4.5 CONTAMINANTS BECOME DISTRIBUTED IN WATER, SOIL, AND AIR

In the environment, contaminants always contact water, air, and soil. No matter where it originated, a contaminant moves across the interfaces between water, soil, and air to become distributed, to different degrees, into every phase it contacts. Partitioning of a pollutant from one phase into other phases serves to deplete the concentration in the original phase and increase it in the other phases. The movement of contaminants through soil is a process of continuous redistribution among the different phases it encounters. It is a process controlled by gravity, capillarity, sorption to surfaces, miscibility with water, and volatility.

VOLATILIZATION

The main partitioning process from liquids and solids to air is *volatilization* which moves a contaminant across the liquid-air or solid-air interface into the atmosphere or into air in soil pore spaces. Volatilization is an important partitioning mechanism for compounds with high vapor pressures. For contaminant mixtures such as gasoline, the most volatile components are lost first causing the composition and properties of the remaining liquid mixture to change over time. For example, the most volatile components of gasoline are also the smallest molecules in the mixture.

Hence, as gasoline "weathers" and loses these smaller molecules by volatilization, its vapor pressure decreases and its viscosity and density increase.

SORPTION

The main partitioning process from liquids and air to solids is *sorption* which moves a contaminant across the liquid-solid or air-solid interface to organic or mineral solid surfaces. For example, contaminants dissolved in stream water may become bound to suspended and bottom sediments. Sorption from the water phase is most important for compounds of low solubility. Once a contaminant is sorbed to a surface, it undergoes chemical and biological transformations at different rates and by different pathways than if it were dissolved.

Example 4.1

Estimating relative air/water partitioning behavior

Suppose you need to compare the air/water partitioning behavior of the compounds tabulated below but are able to find only melting point data. Estimate their relative vapor pressures and water solubilities based on their structures.

Compound	Structure
Phenol Melting temperature = 43.0°C	*(structure: phenol — benzene ring with OH)*
1,2,3,5–tetrachlorobenzene Melting temperature = 54.5°C	*(structure: benzene ring with four Cl substituents)*
1,2,4,5–tetrachlorobenzene Melting temperature = 140°C	*(structure: benzene ring with four Cl substituents)*

Answer: Solubility will vary with polarity. The more polar the molecule, the more soluble it will be, because of stronger attraction to polar water molecules. It also varies with molecular weight. Higher molecular weight tends to decrease solubility because London dispersion forces are stronger, attracting the compound molecules to one another more strongly. Vapor pressure will vary inversely with the melting point because a high melting point indicates strong intermolecular attractive forces, and vice versa.

TABLE 4.1
Measured Values for Melting Point, Vapor Pressure, and Solubility

Compound	T_m (°C)	P_v (atm)	S_w (mol/L)
Phenol	43.0	2.6×10^{-4}	6.3×10^{-1}
1,2,3–tetrachlorobenzene	54.5	1.9×10^{-4}	1.6×10^{-5}
1,2,4–tetrachlorobenzene	140.0	3.0×10^{-5}	2.5×10^{-6}

Note: T_m = melting point; P_v = vapor pressure; S_w = solubility in water.

Vapor pressure: (Lowest to highest vapor pressure is in the order of highest to lowest melting point.)

1,2,4,5–tetrachlorobenzene < 1,2,3,5–tetrachlorobenzene < phenol.

Solubility: (Lowest to highest solubility is in the order of lowest to highest polarity and highest to lowest molecular weight.)

1,2,4,5–tetrachlorobenzene (nonpolar because of symmetry) < 1,2,3,5–tetrachlorobenzene (less symmetrical, more polar) < phenol (most polar and lightest molecular weight of all).

The measured values in Table 4.1 confirm these relative vapor pressure and solubility estimates.

Example 4.2

Rank the four compounds below in order of increasing tendency to partition from water into air.

I.   II.   III.   IV.

Answer: All the compounds are similar in molecular weight and differ only in the top functional group. The molecule having the group with the weakest attractive force to water will most readily partition from water into air. So, we want to rank them by their relative attractions to water.

I: Has no oxygen for hydrogen-bonding to water and is the least polar. It will most readily volatilize.

IV: Has an oxygen, but no hydrogen is attached to it for hydrogen-bonding to water. It will be next in volatility from water to air.

II and III: Can both hydrogen-bond. II can form one H-bond, while III can form two H-bonds. So, II is third in volatility and III is the least volatile in water.

Tendency to volatilize from water: I > IV > II > III.

Their Henry's Law constants are

I: K_H = 1.38; II: K_H = 0.0034; III: K_H = 0.0015; IV: K_H = 1.17.

A larger Henry's Law constant means greater volatility (see Section 4.6).

4.6 PARTITION COEFFICIENTS

The tendency for a pollutant to move from one phase to another is often quantified by the use of a *partition coefficient,* also called a *distribution coefficient.* Partition coefficients are chemical specific. They can be measured directly or, in some cases, estimated from other properties of the chemical. The simplest form of a partition coefficient is the ratio of the pollutant concentration in phase 1 to its concentration in phase 2:

$$K_{1,2} = \frac{\text{concentration in phase 1}}{\text{concentration in phase 2}} = \frac{C_1}{C_2}. \tag{4.6}$$

This expression assumes that a linear relation exists between the concentrations of a substance in different phases and is often satisfactory for low to moderate concentrations. Using the water phase as the reference phase, a linear relation gives Equations 4.7–4.9 for partitioning between water and air, water and a pollutant free product, and water and soil.

$$C_a = K_H C_w, \tag{4.7}$$

K_H is the air-water partition coefficient, also known as Henry's Law constant. C_a and C_w are the pollutant concentrations in air and water, respectively.

$$C_p = K_p C_w, \tag{4.8}$$

K_p is the bulk pollutant-water partition coefficient. C_p and C_w are the pollutant concentrations in bulk pollutant and water, respectively. A bulk pollutant is the portion of a contaminant that remains in its original form, such as a layer of oil floating on a river or above the groundwater table.

$$C_s = K_d C_w, \tag{4.9}$$

K_d is the soil-water partition coefficient. C_s and C_w are the pollutant concentrations sorbed on soil and dissolved in water, respectively.

Each value of K depends on the properties of the particular pollutant and the temperature. K_d also depends on the type of soil.

AIR-WATER PARTITION COEFFICIENT

Example 4.3

Using Henry's Law

Henry's Law, $C_a = K_H C_w$, describes how a substance distributes itself at equilibrium between water and air. The units of Henry's Law constant, $K_H = \dfrac{C_a}{C_w}$, depends on what units are used to express concentrations in air and water.* For the case of oxygen gas, O_2, at 20°C

- When air and water concentrations both have the same units,

$$K_H(O_2, 20°C) = 26 \text{ (unitless)}. \tag{4.10}$$

- For water concentration in mol/L or mol/m³, and air in atmospheres,

$$K_H(O_2, 20°C) = 635 \text{ L·atm/mol} = 0.635 \text{ atm·m}^3/\text{mol}. \tag{4.11}$$

- For water concentration in mg/L and air in atmospheres,

$$K_H(O_2, 20°C) = 0.0198 \text{ L·atm/mg}. \tag{4.12}$$

If soil pore water is measured to contain 3.2 mg/L of oxygen at 20°C, what is the concentration, in mg/L and in atmospheres, of oxygen in the air of the soil pore space?

Answer:

$$\text{For } O_2 \text{ at } 20°C, K_H = 26 = \frac{C_a}{3.2 \text{ mg/L}}; Ca = (26)(3.2 \text{ mg/L}) = \textit{83.2 mg/L.}$$

Also,

$$K_H = 0.0198 \text{ L·atm/mg} = \frac{C_a}{3.2 \text{ mg/L}}; Ca = (0.0198 \text{ L·atm/mg})(3.2 \text{ mg/L}) = \textit{0.063 atm.}$$

Since the normal atmospheric partial pressure of oxygen at sea level is about 0.2 atm, this result indicates the presence in the soil of microbial activity that has consumed oxygen.

Example 4.4

BOD and Henry's Law

A certain sewage treatment plant located on a river typically removes 100,000 lbs (4.54×10^7 g) of biodegradable organic waste each day. If there were a plant upset and it became necessary to

---

* Henry's Law constants are tabulated in many references, such as *Handbook of Chemistry and Physics*, Howard, P.H., Ed., CRC Press, Boca Raton, 1991; *Handbook of Environmental Fate and Exposure Data for Organic Chemicals*, Vols. I-III, Lewis Publishers, Chelsea, MI; Lyman, W.J., Reehl, W.F., and Rosenblatt, D.H., *Handbook of Chemical Property Estimation Methods*, 2nd printing, American Chemical Society, Washington, D.C., 1990; Mackay, D. and Shiu, W.Y., A critical review of Henry's Law constants for chemicals of environmental interest, *J. Phys. Chem. Ref. Data*, 10(4): 1175–1199, 1981.
There are also computer programs that calculate Henry's Law constant from other chemical properties.

release one day's waste into the receiving river, how many liters of river water could potentially be contaminated to the extent of totally depleting the water of all oxygen?

Answer: An approximate chemical equation we have used before as being suitable for biodegradation of organic matter is

$$\{CH_2O\} + O_2 \rightarrow CO_2 + H_2O. \tag{4.2}$$

Assume the river water is saturated with oxygen from the air at 20°C and that no additional oxygen dissolves from the atmosphere, a worst case.

Necessary data

Atmospheric pressure at the treatment plant = 0.82 atm.
Vapor pressure of water at 20°C = 0.023 atm.
Percent O_2 in dry air = 21%.
From Equation 4.11, $K_H(O_2)$ = 635 L·atm/mol.

Calculation

Organic matter is biodegraded, consuming oxygen by Equation 4.2. This chemical equation shows that one mole of O_2 is consumed for each mole of CH_2O biodegraded. The molecular weight of CH_2O is 30 g/mol. Therefore,

$$\text{moles of } CH_2O \text{ in sewage} = \frac{4.54 \times 10^7 \text{ g}}{30 \text{ g/mol}} = 1.5 \times 10^6 \text{ mol} = \text{moles of } O_2 \text{ consumed.}$$

$$\text{Atmospheric pressure} = P_{total} = 0.82 \text{ atm} = P_{O_2} + P_{N_2} + P_{H_2O}.$$

$$P_{dry\ air} = \text{atmospheric pressure} - \text{partial pressure of water vapor} = P_{total} - P_{H_2O}.$$

Therefore, $P_{O_2} = (0.21)(P_{total} - P_{H_2O}) = (0.21)(0.82 \text{ atm} - 0.023 \text{ atm}) = 0.17 \text{ atm}.$

Use Henry's Law to find the concentration of dissolved O_2 in the river:

$$K_H(O_2) = 635 \text{ L·atm/mol} = \frac{C_a}{C_w} = \frac{0.17 \text{ atm}}{C_w}.$$

$$C_w = [O_2(aq)] = \frac{0.17 \text{ atm}}{635 \text{ atm·L/mol}} = 2.7 \times 10^{-4} \text{ mol/L.}$$

or $[O_2(aq)] = (2.7 \times 10^{-4} \text{ mol/L}) \times (32 \text{ g/mol}) = 8.6 \text{ mg/L.}$

In saturated water at 20°C and 0.82 atm total pressure, $[O_2, aq] = 2.7 \times 10^{-4}$ mol/L.

$$\text{Liters of river water depleted of } O_2 = \frac{1.5 \times 10^6 \text{ mol } O_2 \text{ consumed}}{2.7 \times 10^{-4} \text{ mol } O_2/L \text{ in river}} = 5.6 \times 10^9 \text{ L.}$$

Note that both vapor pressure and solubility of a pure solid or liquid generally increase with temperature, but vapor pressure always increases faster. Therefore, the value of K_H increases with temperature, indicating that, for a gas partitioning between air and water, the atmospheric portion increases and the dissolved portion decreases when the temperature rises. This is consistent with the observation that the water solubility of gases decreases with increasing temperature.

Rule of Thumb

Estimating K_H:
If a tabulated value for K_H cannot be found, it may be estimated roughly by dividing the *vapor pressure* of a compound by its *aqueous solubility*. For some compounds, tabulated values of vapor pressure and solubility may be easier to find than K_H values.

$$K_H = \frac{C_a(partial\ pressure\ in\ atmosphere)}{C_w(mol/L)} \approx \frac{vapor\ pressure\ (atm)}{aqueous\ solubility\ (mol/L)}. \qquad (4.13)$$

In this case, the units of K_H are *atm·L·mol^{-1}*.

Example 4.5

Estimate Henry's Law constants for *chlorobenzene* and *bromomethane* using vapor pressure and aqueous solubility.

Answer:

Chlorobenzene: P_v (25°C) = 1.6×10^{-2} atm; C_w (25°C) = 4.5×10^{-3} mol/L.

$$K_H\ (25°C) \approx P_v/C_w = 1.6 \times 10^{-2}\ atm/4.5 \times 10^{-3}\ mol/L = 3.6\ L·atm/mol.$$

This happens to match exactly the experimental value. To put K_H into dimensionless form, divide by RT (R = universal gas constant, T = temperature in degrees Kelvin), equivalent to multiplying by 0.041 mol/L·atm

$$K_H' = \frac{3.6\ L·atm/mol}{(0.0821\ L·atm/mol\ K)(298\ K)} = 0.15.$$

Bromomethane: P_v (liq, 25°C) = 1.8 atm; C_w (1 atm, 25°C) = 0.16 mol/L
At 25°C, bromomethane has a vapor pressure > 1 atm, so it is a gas. Since the solubility is given at 1 atm partial pressure

$$K_H\ (25°C) \approx P_v/C_w = 1\ atm/0.16\ mol/L = 6.3\ L·atm/mol.$$

$$K_H' = \frac{6.3\ L·atm/mol}{(0.0821\ L·atm/mol\ K)(298\ K)} = 0.26.$$

SOIL-WATER PARTITION COEFFICIENT

The partitioning of a compound between water and soil may deviate from linearity. This is particularly true for organic compounds. To account for this, the corresponding partition coefficient is often written in a modified form called the *Freundlich isotherm*. The modification consists of introducing an empirically determined exponent to the C_w term.

$$C_s = K_d C_w^{\ n}, \qquad (4.14)$$

where

C_s = concentration of sorbed organic compound in solid phase (mg/kg).
C_w = concentration of dissolved organic compound in water phase (mg/L).
$K_d = C_s/C_w^n$ = partition coefficient for sorption.
n = empirically determined exponential factor.

When $C_w \ll C_s$ (the common case of organic compounds of low water solubility), then n is close to unity for most organics. However, n typically is temperature dependent.

Equation 4.14 can be written as an equation for a straight line by taking the logarithm of both sides, as in

$$\log C_s = \log K_d + n \log C_w. \tag{4.15}$$

It is important not to extrapolate the Freundlich isotherm too far beyond the range of experimental data.

Example 4.6

Use of the Freundlich isotherm

A power company planned to discharge its power plant cooling water into a small lake. Before purchasing the lake, it tested the water and found the pesticide 2,4-D (2,4-dichlorophenoxyacetic acid) at 0.8 ppt (0.8 parts per trillion, or 0.8×10^{-9} g/L) just a little below the permitted limit of 1 ppt. The company calculated that its operation would raise the water temperature in the mixing zone near its discharge from 5°C to 25°C. Should the company anticipate a problem?

Answer: 2,4-D has low solubility and is denser than water. It will sink in the lake and become sorbed on the bottom sediments. The potential problem is whether or not the expected increase in temperature will cause the 2,4-D limit to be exceeded because of additional 2,4-D partitioning into the water from the bottom sediments. It is a case for the Freundlich isotherm because the constant n is a function of temperature and will cause a change in K_d when the temperature changes. A study[8] measured Freundlich isotherm values for 2,4-D at 5°C and 25°C, as given in Table 4.2.

TABLE 4.2
Values for Freundlich Isotherm Parameters of 2,4-D

Temperature (°C)	n	K_d	$\log K_d$
5	0.76	6.53	0.815
25	0.83	5.20	0.716

$$C_s = K_d C_w^n. \tag{4.14}$$

At 5°C: $C_s = (6.53)(0.8 \times 10^{-9}$ g/L$)^{0.76} = (6.53)(1.22 \times 10^{-7}) = 797 \times 10^{-9}$ g/kg.

There are 797 ppt of 2,4-D sorbed on the sediments at 5°C.

Note that the concentration of 2,4-D sorbed to sediments is about 1000 times greater than the concentration dissolved. Even if a temperature rise to 25°C causes a large percentage increase in the dissolved portion, the percentage loss from the sediment fraction will be one thousand times smaller. Assume that the sediment concentration at 25°C is essentially the same as at 5°C. This allows an approximate calculation of C_w at 25°C.

$$\text{At } 25°C: 797 \times 10^{-9} \text{ g/kg} = (5.20)(C_w)^{0.83}.$$

$$C_w(25°C) = \left(\frac{797 \times 10^{-9}}{5.20} \right)^{\frac{1}{0.83}} = 6.2 \times 10^{-9} \text{ g/L} = 6.2 \text{ ppt}.$$

The expected temperature rise will cause 2,4-D to desorb from the bottom sediments and raise its water concentration well over the permitted limit.

DETERMINING K_d EXPERIMENTALLY

The Freundlich Equation 4.15 can be used to determine K_d experimentally, as follows:

1. Prepare samples having several different concentrations of dissolved contaminant in equilibrium with soil from the site of interest.
2. Measure the contaminant concentrations in water, C_w, in each sample.
3. Measure the corresponding contaminant concentrations sorbed to soil, C_s.
4. Plot log C_s vs. log C_w to get a straight line with slope = n and intercept = log K_d.

Example 4.7

Prepared water samples containing different concentrations of benzene were equilibrated with soil from a site under study. Equilibrium concentrations of dissolved and sorbed benzene are shown in Table 4.3. Find K_d for benzene in this soil.

TABLE 4.3
Benzene Partitioning Data for
Soil and Water in Equilibrium

Dissolved Benzene C_w (mg/L)	Sorbed Benzene C_s (mg/kg)
6.59	2.2
10.00	3.1
33.28	8.1
34.57	9.6
68.31	15.00
88.89	26.00
183.74	44.00
340.54	89.00
452.30	119.00
674.79	130.00
819.56	188.00
955.95	247.00

Answer:

1. Determine the base-10 logarithms of all the concentration values. The logarithmic values are given in Table 4.4.

TABLE 4.4
Logarithms of C_w and C_s

log C_w	log C_s
0.819	0.342
1.000	0.491
1.522	0.908
1.539	0.982
1.834	1.176
1.949	1.415
2.264	1.643
2.532	1.949
2.655	2.076
2.829	2.114
2.914	2.274
2.980	2.393

2. Plot log C_w vs. log C_s and fit a straight line through the points. The formula of the line is log C_s = log K_d + n log C_w. Therefore, the slope of the line is equal to n and the y-axis intercept is equal to log K_d. The resulting plot is shown in Figure 4.5.

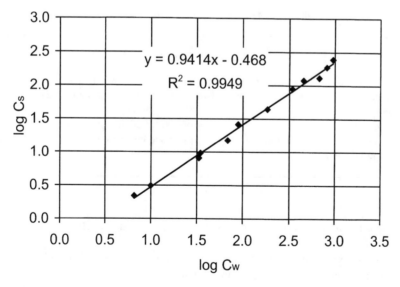

FIGURE 4.5 Freundlich isotherm for benzene partitioning between water and soil.

The equation of the least-squares fitted line is log C_s = 0.941 log C_w – 0.468. Therefore,

$$n = 0.941$$

$$\log K_d = -0.468$$

$$K_d = 0.340 \ L/kg.$$

The Role of Soil Organic Matter

Values for K_d are extremely site and chemical specific because the extent of sorption depends on several physical and chemical properties of both the soil and the sorbed chemical. For dissolved neutral organic molecules, such as fuel hydrocarbons, sorption to soil is controlled mostly by interactions with the organic portion of the soil. Therefore, the value of the soil-water partition coefficient, K_d, depends on the amount of organic matter in the soil. Expressing the soil-water partition coefficient in terms of soil organic carbon (K_{oc}), rather than total soil mass (K_d), can eliminate a large part of the site variability in K_d.

The amount of organic matter in soil is usually expressed as either the weight fraction of *organic carbon*, f_{oc}, or the weight fraction of *organic matter*, f_{om}. The amount of organic matter in typical mineral soils is generally between about 1 and 10% but is typically less than 5%. In wetlands and peat-soils, it can approach 100%. Since soil organic matter is approximately 58% carbon, f_{oc} typically ranges between 0.006 and 0.06. Some characteristic values for f_{oc} for a range of different soil types are given in Table 4.5.

There is a critical fraction of organic carbon in the soil, f_{oc}^* (dependent on the dissolved organic compound), below which sorption to inorganic matter becomes dominant. Typical values for f_{oc}^* are between about 10^{-3} and 10^{-4} (0.1% to 0.01%).

If we define $K_{oc} = \dfrac{C_{oc}}{C_w}$, where C_{oc} is the concentration of contaminant sorbed to organic carbon and C_w is the concentration of contaminant dissolved in water, then the soil/water partition coefficient becomes

$$K_d = K_{oc}f_{oc}. \tag{4.16}$$

This relation is useful as long as f_{oc} is greater by about 0.001.

Rules of Thumb

1. It is often found that typical soil organic matter is about 58% carbon. Using this value to convert between the fraction of organic carbon and the fraction of organic matter leads to the relation $f_{oc} \approx 0.58\, f_{om}$.
2. Soil organic carbon content can vary by a factor of 100 for similar soils. Although there have been tabulations of f_{oc} values according to soil type,[7] it is much better to use values measured at the site for use in Equation 4.16.

Example 4.8

Find K_{oc} for benzene in the soil of Example 4.7. The soil contained 1.1% organic matter.

Answer: The soil's organic matter was measured to be 1.1%, or $f_{om} = 0.011$. This may be converted to fraction of organic carbon (f_{oc}) by the approximate rule of thumb

$$f_{oc} = 0.58\, f_{om}$$

Therefore, $f_{oc} = 0.58 \times 0.011 = 0.0064$. Since $K_d = K_{oc}f_{oc}$, we have:

$$K_{oc}(\text{benzene}) = 0.340/0.0064 = 53 \; L/kg.$$

TABLE 4.5
Typical Values of Fraction of Organic Carbon in Different Soils[a]

Type of Soil	Typical f_{oc} (wt. fraction)
Coarse soil	0.04
Silty loam	0.05
Silty clayey loam	0.03
Clayey silty loam	0.005
Clayey loam	0.004
Sand	0.0005
Glaciofluvial	0.0001

[a] These typical values were collected from many sources. They should be used only as crude estimates when site-specific measurements cannot be obtained. The range of measured values for a single soil type can span a factor of 100.

THE OCTANOL/WATER PARTITION COEFFICIENT, K_{ow}

For laboratory experiments, the liquid compound *octanol*, an eight carbon organic alcohol, $CH_3(CH_2)_7OH$, is a good surrogate for the organic carbon fraction of soils. The partitioning behavior of organic compounds between octanol and water is similar to that between the organic carbon fraction of soil and water. The basic steps for measuring the octanol/water partition coefficient, K_{ow}, are as follows:

1. Combine octanol and water in a bottle. Octanol forms a separate phase floating on top of the water.
2. Add the organic contaminant (e.g., carbon tetrachloride, CCl_4), shake the mixture, and let the phases separate.
3. Measure the contaminant concentrations in the octanol phase and in the water phase. Then,

$$K_{ow} = \frac{C_{octanol}}{C_{water}}. \qquad (4.17)$$

An empirical equation that relates K_{ow} and organic carbon to K_d is

$$K_d = f_{oc}bK_{ow}{}^a, \text{ or}$$

$$\log K_d = a \log K_{ow} + \log f_{oc} + \log b, \qquad (4.18)$$

where the empirically determined constants a and b depend on the organic compound.

Equation 4.18 defines a relation that allows a calculation of K_{oc} in terms of K_{ow}. Since

$$K_d = \frac{C_{soil}}{C_{water}} = K_{oc}f_{oc}, \text{ then } K_{oc} = bK_{ow}{}^a. \qquad (4.19)$$

Consider the pesticide DDT. Its $K_{ow} = 3.4 \times 10^6$, which is well into the high value range for K_{ow}. Using the rules of thumb for K_{ow}, one can predict that DDT has low water solubility (solubility = 0.025 mg/L), is slowly biodegraded, persistent in the environment (low mobility and biodegradation), strongly adsorbed to soil, and strongly bioaccumulated.

On the other hand, phenol has K_{ow} = 30.2, a low value. Phenol has high water solubility (solubility = 8.3×10^4 mg/L), biodegrades rapidly, is not persistent, is weakly sorbed to soil, highly mobile, and weakly bioaccumulated.

Rules of Thumb for K_{ow}

(High K_{ow} = >1000)
The higher K_{ow} is for a compound,

- The higher is the sorption to soil
- The higher is the bioaccumulation
- The lower is the biodegradation rate
- The lower is the water solubility
- The lower is the mobility

(Low K_{ow} = <500)
The lower K_{ow} is for a compound,

- The lower is the sorption to soil
- The lower is the bioaccumulation
- The higher is the biodegradation rate
- The higher is the solubility
- The higher is the mobility

ESTIMATING K_d USING SOLUBILITY OR K_{ow}

Literature values for K_d measurements vary considerably because they are so site specific. Considerable effort has been expended in finding more consistent approaches to soil sorption. EPA has recently published a comprehensive evaluation of the soil-water partition coefficient, K_d.[12]

The following empirical observations have led to methods for estimating K_d from more easily measured parameters:

- Water solubility is inversely related to K_d; the lower the solubility, the greater is K_d.
- Since molecular polarity correlates with solubility, molecules whose structure indicates low polarity (hence, low solubility) may be expected to have a high K_d.
- For compounds of low solubility, such as fuel hydrocarbons, sorption is controlled primarily by interactions with the organic portion of the solid sorbent.
- The surface area of the solid is important. The larger the surface area, the larger will be K_d.
- A simple way to estimate the tendency for a compound to partition between water and organic solids in the soil is to measure K_{ow}, the partition coefficient for the compound between water and octanol. The larger is K_{ow}, the larger is K_d.

Because K_d is site specific, it is generally preferable to calculate it from the more easily obtained quantities K_{ow}, K_{oc}, f_{oc}, and solubility. K_{ow} and solubility (S) are easily measured in a laboratory and K_{oc} may be normalized to f_{oc}, percent organic carbon in soil, which is an easily measured site parameter. There are linear relationships between log K_{oc}, log K_{ow}, and S that vary according to the class of compounds tested[6a] (e.g., chlorinated compounds, aromatic hydrocarbons, ionizable organic acids, pesticides, etc.). These relations can be used to calculate K_{oc} from K_{ow} or S in the absence of measured K_{oc} data. The EPA has reviewed the soil-water partitioning literature and selected or calculated the most reliable values in their judgment. In these calculations, the EPA used the following relations.

- For nonionizable, semivolatile organic compounds:

$$\log K_{oc} = 0.983 \log K_{ow} + 0.00028. \qquad (4.20)$$

- For nonionizable, volatile organic compounds:

$$\log K_{oc} = 0.7919 \log K_{ow} + 0.0784. \qquad (4.21)$$

In addition, the EPA suggests the use of the following equation for estimating K_{oc} from solubility (S) or bioconcentration factors (BCF):

$$\log K_{oc} = -0.55 \log S + 3.64. \tag{4.22}$$

$$\log K_{oc} = 0.681 \log BCF + 1.963. \tag{4.23}$$

Values for K_{oc}, K_{ow}, S, and BCF for many environmentally important chemicals have been collected in look-up tables for use in the EPA soil screening guidance procedures.[11] Table 4.6 is adapted from these EPA tables.

Example 4.9

Benzene spill

A benzene leak soaked into a patch of soil. To determine how much benzene was in the soil, several soil samples were taken in a grid pattern across the area of maximum contamination. When the samples were analyzed, the average benzene concentration in the soil was 2422 mg/kg (ppm). The soil contained 2.6% organic matter. From Table 4.6, $\log K_{ow}$ (benzene) = 2.13. If rainwater percolates down through the contaminated soil, what concentration of benzene might initially leach from the soil and be found dissolved in the water? Assume the water and soil are in equilibrium with respect to benzene.

Calculation: The general approach to this type of problem is

1. Find a tabulated value of K_{oc} for the chemical of concern, or calculate it from tabulated values for K_{ow} or S.
2. Obtain a measurement of f_{oc} or f_{om} in soil at the site or estimate it from the soil type.
3. Calculate K_d from the above quantities.

Benzene is a volatile, nonionizable organic compound (refer to Group 2 in Table 4.6).

Use $K_d = \dfrac{C_{soil}}{C_w} = K_{oc} f_{oc}$ and Equation 4.21:

$$\log K_{oc} = 0.7919 \log K_{ow} + 0.0784.$$

$$\log K_{oc} = 0.7919(2.13) + 0.0784 = 1.7651.$$

$$K_{oc} = 58.2.$$

Since $f_{oc} \approx 0.58\, f_{om} = 0.58(0.026) = 0.015$, we have

$$K_d = \frac{C_{soil}}{C_w} = K_{oc} \times f_{oc} = 58.2(0.015) = 0.88 \text{ L/kg}.$$

When 1 L of water is in equilibrium with 1 kg of soil, $C_{soil} + C_w = 2422$ ppm.

Therefore, $K_d = \dfrac{2422 - C_w}{C_w} = 0.88.$

$$0.88\, C_w = 2422 - C_w.$$

$$C_w = \frac{2422 \text{ mg/kg}}{1.88 \text{ L/kg}} = 1288 \text{ mg/L} = 1300 \text{ mg/L to 2 significant figures.}$$

Note that a substantial portion of the sorbed benzene partitions into the water, indicating that benzene has sufficient solubility to be mobile in the environment. This is predictable from the rules of thumb for K_{ow}, which classify K_{ow}(benzene) = 2.13 (from Table 4.6) as a low value giving benzene correspondingly high solubility and mobility.

4.7   MOBILITY OF CONTAMINANTS IN THE SUBSURFACE

The sorption of a contaminant from a liquid to a solid is a reversible reaction. Just as a contaminant has some probability of sorbing from water to a surface it comes in contact with, a sorbed contaminant also has a probability of desorbing from the surface back into the water. For strongly sorbed contaminants, the probability of sorption is much greater than the probability of desorption, but both processes continually take place. The rates of sorption and desorption depend on the strength of the bond holding the sorbed compound to the surface, and on the concentrations of the dissolved and sorbed contaminant.

$$\text{Rate of sorption} = k_{sorb} \, C_w \qquad\qquad (4.24)$$

$$\text{Rate of desorption} = k_{desorb} \, C_s \qquad\qquad (4.25)$$

where C_w is the contaminant concentration in water, C_s is the contaminant concentration sorbed to soil, and the rate constants, k_{sorb} and k_{desorb}, depend on the binding strength of sorption. For strongly sorbed contaminants, $k_{sorb} \gg k_{desorb}$. For weakly sorbed contaminants, $k_{sorb} \ll k_{desorb}$.

The partition coefficient K_d quantifies the equilibrium condition of sorption, where sorption and desorption occur at the same rate. When both rates are equal,

$$k_{sorb} \, C_w = k_{desorb} \, C_s$$

$$\frac{k_{sorb}}{k_{desorb}} = \frac{C_s}{C_w} = K_d. \qquad\qquad (4.26)$$

Equation 4.26 shows that if $k_{sorb} \gg k_{desorb}$, which is the case of strong sorption, then at equilibrium $C_s \gg C_w$, and more of the contaminant is sorbed than is dissolved, as expected. If $k_{sorb} \ll k_{desorb}$, the reverse is true. Thus, strongly sorbed contaminants accumulate to higher concentrations in the soil than do more weakly sorbed contaminants.

In terms of contaminant mobility, strongly sorbed contaminants remain sorbed to soil surfaces longer than do weakly sorbed contaminants and consequently move more slowly through the subsurface than does the groundwater in which they are dissolved. The movement of contaminants dissolved in groundwater through the subsurface is analogous to the movement of analytes through a chromatograph column. Because each analyte binds to the column wall (stationary phase) with a unique binding strength, different contaminants move through the column at different velocities and eventually become separated in space.

Contaminants dissolved in groundwater are similarly retarded in their downgradient movement relative to the flow of groundwater. Soil serves as the chromatographic stationary phase. The extent of retardation is related to the contaminants' value of K_d. At the front of a groundwater contaminant plume one will find the fastest moving contaminants with the lowest values of K_d (or weakest

TABLE 4.6
Chemical Properties Used for Calculating Partition Coefficients and Retardation Factors[a]

CAS No.	Compound	Chemical Group[b]	S Solubility of Pure Compound (mg/L)	HLC Henry's Law Constant (atm-m³/L)	H Henry's Law Constant (unitless)	log K_{ow}	log K_{oc}	K_{oc} Organic Carbon Partition Coefficient (L/kg)	T_{bp} Normal Boiling Point (°C)	T_{mp} Normal Melting Point[c] (°C)
83-32-9	Acenaphthene	1	4.24E+00	1.55E-04	6.36E-03	3.92	3.85	7.08E+03	288	93
120-12-7	Anthracene	1	4.34E-02	6.50E-05	2.67E-03	4.55	4.47	2.95E+04	324	215
71-43-2	Benzene	2	1.75E+03	5.55E-03	2.28E-01	2.13	1.77	5.89E+01	178	5.5
56-55-3	Benzo(a)anthracene	1	9.40E-03	3.35E-06	1.37E-04	5.70	5.60	3.98E+05	376	84
205-99-2	Benzo(b)fluoranthene	1	1.50E-03	1.11E-04	4.55E-03	6.20	6.09	1.23E+06	380	168
207-08-9	Benzo(k)fluoranthene	1	8.00E-04	8.29E-07	3.40E-05	6.20	6.09	1.23E+06	401	217
50-32-8	Benzo(a)pyrene	1	1.62E-03	1.13E-06	4.63E-05	6.11	6.01	1.02E+06	380	176
117-81-7	Bis(2-ethylhexyl)phthalate	1	3.40E-01	1.02E-07	4.81E-06	7.30	7.18	1.51E+07	347	-55
56-23-5	Carbon tetrachloride	2	7.93E+02	3.04E-02	1.25E+00	2.73	2.24	1.74E+02	177	-23
57-74-9	Chlordane	2	5.60E-02	4.86E-05	1.99E-03	6.32	5.08	1.20E+05	329	106
108-90-7	Chlorobenzene	2	4.72E+02	3.70E-03	1.52E-01	2.86	2.34	2.19E+02	207	-45
67-66-3	Chloroform	2	7.92E+03	3.67E-03	1.50E-01	1.92	1.60	3.98E+01	168	-64
218-01-9	Chrysene	1	1.60E-03	9.46E-05	3.88E-03	5.70	5.60	3.98E+05	379	258
50-29-3	DDT	1	2.50E-02	8.10E-06	3.32E-04	6.53	6.42	2.63E+06	278	109
53-70-3	Dibenzo(a,h)anthracene	1	2.49E-03	1.47E-08	6.03E-07	6.69	6.58	3.80E+06	395	269
84-74-2	Di-n-butylphthalate	1	1.12E+01	9.38E-10	3.85E-08	4.61	4.53	3.39E+04	323	-35
95-50-1	1,2-Dichlorobenzene	2	1.56E+02	1.90E-03	7.79E-02	3.43	2.79	6.17E+02	234	-17
106-46-7	1,4-Dichlorobenzene	2	7.38E+01	2.43E-03	9.96E-02	3.42	2.79	6.17E+02	231	53
75-34-3	1,1-Dichloroethane (1,1-DCA)	2	5.06E+03	5.62E-03	2.30E-01	1.79	1.50	3.16E+01	166	-97
107-06-2	1,2-Dichloroethane (1,2-DCA)	2	8.52E+03	9.79E-04	4.01E-02	1.47	1.24	1.74E+01	181	-36
75-35-4	1,1-Dichloroethlyene	2	2.25E+03	2.61E-02	1.07E+00	2.13	1.77	5.89E+01	152	-123
156-59-2	cis-1,2-Dichloroethylene	2	3.50E+03	4.08E-03	1.67E-01	1.86	1.55	3.55E+01	168	-80
156-60-5	trans-1,2-Dichloroethlyene (DCE)	2	6.30E+03	9.38E-03	3.85E-01	2.07	1.72	5.25E+01	161	-50

TABLE 4.6 (continued)
Chemical Properties Used for Calculating Partition Coefficients and Retardation Factors[a]

CAS No.	Compound	Chemical Group[b]	S Solubility of Pure Compound (mg/L)	HLC Henry's Law Constant (atm-m³/L)	H Henry's Law Constant (unitless)	log K_{ow}	log K_{oc}	K_{oc} Organic Carbon Partition Coefficient (L/kg)	T_{bp} Normal Boiling Point (°C)	T_{mp} Normal Melting Point[c] (°C)
78-87-5	1,2-Dichloropropane	2	2.80E-03	2.80E-03	1.15E-01	1.97	1.64	4.37E+01	187	-70
60-57-1	Dieldrin	2	1.95E-01	1.51E-05	6.19E-04	5.37	4.33	2.14E+04	323	176
121-14-2	2,4-Dinitrotoluene	1	2.70E+02	9.26E-08	3.80E-06	2.01	1.98	9.55E-01	310	71
115-29-7	Endosulfan	2	5.10E-01	1.12E-05	4.59E-04	4.10	3.33	2.14E+03	357	106
72-20-8	Endrin	2	2.50E-01	7.52E-06	3.08E-04	5.06	4.09	1.23E+04	381	200
100-41-4	Ethylbenzene	2	1.69E+02	7.88E-03	3.23E-01	3.14	2.56	3.63E+02	209	-95
206-44-0	Fluoranthene	1	2.06E-01	1.61E-05	6.60E-04	5.12	5.03	1.07E+05	347	108
86-73-7	Fluorene	1	1.98E+00	6.36E-05	2.61E-03	4.21	4.14	1.38E+04	299	115
76-44-8	Heptachlor	1	1.80E-01	1.09E-03	4.47E-02	6.26	6.15	1.41E+06	318	96
58-89-9	γ-HCH (Lindane)	2	6.80E+00	1.40E-05	5.74E-04	3.73	3.03	1.07E+03	314	113
193-39-5	Indeno(1,2,3-cd)pyrene	1	2.20E-05	1.60E-06	6.56E-05	6.65	6.54	3.47E+06	432	162
74-83-9	Methyl bromide	2	1.52E+04	6.24E-03	2.56E-01	1.19	1.02	1.05E+01	136	-94
75-09-2	Methylene chloride	2	1.30E+04	2.19E-03	8.98E-02	1.25	1.07	1.17E+01	156	-96
91-20-3	Naphthalene	1	3.10E+01	4.83E-04	1.98E-02	3.36	3.30	2.00E+03	255	80
108-95-2	Phenol	1	8.28E+04	3.97E-07	1.63E-05	1.48	1.46	2.88E+01	235	41
129-00-0	Pyrene	1	1.35E-01	1.10E-05	4.51E-04	5.11	5.02	1.05E+05	353	151
100-42-5	Styrene	1	3.10E+02	2.75E-03	1.13E-01	2.94	2.89	7.76E+02	215	-33
79-34-5	1,1,2,2-Tetrachloroethane (PCA)	2	2.97E+03	3.45E-04	1.41E-02	2.39	1.97	9.33E+01	215	-44
127-18-4	Tetrachloroethylene (PCE, PERC)	2	2.00E+02	1.84E-02	7.54E-01	2.67	2.19	1.55E+02	201	-22
108-88-3	Toluene	2	5.26E+02	6.64E-03	2.72E-01	2.75	2.26	1.82E+02	195	-95
8001-35-2	Toxaphene	1	7.40E-01	6.00E-06	2.46E-04	5.50	5.41	2.57E+05	347	65 to 90
120-82-1	1,2,4-Trichlorobenzene	2	3.00E+02	1.42E-03	5.82E-02	4.01	3.25	1.78E+03	252	17
71-55-6	1,1,1-Trichloroethane (1,1,1-TCA)	2	1.33E+03	1.72E-02	7.05E-01	2.48	2.04	1.10E+02	175	-30

TABLE 4.6 (continued)
Chemical Properties Used for Calculating Partition Coefficients and Retardation Factors[a]

CAS No.	Compound	Chemical Group[b]	S Solubility of Pure Compound (mg/L)	HLC Henry's Law Constant (atm·m³/L)	H Henry's Law Constant (unitless)	log K_{ow}	log K_{oc}	K_{oc} Organic Carbon Partition Coefficient (L/kg)	T_{bp} Normal Boiling Point (°C)	T_{mp} Normal Melting Point[c] (°C)
79-00-5	1,1,2-Trichloroethane (1,1,2-TCA)	2	4.42E+03	9.13E-04	3.74E-02	2.05	1.70	5.01E+01	197	-37
79-01-6	Trichloroethylene (TCE)	2	1.10E+03	1.03E-02	4.22E-01	2.71	2.22	1.66E+02	183	-85
75-01-4	Vinyl chloride	2	2.76E+03	2.70E-02	1.11E+00	1.50	1.27	1.86E+01	126	-154
108-38-3	m-Xylene	2	1.61E+02	7.34E-03	3.01E-01	3.20	2.61	4.07E+02	211	-48
95-47-6	o-Xylene	2	1.78E+02	5.19E-03	2.13E-01	3.13	2.56	3.63E+02	214	-25
106-42-3	p-Xylene	2	1.85E+02	7.66E-03	3.14E-01	3.17	2.59	3.89E+02	211	13

[a] Adapted from: U.S. EPA, 1996, *Soil Screening Guidance: Technical Background Document*, Office of Emergency and Remedial Response, Washington, D.C., EPA/540/R95/128.

[b] Group 1: Semi-volatile nonionizing organic compounds. Fitted to: log K_{oc} = 0.983 log K_{ow} + 0.00028.
Group 2: Volatile organic compounds, chlorobenzenes, and certain chlorinated pesticides. Fitted to: log K_{oc} = 0.7919 log K_{ow} + 0.0784.

[c] Compounds solid at soil temperature are defined as those with a melting point > 20°C.
Compounds liquid at soil temperature are defined as those with a melting point < 20°C.

sorption strengths), and moving back upgradient through the plume, there are progressively slower moving contaminants with progressively larger values of K_d.

Rules of Thumb

The mobility of dissolved organic compounds in groundwater depends on K_d.

1. If $K_d = 0$, the organic does not sorb to the soils it passes through and moves at the groundwater velocity.
2. If $K_d > 0$, movement of the organic is retarded.

Example 4.10

Consider the case where groundwater contaminated with benzene and toluene is moving through subsurface soils containing 1.6% organic carbon. Compare benzene and toluene qualitatively with respect to their mobility in the subsurface.

Answer: From Table 4.6,

$$K_{oc}(\text{benzene}) = 58.9 \text{ L/kg, and } K_{oc} (\text{toluene}) = 182 \text{ L/kg.}$$

Using $K_d = K_{oc}f_{oc}$,

$$K_d (\text{benzene}) = 58.9 \times 0.016 = 0.942 \text{ and } K_d (\text{toluene}) = 182 \times 0.016 = 2.91.$$

K_d (toluene) is approximately 3 times larger than K_d(benzene), indicating that toluene sorbs more strongly to the soil and will move significantly more slowly than benzene through the subsurface.

RETARDATION FACTOR

A retardation factor for contaminant movement can be calculated using K_d.[3,4]
    The *retardation factor, R,* for a contaminant is defined as

$$R = \frac{\text{average linear velocity of groundwater}}{\text{average linear velocity of contaminant}}. \tag{4.27}$$

Thus, a retardation factor of 10 means that the contaminant moves at one-tenth of the average velocity of the groundwater. The average linear velocity of the contaminant is measured at the point where its concentration is equal to 1/2 of its concentration at the contaminant source.

Assuming a linear partition coefficient, $K_d = \dfrac{C_{soil}}{C_w}$, the retardation factor becomes[3]

$$R = 1 + \frac{\rho K_d}{h}, \tag{4.28}$$

or for a Freundlich partition coefficient, $K_d = \dfrac{C_{soil}}{C_w^n}$,

$$R = 1 + \left(\frac{\rho}{h}\right)\left(\frac{K_d}{n}\right)\left(C_w^{1/(n-1)}\right), \tag{4.29}$$

where

ρ = soil bulk density (g/cm^3), typically 1.5 to 1.9 g/cm^3.

h = effective soil porosity,* typically 0.35 to 0.55.

Some retardation factors and qualitative mobility classifications determined from data in Table 4.6 are in Table 4.7. The table was developed using the linear Equation 4.28 with typical values for soil f_{oc}, porosity, and bulk density.

TABLE 4.7
Calculated Retardation Factors and Mobility Classifications

R	Examples	K_d	Mobility Classification
<3	Methylene Chloride, MTBE, 1,2-DCA	<0.03	Very Mobile
3–9	Benzene, 1,1-DCA, Chloroform	0.03–1.2	Mobile
9–30	Ethylbenzene, Toluene, Xylenes	1.2–4.3	Intermediate
30–100	Styrene, Pyrene, Lindane	4.3–15	Low Mobility
>100	Naphthalene, Dioxin, Heptachlor	>15	Immobile

Note: Assumed values: f_{oc} = 0.01; ρ = 2.0 g/cm^3; h = 0.3.

EFFECT OF BIODEGRADATION ON EFFECTIVE RETARDATION FACTOR

Equations 4.28 and 4.29 are based only on sorption to organic carbon and do not consider the influence of processes like biodegradation, ion-exchange, precipitation, and chemical changes. As shown in Figure 4.6, biodegradation causes additional retardation to plume movement.

Rules of Thumb

1. Dissolved contaminants generally move more slowly than groundwater because of sorption, ion-exchange, precipitation, and biodegradation, but the calculated retardation factor considers sorption only.
2. The movement rate of biodegradable compounds is overestimated by the sorption retardation factor, especially over long periods of time.
3. Sorption retardation factors not only slow the growth of a plume but also the rate of cleanup by pump-and-treat, and they increase the water volume that must be extracted.

Example 4.11

Calculation of a Sorption Retardation Factor

Calculate the retardation factor for 1,2-dichloroethane (1,2-DCA) in a soil with 2.7% TOC (total organic carbon), dry bulk density of 1.7 g/cm^3, and soil porosity of 40%.

---

* Soil *porosity* is the ratio of the volume of empty space (pore volume) to total volume of soil. It can be expressed as a simple ratio or as a percentage. Thus, for a soil sample in which 1/4 of the total volume is empty, or void, space, the porosity may be expressed as 0.25 or 25%. *Effective porosity* is the ratio of the volume of effective void space to the total volume of soil, expressed as a percentage. Effective porosity accounts for the fact that pore spaces that are not connected, or are too small for water to overcome capillarity, do not contribute to fluid movement. The difference between true porosity and effective porosity is most noticeable with clay, where true porosity is very high, 34–60%. Effective porosity, however, is very low, 1–2%, making clay relatively impermeable.

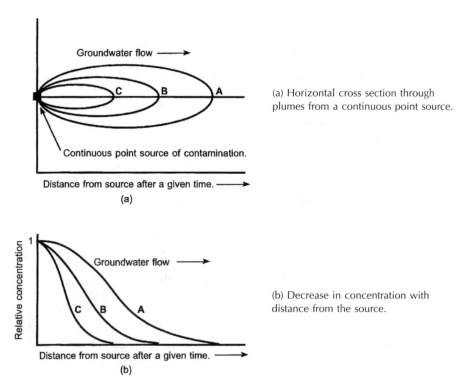

(a) Horizontal cross section through plumes from a continuous point source.

(b) Decrease in concentration with distance from the source.

FIGURE 4.6   Positions of contaminant plume front as affected by different flow conditions. A represents a nonretarded plume, B represents a plume retarded by sorption, and C represents a plume retarded by sorption and biodegradation.

Answer: $K_d = K_{oc}f_{oc}$ where $K_{oc} = 17.4$ (from Table 4.6).

$$R = 1 + \frac{rK_d}{h}.$$

r = dry bulk soil density = 1.7 g/cm³.

h = effective porosity of soil = 0.40.

$f_{oc} = 0.025$.

$K_d = 17.4 \times 0.027 = 0.470$.

$$R = 1 + \frac{1.7 \times 0.470}{0.40} = 3.0.$$

Divide the groundwater velocity by 3.0 to get the velocity with which 1,2-DCA moves through this particular soil.

Example 4.12

Calculate the sorption retardation factor for tetrachloroethene (PERC) in a silty clayey loam soil, using data from Table 4.6. Soil density is 2.5 g/cm³ and porosity is 31%.

Answer: For tetrachloroethene, $K_{oc} = 155$. Since no measured value for f_{oc} is available, a value from Table 4.5 can be used. For a silty clayey loam soil, $f_{oc} \approx 0.03$.

$$K_d = 155 \times 0.03 = 4.7, \text{ and } R = 1 + \frac{2.5 \times 4.7}{0.31} = 38.$$

Tetrachloroethene has a K_{oc} nearly 10 times that of 1,2-DCA, which means that it is much less water-soluble. Thus, it is less mobile and has a higher retardation factor.

A MODEL FOR SORPTION AND RETARDATION

Consider the result of Example 4.11, where the retardation factor for 1,2-DCA was found to be 3.0 in a particular soil. Since the contaminant is slowed by a factor of 3 relative to the groundwater velocity, it might appear that flushing 3 pore volumes of water through the impacted soil would completely desorb 1,2-DCA from that part of the subsurface. In other words, the retardation factor might be erroneously interpreted as the number of groundwater pore volumes that must be flushed through the contaminated zone to desorb a contaminant from the impacted soil.

In fact, however, most low solubility organic contaminants can never be completely flushed from the soil because normally there is some fraction of the contaminant that becomes almost irreversibly bound to the organic matter in soil. This part cannot be removed in any reasonable time by water flushing. It can remain in the soil as a slowly diminishing source of groundwater contamination for tens, or even hundreds, of years.

There is an aging process that causes the fraction of a contaminant that is desorbed by flushing to decrease with time. This is why pump-and-treat remediation methods are seldom successful at removing the last part of contamination from the soil. When pump-and-treat methods become ineffective, other approaches, such as bioremediation, are needed to achieve required cleanup levels.

A conceptual model for sorption has been proposed[1] that is consistent with time-dependent irreversible sorption and other observations, such as a decrease in biodegradation rates with time.

Model of Sorption

- A compound with large K_d is initially adsorbed rapidly from water to the external surfaces of soil particles.
- The initially adsorbed fraction can be rapidly desorbed again to water, if the elapsed time is not too long. It is also available to microorganisms for biodegradation, and to organisms that might be susceptible to toxic effects such as fish and humans.
- As time passes, a portion of the surface-adsorbed compound begins to diffuse into micropores in the solid surface, moving away from the surface into the particle interior. Here, the compound becomes sequestered within the soil in locations remote from the surface where the compound is less available for desorption.
- Thus, aging appears to be associated with continuous diffusion into more remote sites on the solid particle where the molecules are retained and rendered less accessible to biological, chemical, and physical changes. After 1 to 10 years, depending on site-specific soil and pollutant conditions, a large fraction of the remaining sorbed pollutant will not desorb from the soil.

Another limitation of the retardation factor is that the equations unrealistically assume that the soil matrix is homogeneous, which is rarely the case. However, the results are still useful for rough estimates and for estimating relative mobilities of dissolved contaminants.

SOIL PROPERTIES

The subsurface environment contains inorganic minerals, organic humic materials, air, and water. Also found are plant roots, microorganisms, and burrowing animals, not to mention building

TABLE 4.8
**Representative Values of Effective Porosity
for Some Soil Types**

Soil Type	Effective Porosity (%)
Well-sorted sand or gravel	25–40
Sand and gravel, mixed	20–35
Medium sand	15–30
Glacial sediments	5–20
Silt	1–20
Clay	1–2

TABLE 4.9
Soil Particle Size Range for Some Soil Types

Soil type	Particle size range (mm)
Clay	<0.002
Silt	0.002–0.04
Very fine sand	0.04–0.10
Fine sand	0.10–0.20
Medium sand	0.20–0.40
Coarse sand	0.40–0.9
Very coarse sand	0.9–2.0
Fine gravel	2.0–10.0
Medium gravel	10.0–20
Coarse gravel	20–40
Very coarse gravel	40–80

foundations, utility service lines, and other man-made structures. All these can affect the movement of contaminants through the subsurface.

The physical properties of soil that have the greatest effect on water and contaminant movement are *effective porosity, particle size range*, and *hydraulic conductivity*. Representative values for these properties are given in Table 4.8, 4.9, and 4.10.

- *Effective porosity* is the ratio of the volume of effective void space to the total volume of material, expressed as a percentage. Effective porosity accounts for the fact that pore spaces that are not connected, or are too small for water to overcome capillarity, do not contribute to fluid movement. The difference between true porosity and effective porosity is most noticeable with clay, where true porosity is very high, 34–60%. Effective porosity, however, is very low, 1–2%, making clay relatively impermeable.
- *Soil particle range* determines the average soil pore size and strongly influences the capillary attraction of soil particles to moving liquids. The smaller the pore size, the stronger is capillary attraction and the greater is the total particle surface area within a given volume of soil. The larger the surface area, the larger is the volume of liquid immobilized by sorption to soil surfaces. Where pore size is small, the distance between adjacent particle surfaces also is small and capillary attraction can extend across significant fractions of the pore volume. Capillary attractions retard the movement of liquids through soil and can also immobilize a fraction of the liquid. Thus, silt retards the movement of liquids more than coarse sand. Silt also will immobilize a larger quantity of liquid than will an equal volume of coarse sand.

- *Hydraulic conductivity* indicates the ability of subsurface materials to transmit a particular fluid. For example, the hydraulic conductivity of a given soil to transmit water is greater than it would be to transmit more viscous fluids, such as diesel fuel. The hydraulic conductivity of a very porous soil, such as coarse sand, is greater than for a less porous soil, such as fine silt.

TABLE 4.10
Representative Values of Water Hydraulic Conductivity for Some Soils

Soil Type	Hydraulic Conductivity for Water: Typical Range (cm/sec)
Clay	10^{-6} to 10^{-9}
Silt	10^{-3} to 10^{-7}
Fine sand	10^{-2} to 10^{-5}
Medium sand	10^{-1} to 10^{-4}
Coarse sand	1 to 10^{-4}
Gravel	10^{2} to 10^{-1}

4.8 PARTICULATE TRANSPORT IN GROUNDWATER: COLLOIDS

Contaminants move in groundwater systems as dissolved species in the water, as flowing free-phase liquids, or combined with moving particulates. Particulates that can move through soils with groundwater must be small enough to move through the soil pore spaces. Such particulates are generally less than 2.0 μm (micron) in diameter and are called *colloids*. Colloids are a special class of matter with properties that lie between those of the dissolved state and the solid or immiscible liquid states.

Colloids have a high surface area to mass ratio due to their very small size. Groundwater concentrations of colloidal materials can be as high as 75 mg/L, corresponding to as many as 10^{12} particles/L. This represents a large surface area available for transporting sorbed contaminants.

There are many sources of colloidal material in groundwater. Colloids are formed in soil when fragments of soil, mineral, or contaminant particles become detached from their parent solid because of weathering. Then they are carried to the groundwater when water from irrigation or precipitation percolates downward through the soil. Colloids also form as fine precipitates when dissolved minerals in groundwater undergo pH or redox potential changes. Colloids are often introduced directly into groundwater from landfills, and they can form as emulsions of small droplets from free-phase hydrocarbons or other immiscible liquids.

COLLOID PARTICLE SIZE AND SURFACE AREA

When particle size is reduced to 1–2 μm or smaller, surface forces arising from surface charge or London force attractions begin to exert a significant influence on particle behavior. Consider the effect of reducing the particle size of a given mass of solid. A cube that is 10 mm on a side has a surface area of 6.0×10^{-4} m^2. Cut it in half in each of the 3 directions to get 8 cubes, which are each 5 mm on a side. The surface area is now 12.0×10^{-4} m^2. Continue subdividing until the cubes are 1 μm on an edge. The total surface area is now 6.0 m^2 — an increase of 10,000 times over the original cube.

Montmorillonite, a clay mineral, in the dispersed state may break down into platy particles only one unit cell in thickness, about 10^{-9} meters. Its specific surface area then is about 800 m^2/g. A monolayer of 10 g of this material would cover a football field.

PARTICLE TRANSPORT PROPERTIES

Contaminants of low solubility can be transported as colloids or associated by adsorption or occlusion with colloids, resulting in the unexpected mobility of a low-solubility material. When contaminants are sorbed to colloids, their transport behavior is determined by the properties of the colloid, not the sorbed contaminant. In the low velocity flow conditions of groundwater, particles larger than 2 μm tend to settle by gravity. Particles smaller than 0.1 μm tend to sorb readily to larger soil particles, becoming retarded or immobilized. Thus, particles in the range 0.1 to 2.0 μm are the most mobile in groundwater.

Adsorptive Processes

Colloid adsorptive behavior is influenced by

1. Forces acting on colloidal particles
    a) Electrostatic attraction and repulsion
    b) London attractions
    c) Brownian motion
2. Properties of the groundwater
    a) Ionic strength (related to TDS and conductivity)
    b) Ionic composition
    c) Flow velocity
3. Properties of the colloids
    a) Size
    b) Chemical nature
    c) Concentration
4. Properties of the soil matrix
    a) Geologic composition
    b) Particle size distribution
    c) Soil surface area

Rules of Thumb

1. Usually, the most important factor governing colloid behavior is groundwater chemistry, and the least important is flow velocity.
2. Generally, colloids are more mobile when the dissolved ion concentration is *low*, and less mobile in solutions with high salt content.

ELECTRICAL CHARGES ON COLLOIDS AND SOIL SURFACES

The explanation for how water chemistry affects colloidal behavior lies in the behavior of charged particles in water solutions. Soil surfaces in an aquifer generally have a net negative charge due to the dominance of silicates in the minerals which have exposed electronegative oxygen atoms. Colloidal particles also are usually charged.

- Metal oxide colloids tend to be positively charged.
- Sulfur and the noble metals tend to be negatively charged.
- Organic macromolecules, such as humic materials, proteins, or flocculating resins, acquire a charge that depends on the pH. At a low pH, protons bind to the molecules and make the colloid positive. At a high pH, the binding of hydroxyl ions makes the charge negative.

The Electrical Double Layer

The charge on colloids and on soil surfaces attracts dissolved ions into a configuration known as an *electric double layer*. Consider the double layer that forms adjacent to a negatively charged surface. The first layer forms when dissolved cations (positive ions) are attracted to the oppositely charged surfaces. Then anions (negative ions) are attracted to the positive ion region of the first layer to form a second, more diffuse, oppositely charged layer surrounding the first. Positively charged surfaces acquire an inner layer of anions and an outer layer of cations. The inner layer effectively neutralizes the surface charge, and the second layer effectively neutralizes the inner layer. Subsequent layers can form in principle, but they are too diffuse to have any effect.

Adsorption and Coagulation

If two colloid particles can come close enough together, London attractive forces will pull them together into a larger particle starting the process of coagulation. The same thing happens if a colloidal particle comes close enough to a soil grain. London attractive forces will pull them together and the colloid will become sorbed to the soil surface.

Two colloid particles of the same material have the same sign of charge in the outer part of their double layer and, therefore, repel one another. For two colloid particles to coagulate, they must collide with enough energy to force past their repulsive double layers and approach close enough for London attractive forces to be effective. The same is true for adsorption to soil particle surfaces.

Brownian motion provides the collisional energy for overcoming the electrical repulsion of the double layer. In high-energy collisions, particle momentum overcomes charge repulsion and allows particles to approach close enough to enter the zone of London attraction where adsorption and coagulation can occur.

- *At high ionic strength* (high dissolved ion concentrations; high TDS), the double layer is thin because the ion atmosphere is dense. The higher ion charge density neutralizes the net particle charge with a thinner layer.
- *At low ionic strength* (low TDS), a thicker ionic charge layer is required to neutralize the charge on colloidal particles.

Thus, colloidal particles can approach surfaces and other particles more closely at high TDS than at low TDS, allowing London attraction to be more effective. There are several familiar illustrations of this principle:

- When river water carrying colloidal clay reaches the ocean, the salt water induces coagulation. This is a major cause of silting in estuaries.
- Applying a styptic pencil stops bleeding from small cuts. The styptic pencil contains aluminum salts. Dissolving high concentrations of Al^{3+} into the wound initiates coagulation of colloidal proteins in the blood.
- Colloidal material can be "salted out" by dissolving salts into the solution.
- The high TDS of landfill leachate provides optimal conditions for immobilizing entrained colloids by sorption and coagulation.

Rule of Thumb

Coagulation and adsorption of colloids are more efficient under high TDS conditions.

4.9   BIODEGRADATION

For organic contaminants of low solubility and low volatility, biodegradation is the ultimate form of contaminant removal and soil and groundwater cleanup. Because some fraction of these contaminants always becomes nearly irreversibly sorbed to soil and cannot readily be removed mechanically, it can remain in place for many years continually serving as a source of groundwater contamination. There are only three practical approaches to achieving a complete remediation in such cases:

1. Excavation of the contaminated soil and treating it on the surface (e.g., by incineration), or sending it to a landfill.
2. Isolating the contaminated soil by capping its surface with a clay or membrane liner, and diverting surface and groundwater away from the area.
3. Allowing biodegradation to transform the contaminants into nonhazardous substances.

Biodegradation is often the most economical and practical approach. If the rate of natural biodegradation (intrinsic bioremediation) is too slow, adding nutrients, oxygen, and appropriate microbes can often accelerate it (engineered bioremediation). Much progress has been made in recent years in understanding the many different processes of biodegradation and adapting them to successfully degrading types of organic compounds once thought resistant to biodegradation such as chlorinated hydrocarbons. The EPA issues many bulletins related to this new technology. The EPA's Web site is a good place to find the latest references. An extensive and excellent reference for the application of intrinsic bioremediation to fuel hydrocarbon contaminants is Wiedemeier, et al., 1995 (or Reference 13).

BASIC REQUIREMENTS FOR BIODEGRADATION

Biodegradation is an oxidation-reduction process. This means that energy for biodegradation arises from electron-transfer reactions. There are six basic requirements that must be present for biodegradation to occur:

1. Suitable degrading organisms, generally bacteria or fungi.
2. An energy source, generally organic carbon, which is mostly oxidized to CO_2. The energy source is the electron-donor.
3. Electron acceptors, generally O_2, NO_3^-, SO_4^{2-}, Fe^{3+}, and CO_2.
4. Carbon for cell growth, which also comes from organic carbon. About 50% of bacterial dry mass is carbon.
5. Nutrients, including nitrogen, phosphorus, calcium, magnesium, iron, and trace elements. Nitrogen and phosphorus are needed in the largest quantities. Bacterial dry mass is about 12% nitrogen and about 2–3% phosphorus.
6. Acceptable environmental conditions: pH, salinity, hydrostatic pressure, solar radiation, toxic substances, oxygen, etc., must all be within acceptable limits for the particular bioprocesses.

Upon accepting electrons from an energy source, electron acceptors are converted to the products indicated in Equations 4.30–4.34.

$$e^- + O_2 \rightarrow H_2O. \tag{4.30}$$

$$e^- + NO_3^- \rightarrow N_2. \tag{4.31}$$

$$e^- + SO_4^{2-} \rightarrow H_2S. \tag{4.32}$$

$$e^- + Fe^{3+} \rightarrow Fe^{2+}. \tag{4.33}$$

$$e^- + CO_2 \rightarrow CH_4. \tag{4.34}$$

Microbes capable of degrading petroleum hydrocarbons generally prefer a pH between 6 and 8, and temperatures between 5°C and 25°C. Temperatures lower than 5°C tend to inhibit biodegradation in general. Microbial reactions can be grouped into two classes, aerobic and anaerobic. Aerobic microbial reactions require the presence of oxygen as an electron acceptor. Anaerobic microbial reactions utilize electron acceptors other than oxygen. When sufficient oxygen is present, it will be utilized in preference to alternative electron acceptors. Aerobic and anaerobic reactions require different kinds of bacteria, with aerobic reactions being considerably faster than anaerobic reactions.

NATURAL AEROBIC BIODEGRADATION OF NAPL HYDROCARBONS

Hydrocarbons of low solubility are called *nonaqueous phase liquids,* or NAPL. NAPL is further divided into hydrocarbons that are less dense than water, *light nonaqueous phase liquids,* or LNAPL, and hydrocarbons that are denser than water, *dense nonaqueous phase liquids,* or DNAPL. If nonaqueous phase liquids are mixed with water, they separate from water into a separate immiscible liquid phase. LNAPL floats on the water surface and DNAPL sinks to the bottom of the water. Generally, most of the NAPL (>90%) is in the immiscible phase and only a small fraction dissolves into the water (<10%).

When natural biodegradation of NAPL (oils, many solvents, gasoline, etc.) occurs in the saturated zone, indigenous aerobic bacteria react with dissolved oxygen to consume some of the immiscible-phase NAPL directly. These bacteria also form a biosurfactant that helps to increase the rate of NAPL dissolution into groundwater.

Rules of Thumb

1. Within the unsaturated zone, *in situ* aerobic biodegradation of NAPL is often nutrient-limited by available nitrogen. Addition of nitrogen, as nitrate or ammonia, usually enhances biodegradation.
2. Within the saturated zone, dissolved oxygen is usually the limiting factor for *in situ* aerobic biodegradation of NAPL.
3. For fuels, the contaminants of regulatory interest are usually the most toxic and soluble components known as BTEX: benzene, toluene, ethylbenzene, and the xylene isomers.
4. It takes about 1 mg of O_2 to biodegrade 0.32 mg of BTEX.
5. About 95% of the easily biodegradable hydrocarbons are converted to CO_2 and water in a few months. These are low molecular weight unbranched alkanes (smaller than C-30 to C-40) and aromatics (smaller than C-10).
6. The remainder — unbranched alkanes larger than C–40, branched alkanes, alkenes, and aromatics larger than C-10 — can last for many years.
7. Polar hydrocarbons containing S, O, N, Cl, or Br may be resistant to biodegradation.
8. Chemicals that are highly water soluble biodegrade more readily than those with low water solubility.
9. Chemicals that adsorb weakly to soils biodegrade more readily than those that adsorb strongly.
10. Chemicals with low K_{ow} values biodegrade more readily than those with high K_{ow} values.
11. Chemicals that leach easily from soils biodegrade more readily than those that are not easily leached.
12. Rates of hydrocarbon biodegradation roughly double for every 10°C increase in groundwater temperature, over the range 5 to 25°C.

Example 4.13

How long will it take to naturally biodegrade BTEX from a 100-gallon gasoline (an LNAPL) release (about 250 kg) in the saturated zone, given the following conditions:

Depth of LNAPL penetration into saturated zone = 2 m.
Width of LNAPL release in saturated zone = 10 m.
Groundwater Darcy velocity = 1 m/day.
Background DO concentration = 5 mg/L.
Oxygen-hydrocarbon consumption ratio = 1 mg O_2/0.32 mg BTEX (from rules of thumb).

   Assume the gasoline LNAPL is immobilized by sorption in the soil matrix and that BTEX is 25% of the LNAPL weight. Also assume that aerobic biodegradation is instantaneous compared to normal groundwater movement. In the presence of excess oxygen, aerobic bacteria can degrade 1 mg/L of benzene in about 8 days, essentially instantaneous compared to the years required for flowing groundwater to replenish a plume area with oxygen. Under these conditions, the rate of biodegradation is equal to the rate at which sufficient dissolved oxygen can be brought into the residual gasoline LNAPL plume by groundwater flow. Approach the problem by calculating how long it will take enough water to pass through the plume cross-sectional area to supply 1 mg O_2 per 0.32 mg of BTEX.

Answer:

$$\text{Time to degrade} = (250 \text{ kg NAPL}) \times \left(\frac{0.25 \text{ g BTEX}}{1 \text{ g NAPL}} \right) \times (10^6 \text{ mg/kg}) \times \left(\frac{1 \text{ mg } O_2}{0.32 \text{ mg BTEX}} \right).$$

$$\times \left(\frac{1 \text{ L}}{5 \text{ mg } O_2} \right) \times \left(\frac{1}{(2 \times 10) \text{ m}^2} \right) \times \left(\frac{1 \text{ day}}{1 \text{ m}} \right) \times \left(\frac{1 \text{ m}^3}{10^3 \text{ L}} \right).$$

   Time to degrade = *1950 days or about 5.4 years.*

   This approach of calculating how rapidly dissolved oxygen can be supplied to the plume area does not work for DNAPL composed of chlorinated compounds because they are resistant to aerobic biodegradation.

4.10   BIODEGRADATION PROCESSES

The most important metabolic processes in biodegradation are redox reactions.

- Aerobic respiration:
  — Microbes use oxygen electron acceptors to transform organic carbon to carbon dioxide.
  — Electrons are transferred from the contaminant to oxygen. The contaminant is oxidized and the oxygen is reduced, forming water and carbon dioxide.
  — The key requirement is adequate oxygen.

- Anaerobic respiration:
  — Microbes use an "oxygen substitute" to serve as electron acceptors (usually nitrate, sulfate, Fe^{3+}, or carbon dioxide). Organic carbon is chemically transformed, often to carbon dioxide, but sometimes to methane.
  — Electrons are transferred from the contaminant (oxidizing it) to an electron acceptor (reducing it). Different electron acceptors form different products:

$$CO_2 \rightarrow CH_4; \; NO_3^- \rightarrow NO_2^-, \; NH_3, \text{ or } N_2; \; SO_4^{2-} \rightarrow H_2S \text{ or } S; \; Fe^{3+} \rightarrow Fe^{2+}.$$

   The key requirement is an adequate supply of electron acceptors.

- Cometabolism:
  — Enzymes produced by microbes during the degradation of some organic matter fortuitously react chemically to transform a contaminant that resists biodegradation. For example, during biodegradation of methane, some bacteria produce an enzyme that breaks down chlorinated solvents, such as trichloroethylene.
  — The key requirement is the presence of a substance that, when metabolized by microbes, produces the right enzymes to transform the contaminants.

Each of these three biodegradation processes requires an electron donor that is oxidized (usually the contaminant, except in cometabolism), and an electron acceptor that is reduced. The overall electron transfer process provides metabolic energy for the microbes. Favorable conditions for biodegrading various organic compounds are listed in Table 4.11.

TABLE 4.11
Biodegradation Processes for Some Organic Compounds

Hydrocarbons (HCs)

Gasoline, diesel, fuel oil	Readily biodegradable under aerobic conditions; more slowly degradable under anaerobic conditions.
PAHs	Aerobically biodegradable under a narrow range of conditions.
Creosote	Readily biodegradable under aerobic conditions.
Alcohols, ketones, esters	Readily biodegradable under aerobic conditions.
Ethers	Biodegradable under a narrow range of conditions using aerobic or nitrate-reducing microbes.

Chlorinated aliphatic HCs

Highly chlorinated (e.g., tetrachloroethene)	Cometabolized by anaerobic microbes; cometabolized by aerobic microbes in special cases.
Less chlorinated (e.g., dichloroethene)	Aerobically biodegradable under a narrow range of conditions; cometabolized by anaerobic microbes.

Chlorinated aromatic HCs

Highly chlorinated (e.g., pentachlorophenol)	Aerobically biodegradable under a narrow range of conditions; cometabolized by anaerobic microbes.
Less chlorinated (e.g., chlorobenzene)	Readily biodegradable under aerobic conditions.

PCBs

Highly chlorinated	Cometabolized by anaerobic microbes.
Less chlorinated	Aerobically biodegradable under a narrow range of conditions.

Nitroaromatics

TNT, nitrobenzene, etc.	Aerobically biodegradable; converted to innocuous volatile organic acids under anaerobic conditions.

Metals

Cr, Cu, Ni, Hg, Cd, Zn, etc.	Solubility and reactivity can be changed by a variety of microbial processes.

4.11 CALIFORNIA STUDY

Sometimes the best approach to treating soil that is contaminated with fuel hydrocarbons is to do nothing. If intrinsic biodegradation rates are high enough, a pollutant plume may shrink or become immobile.

Passive or intrinsic bioremediation is becoming increasingly acceptable as a treatment alternative, especially for fuel hydrocarbons (LNAPL) if there is no immediate threat to water uses,

TABLE 4.12
Hydrogeologic Factors Favoring *In Situ* Bioremediation

Type of Bioremediation	Important Site Characteristic	Favorable Indicators
Engineered	Transmissivity of subsurface to fluids	Hydraulic conductivity > 10^{-4} cm/s (if system circulates water)
		Intrinsic permeability > 10^{-9} cm^2 (if system circulates air)
	Relative uniformity of subsurface medium	Common in river delta deposits, floodplains of large rivers, and glacial outwash aquifers
	Low residual concentrations of NAPL contaminants on subsurface solids	NAPL conc. < 10,000 mg/kg
Intrinsic/Passive	Consistent groundwater flow (velocity and direction)	Seasonal variation in depth to water table < 1 meter
		Seasonal variation in regional flow trajectory < 25°
	Presence of pH buffers	Carbonate minerals (limestone, dolomite, and shell material)
	High concentration of electron acceptors	Oxygen, nitrate, sulfate, Fe^{3+}
	Presence of elemental nutrients	Nitrogen and phosphorus

and if natural site conditions are favorable (see Table 4.12). With favorable site conditions, passive remediation may be expected to stabilize a fuel contaminant plume's length and mass, even if an active source is present, such as a free product on the water table that continues to dissolve hydrocarbons into the plume. If all active sources are removed to the point of residual saturation, biodegradation is likely to eventually reduce the plume mass to the point of completing the cleanup.

These are conclusions from a study by Lawrence Livermore National Laboratory, commissioned in 1994 by the Underground Storage Tank (UST) Program of the California State Water Resources Control Board.[9,10] In a review of 200 LUFT (leaking underground fuel tanks) sites in Napa Valley, it was found that

- 57 sites had fuel-impacted groundwater.
- 51 of these sites had TPH (total petroleum hydrocarbons) and gasoline concentrations in the pollutant plume that were greater than 50 ppb.
- In 90% of these sites, the nondetect perimeter of the plume was less than 60 m from the source.
- After removing contaminant sources, degradation removed 50–60% of the remaining pollutant mass per year.

A detailed analysis of 271 LUFT cases showed that

- In general, plume lengths change slowly and tend to stabilize at short distances from the source.
- Plume boundaries, defined by a concentration of 10 ppb, extended no farther than 76 m in 90% of the cases.
- Plume mass decreased more rapidly than plume length.
- Residual fuel in the soil degraded more slowly than dissolved fuel and continued to dissolve contaminants into the groundwater. Thus, the length of the plume is defined mainly by the extent of soils containing residual adsorbed fuel.
- In 50% of sites with no actively engineered remediation, groundwater benzene concentrations decreased about 70% as fast as where pump-and-treat plus excavation treatment was applied.

> **Rule of Thumb From This Study**
>
> Once a fuel hydrocarbon source is removed, passive remediation requires 1 to 3 years to reduce the dissolved plume mass by a factor of 10.

4.12   DETERMINING THE EXTENT OF BIOREMEDIATION OF LNAPL

Spilled fuel LNAPL (light nonaqueous phase liquids) is the largest single type of subsurface contamination. LNAPL is present in the subsurface as

- Mobile LNAPL free product, which will drain from the surrounding soil into a well by gravity
- Residual LNAPL held by adsorption and capillarity in soil pore spaces, which is immobile and unable to drain into a well by gravity
- Dissolved LNAPL compounds in water
- Volatile LNAPL vapors

In an LNAPL spill, the most soluble and volatile components are lost first from the LNAPL free product. Nevertheless, the remaining LNAPL free product still contains nearly

- 90% of the benzene
- 99% of the total BTEX (benzene, toluene, ethylbenzene, and xylenes)
- 99.9% of the TPH (total petroleum hydrocarbons)

It seems clear that the first remediation step is to physically remove as much LNAPL as possible. However, frequently less than 10% of the total LNAPL can be removed by recovery of mobile LNAPL. The remaining part stays trapped in the soil by sorption and capillarity. For this remaining part of the contamination, the best choice for corrective action is bioremediation. The next remediation step should be to determine if intrinsic bioremediation is occurring at a sufficient rate such that no other action is required.

The U.S. Air Force Center for Environmental Excellence has developed a technical document that describes a protocol for data collection and analysis that can be used for judging whether intrinsic bioremediation is occurring at a useful rate.[13] This report is notable for comprehensively discussing the current state of knowledge and for its thorough list of references. Much of the following material is adapted from the Wiedemeier report.[13]

USING CHEMICAL INDICATORS OF THE RATE OF INTRINSIC BIOREMEDIATION

Certain water quality parameters change as a result of biodegradation. By measuring how these parameters change with time, the occurrence and rate of active biodegradation can be determined. The most important parameters that change are presented in Table 4.13. The first step in determining whether biodegradation is occurring at a rate fast enough to be useful for remediation is to measure the concentration of time and distance behavior of these chemical indicators.

HYDROCARBON CONTAMINANT INDICATOR

If significant biodegradation is occurring, hydrocarbon concentrations must diminish with time and distance from the spill source. However, because this could occur due to dilution, confirmatory evidence from the other indicators is always required. Seek confirmatory evidence by measuring hydrocarbon concentrations in the groundwater plume near the spill source. Analyze the ground-

TABLE 4.13
Water Quality Parameters That Indicate Biodegradation Activity

Chemical Indicators in Groundwater for Biodegradation	Trend in Indicator Concentration During Biodegradation	Processes Responsible for Trend
Hydrocarbon Concentrations.	Decreases.	Biodegradation.
Dissolved Oxygen.	Decreases.	Aerobic Respiration.
Nitrate.	Decreases.	Denitrification.
Iron (II).	Increases.	Iron (III) Reduction.
Sulfate.	Decreases.	Sulfate Reduction.
Methane.	Increases.	Methanogenesis.
Alkalinity.	Increases.	Increased by aerobic respiration, denitrification, iron (III) reduction, and sulfate reduction; not affected much by methanogenesis.
Oxidation/Reduction Potential.	Generally decreases toward plume center.	Serves as a crude indicator of which redox reactions may be operating at a given time.
Volatile Fatty Acids.	Increases.	Metabolic byproducts of biodegradation.

water for hydrocarbon compounds of regulatory concern. These are usually BTEX and trimethyl-benzenes, but sometimes TVH, TEH, and/or PAHs should also be measured. The determination of whether biodegradation is occurring at a useful rate hinges on the rate at which these parameters are disappearing.

Based on solubilities, the highest combined dissolved concentrations of BTEX plus trimethyl-benzenes should not be greater than about 30 mg/L for JP–4 jet or diesel fuel, or about 135 mg/L for gasoline. If these concentrations are exceeded, sampling errors have probably occurred, such as collecting some free product. This error is likely if emulsification of LNAPL has occurred in the water sample.

Figures 4.7a and 4.7b are idealized representations of an LNAPL groundwater plume that is caused by a gasoline spill and measured twice, one year apart.

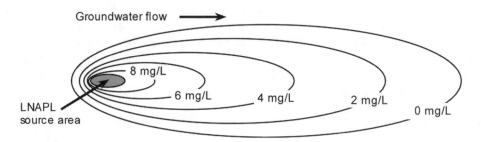

FIGURE 4.7a   Initial measurement of total BTEX isopleths for groundwater plume. Contour interval is 2 mg/L. Indicated isopleth values are mg/L of total BTEX.

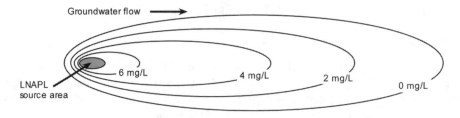

FIGURE 4.7b   Measurement of total BTEX isopleths, one year later.

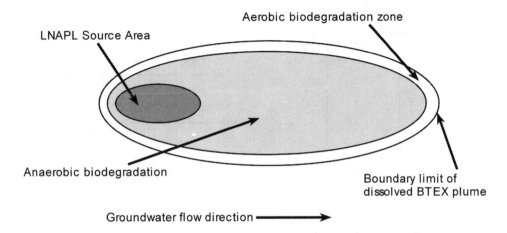

FIGURE 4.8 Aerobic and anaerobic zones of biodegradation in an LNAPL plume.

Comparing total BTEX plumes at the start of the study and one year later shows that, although plume area had increased a little, total mass of BTEX in the plume had decreased. Data from other chemical indicators may provide evidence that the decrease is primarily due to intrinsic bioremediation and not dilution.

ELECTRON ACCEPTOR INDICATORS

Subsurface microbes utilize different oxidation-reduction reactions in the order of decreasing energy-yielding value. If oxygen is available, using it as an electron acceptor will always yield the greatest energy. Therefore, aerobic biodegradation reactions always occur first if oxygen is available. Anaerobic denitrification of nitrate is the second highest energy-yielding process and will occur next. However, if dissolved oxygen (DO) is present in concentrations greater than 0.5 mg/L, it is toxic to anaerobic-only (obligate) bacteria. Therefore, nitrate denitrification cannot begin until most of the DO has been consumed.

The electron acceptors that are available in the groundwater determine the sequence in which biodegradation reactions will occur. As the oxidation-reduction potential changes from positive to increasingly negative values, different electron-acceptors are used in the biodegradation process. As reduction of electron acceptors progresses, the oxidation-reduction potential (E_h) of the groundwater becomes increasingly negative, and the energy obtained per electron transfer decreases.

Figure 4.8 shows how aerobic and anaerobic biodegradation zones develop in an LNAPL plume undergoing active biodegradation. Dissolved oxygen is rapidly depleted where LNAPL concentrations are high; therefore, anaerobic redox reactions occur in most of the plume. Water carrying dissolved oxygen diffuses into the plume from outside the plume boundary, enabling aerobic redox reactions to occur around the plume periphery. It is likely that biodegradation is inhibited in the LNAPL source area because of pollutant concentrations that are high enough to be toxic to microorganisms.

Figure 4.9 illustrates the normal sequence of the most important biodegradation redox reactions and the potential at which they are initiated. Table 4.14 gives the stoichiometry of the redox reactions. Each electron acceptor indicator is discussed in detail below.

DISSOLVED OXYGEN (DO)

If significant biodegradation is occurring, dissolved oxygen must diminish with time in the plume zone as indicated in Figures 4.10a and 4.10b. DO will be consumed first, before other electron acceptors are used. Each 1.0 mg/L of DO consumed by microbes will destroy approximately 0.32 mg/L of BTEX, based only on production of CO_2 and H_2O. This is a conservative estimate that

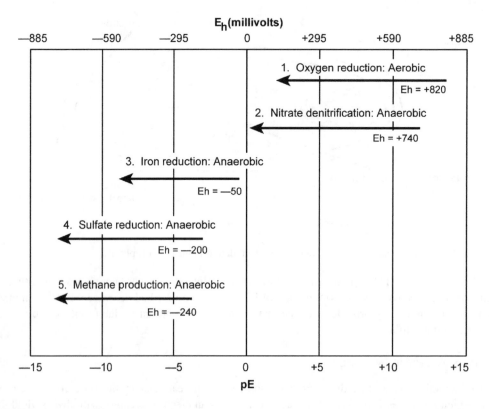

FIGURE 4.9 Order in which successive microbially mediated oxidation-reduction reactions occur, based on increasing negative redox potentials. The more positive the redox potential, the more energy is released per electron transferred. Redox potential (millivolts) = E_h = 59 pE.[13]

TABLE 4.14
Normal Sequence of Biodegradation Reactions

Oxygen Reduction	$\{CH_2O\} + O_2 \rightarrow CO_2 + H_2O$	(aerobic)
↓		
NO_3^- Reduction to N_2 (denitrification)	$5 \{CH_2O\} + 4 NO_3^- + 4 H^+ \rightarrow 5 CO_2 + 2 N_2 + 7 H_2O$	(anaerobic)
↓		
Fe^{3+} Reduction	$\{CH_2O\} + 4 FeOOH(s) + 8 H^+ \rightarrow CO_2 + 4 Fe^{2+} + 7 H_2O$	(anaerobic)
↓		
SO_4^{2-} Reduction	$2 \{CH_2O\} + SO_4^{2-} + H^+ \rightarrow 2 CO_2 + HS^- + 2 H_2O$	(anaerobic)
↓		
Methane Production	$2 \{CH_2O\} \rightarrow CH_4 + CO_2$	(anaerobic)

ignores the conversion of carbon to cell mass. If cell mass production is included, each 1.0 mg/L of DO consumed by microbes can destroy as much as 0.97 mg/L of BTEX under ideal conditions. Because other factors such as nutrient availability affect respiration, it is best to use the more conservative 0.32 mg/L value as the amount of BTEX destroyed per 1.0 mg/L of DO consumed. The overall redox reaction with benzene is

$$C_6H_6 + 7.5 \ O_2 \rightarrow 6 \ CO_2 + 3 \ H_2O. \tag{4.35}$$

Example 4.14

Suppose the background DO in groundwater at a remediation site is 6 mg/L. Excluding cell mass production, the groundwater conservatively has the capacity to biodegrade:

$$6\text{mg/L DO} \times \frac{0.32 \text{ mg/L BTEX}}{1.0 \text{ mg/L DO}} = 1.9 mg/L \; BTEX.$$

Field Test for DO Consumption by Microbes

- Analyze groundwater for DO at different locations in the BTEX plume.
- Areas with elevated BTEX concentrations should have depleted (relative to background) or zero DO concentrations.
- This is strong evidence for the occurrence of aerobic biodegradation.

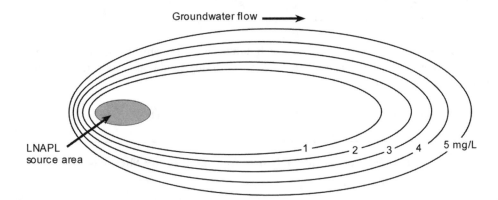

FIGURE 4.10a Initial measurements of dissolved oxygen isopleths show that dissolved oxygen levels are diminished relative to background within the BTEX plume.

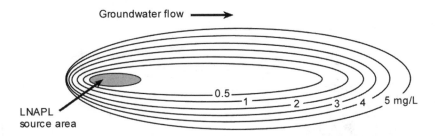

FIGURE 4.10b Dissolved oxygen isopleths one year later. The oxygen-depleted zone has become smaller, as the BTEX concentrations decreased. Reducing conditions have been established in the center of the plume.

NITRATE + NITRITE DENITRIFICATION

If significant biodegradation is occurring and DO has been depleted, nitrate and nitrite concentrations must diminish with time in the plume zone, as indicated in Figures 4.11a and 4.11b. After DO is largely consumed, anaerobic bioreactions can begin by using available nitrate (NO_3^-) and nitrite (NO_2^-) as electron acceptors. Each 1.0 mg/L of dissolved NO_3^- will destroy about 0.21 mg/L of BTEX. The final reaction products are CO_2, H_2O, and N_2.

The denitrification steps are

$$NO_3^- \rightarrow NO_2^- \rightarrow NO \rightarrow N_2O \rightarrow NH_4^+ \rightarrow N_2. \tag{4.36}$$

The overall redox reaction with benzene is

$$6\ NO_3^- + 6\ H^+ + C_6H_6 \rightarrow 6\ CO_2 + 6\ H_2O + 3\ N_2. \tag{4.37}$$

Requirements for denitrification are

- Dissolved nitrate/nitrite
- Organic carbon
- Denitrifying bacteria
- Reducing conditions (DO < 0.5 mg/L)

Denitrification is favored when

- $6.2 < pH < 10.2$
- -200 mV < redox potential (E_h) < $+665$ mV

Nitrate reduction is rapid. The rate at which nitrate and nitrite are supplied by groundwater to the reduction zone limits the reaction rate. Under denitrifying conditions, biodegradation of BTEX occurs in the following order:

toluene > p-xylene > m-xylene > ethylbenzene > o-xylene >> benzene.

Field Test for Nitrate + Nitrite Consumption by Microbes

- Analyze groundwater for nitrate + nitrite.
- Areas with elevated BTEX concentrations should have depleted (relative to background) or zero nitrate + nitrite concentrations.

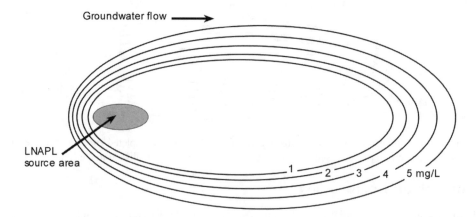

FIGURE 4.11a Initial measurements of nitrate + nitrite isopleths show that nitrate + nitrite levels are diminished relative to background, within the BTEX plume.

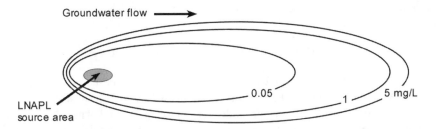

FIGURE 4.11b Nitrate + nitrite isopleths one year later show that the nitrate + nitrite depleted zone has become smaller. As the nitrate + nitrite concentrations decrease and oxidation potentials become lower, other electron acceptors are used.

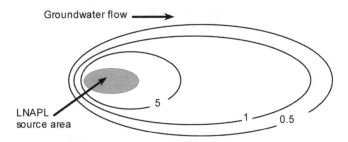

FIGURE 4.12a Initial measurements of ferrous iron (Fe^{2+}) isopleths show that ferrous iron levels are increased relative to background within the BTEX plume.

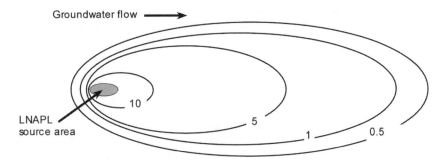

FIGURE 4.12b One year later, ferrous iron isopleths show that the Fe^{2+} enhanced zone has become larger as the BTEX and nitrate/nitrite concentrations decreased.

IRON (III) REDUCTION TO IRON (II)

If significant biodegradation is occurring and DO and nitrate/nitrite have been depleted, dissolved iron (II) concentrations must increase with time in the plume zone, as indicated in Figures 4.12a and 4.12b. After available DO and nitrate are consumed, available forms of iron (III) minerals can be used as electron acceptors, forming dissolved iron (II). Aquifer sediments often contain large quantities of iron (III), frequently in the form of amorphous iron oxyhydroxides, so the reduction reaction in Equation 4.38 is a common occurrence:

$$Fe^{3+} \text{ (insoluble)} \rightarrow Fe^{2+} \text{ (soluble)}. \tag{4.38}$$

Fe^{3+} is not a good indicator to measure because its bioavailability depends on the degree of crystallinity. Amorphous Fe^{3+} minerals or Fe^{3+} oxyhydroxides are more bioavailable than are highly crystallized forms. Also, soil samples are inherently more difficult to analyze and interpret than water samples. Therefore, dissolved Fe^{2+} is the preferred indicator. The overall redox reaction with benzene is

$$60 \ H^+ + 30 \ Fe(OH)_3(amorphous) + C_6H_6 \rightarrow 6 \ CO_2 + 30 \ Fe^{2+} + 78 \ H_2O. \qquad (4.39)$$

The degradation of 1 mg/L of BTEX by Fe^{3+} reduction results in the production of about 22 mg/L of Fe^{2+}. Most of the soluble Fe^{2+} produced in the BTEX plume subsequently precipitates again as Fe^{3+} when it is carried with the groundwater into oxygenated downgradient regions of the aquifer. Some research indicates that reduction of Fe^{3+} *requires* microbial remediation.[2,5,6] Thus, the presence of increasing concentrations of Fe^{2+} within the BTEX plume is strong evidence for anaerobic biodegradation reactions.

Field Test for Iron (III) Reduction to Iron (II) by Microbes

- Analyze groundwater for Fe^{2+}.
- Areas with elevated BTEX concentrations should have elevated (relative to background) Fe^{2+} concentrations.

SULFATE REDUCTION

After available DO, nitrate, and iron (III) are consumed, available sulfate can be used as an electron acceptor. Sulfate reduction to sulfide is favored at a pH of 7 and E_h of -200 mV. Sulfate-reducing microorganisms are sensitive to temperature, inorganic nutrients, pH, and oxidation potential. Small imbalances in environmental conditions can severely limit the rate of BTEX degradation via sulfate reduction. Each 1.0 mg/L of BTEX that is biodegraded requires the reduction of about 4.7 mg/L of sulfate. The overall reaction for benzene oxidation by sulfate reduction is

$$7.5 \ H^+ + 3.75 \ SO_4^{2-} + C_6H_6 \rightarrow 6 \ CO_2 + 3.75 \ H_2S + 3 \ H_2O. \qquad (4.40)$$

Field Test for Sulfate Reduction by Microbes

- Analyze groundwater for sulfate (SO_4^{2-}), and perhaps sulfide (S^{2-}).
- Depleted sulfate concentrations and increased sulfide concentrations (relative to background) within the BTEX plume indicates active biodegradation by sulfate-reducing bacteria.

METHANOGENESIS (METHANE FORMATION)

After available DO, nitrate, iron (III), and sulfate are consumed, the redox reactions of organic carbon to form CO_2 (C^{4+}) and CH_4 (C^{4-}) can be used to biodegrade BTEX.

Methanogenesis generates less energy for microbes than the other reducing reactions and always occurs last, after other electron acceptors have been depleted. Methanogenesis causes the oxidation potential to fall below -200 mV at pH 7. The presence of elevated levels of methane in the presence of elevated levels of BTEX indicates that BTEX biodegradation is occurring as a result of methanogenesis.

Because methane is not present in LNAPL fuels, the presence of methane above the background in groundwater adjoining LNAPL fuels indicates microbial degradation of fuel hydrocarbons. The overall reaction for benzene oxidation by methanogenesis is

$$C_6H_6 + 4.5 \ H_2O \rightarrow 2.25 \ CO_2 + 3.75 \ CH_4. \qquad (4.41)$$

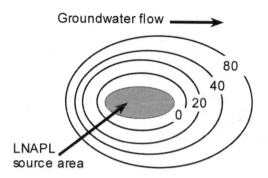

FIGURE 4.13a   Initial measurements of sulfate isopleths show that sulfate levels are diminished relative to background within the BTEX plume. Note that sulfate reduction is confined to a small portion of the plume where BTEX concentrations are the highest. Alternative electron acceptors have been depleted in this region.

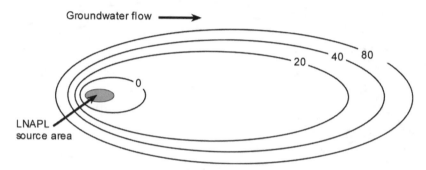

FIGURE 4.13b   One year later, sulfate isopleths show that the sulfate-depleted zone has become much larger as the concentrations of alternative electron acceptors were depleted.

This reaction occurs in at least four steps, at least one of which involves CO_2 accepting electrons and reacting with H^+ to form CH_4. In the process, C_6H_6 is oxidized to form additional CO_2. The biodegradation of 1 mg/L of BTEX by methanogenesis produces about 0.78 mg/L of methane.

Field Test for Methanogenesis

- Analyze groundwater for methane (CH_4).
- High methane concentrations (relative to background) within the BTEX plume indicate active biodegradation by methanogenesis.

REDOX POTENTIAL AND ALKALINITY AS BIODEGRADATION INDICATORS

Groundwater redox potential and alkalinity undergo measurable changes in regions where significant biodegradation of fuel hydrocarbons is occurring.

Using Redox Potentials to Locate Anaerobic Biodegradation Within the Plume

Each biodegradation redox reaction involving the succession of electron acceptors from dissolved oxygen to methane serves to lower the redox potential of the groundwater in which it occurs. Thus, the change with time of groundwater redox potential should serve as an indication of biodegradation activity.

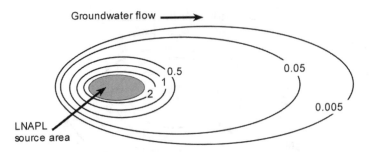

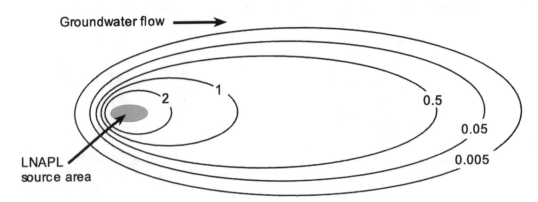

FIGURE 4.14a Initial measurements of methane isopleths show that methane levels are increased relative to background within the BTEX plume. The methane buildup is limited to the central part of the BTEX plume where alternative electron acceptors are depleted.

FIGURE 4.14b One year later, methane isopleths show that the methane enhanced zone has become larger, as the concentrations of alternative electron acceptors decreased.

Field Test for Occurrence of Redox Reactions

- Map groundwater redox potentials at the site. Include at least one location upgradient of the plume. It is important to avoid aeration of well samples for these measurements.
- Locations where groundwater redox potentials are lower than background are where electron acceptor species are being reduced, a sign of biodegradation.
- Redox potentials within the plume can help indicate which electron acceptors are active in different locations.
- In regions of biodegradation activity, the zone of low oxidation potential (reducing zone) will become larger with time, as diminishing DO concentrations move farther and farther from the spill region.

Using Alkalinity to Locate Anaerobic Biodegradation Within the Plume

Groundwater alkalinity increases during aerobic respiration, denitrification, iron (III) reduction, and sulfate reduction and is unchanged during methanogenesis. The two main processes that increase alkalinity are

1. Production of carbon dioxide in the biodegradation of fuel hydrocarbons, which increases the total carbonate and alkalinity of the groundwater.
2. Redox reactions involving nitrate, iron (III), and sulfate as electron acceptors all consume acidity as H^+ (see Equations 4.37, 4.39, and 4.40). Thus, alkalinity increases in groundwater where these reactions are taking place.

A measurement of alkalinity within a hydrocarbon plume can be used to infer the amount of petroleum hydrocarbons destroyed. For every 1 mg/L of alkalinity (as $CaCO_3$) produced, 0.13 mg/L of BTEX is destroyed.[13]

Field Test for Using Alkalinity as an Indicator of Biodegradation

1. Map groundwater alkalinity concentrations at the site. Include at least one location upgradient of the plume.
2. Locations where groundwater alkalinity is higher than background are where CO_2 is being produced and H^+ is being consumed, which are signs of biodegradation.
3. Alkalinity levels within the plume can help indicate the amount of petroleum hydrocarbons that have been destroyed.
4. In regions of biodegradation activity, the zone of higher alkalinity will become larger with time.
5. Increased alkalinity levels within the plume can be used to infer the extent of biodegradation occurring.

REFERENCES

1. Alexander, M., How toxic are toxic chemicals in soil? *Environmental Science and Technology*, Vol. 29, No. 11, pp. 2713–2717, 1995.
2. Chapelle, F.H., *Groundwater Microbiology and Geochemistry*, John Wiley & Sons, Inc., New York, 1993, 424.
3. Fetter, C.W., *Contaminant Hydrogeology*, Macmillan, New York, 1993, p. 119.
4. Freeze, R.A. and Cherry J.A., *Groundwater*, Prentice-Hall, Inc., Englewood Cliffs, NJ, 1979.
5. Lovely, D.R., Dissimilatory Fe(III) and Mn(IV) reduction, *Microbiological Reviews*, pp. 259–287, June 1991.
6. Lovely, D.R. and Phillips E.J.P., Competitive mechanisms for inhibition of sulfate reduction and methane production in the zone of ferric iron reduction in sediments, *Appl. Environ. Microbiol.*, 53, pp. 2636–2641, 1987.
6a. Lyman, W.J., Rheel W.F., and Rosenblatt, D.H., *Handbook of Chemical Property Estimation Methods*, American Chemical Society, Washington, D.C., 1990.
7. Maidment, D.R., Ed., *Handbook of Hydrology*, McGraw-Hill, New York, 1993, p. 16.22.
8. Means, J.C. and Wijayratne R., Role of natural colloids in the transport of hydrophobic pollutants, *Science*, 215, pp. 968–970,
9. Rice, D.W., Dooher B.P., Cullen S.J., Everett L.E., Kastenberg W.E., Grose R.D., and Marino M.A., 1995a, Recommendations to improve the cleanup process for California's leaking underground fuel tanks (LUFTs), Lawrence Livermore National Laboratory, Livermore, CA (UCRL-AR-121762).
10. Rice, D.W., Grose R.D., Michaelsen J.C., Dooher B.P., MacQueen D.H., Cullen S.J., Kastenberg W.E., Everett L.E., and Marino M.A., 1995b, *California Leaking Underground Fuel Tanks (LUFTs)*, Lawrence Livermore National Laboratory, Livermore, CA (UCRL-AR-122207).
11. U.S. EPA, 1996, *Soil Screening Guidance: Technical Background Document*, Office of Emergency and Remedial Response, Washington, D.C., EPA/540/R95/128.
12. U.S. EPA, *Understanding Variation in Partition Coefficient, K_d, Values*, Vol. 1 (EPA 402-R-99-004A) and Vol. 2 (EPA 402-R-99-004B), August, 1999.
13. Wiedemeier, T., Wilson J.T., Kampbell D.H., Miller R.N., and Hanson J.E., *Technical Protocol for Implementing Intrinsic Remediation with Long-term Monitoring for Natural Attenuation of Fuel Contamination Dissolved in Ground Water*, Volumes I and II, Air Force Center for Environmental Excellence, Technology Transfer Division, Brooks AFB, San Antonio, Texas, Revision 0, 11/11/95.

5 Petroleum Releases to the Subsurface

CONTENTS

5.1 THE PROBLEM

A major federal law governing pollution from underground storage tanks is described in Subtitles I and C of the Resource Conservation and Recovery Act (RCRA). Spills to any navigable waters are regulated under the Federal Clean Water Act. One EPA estimate puts leaks from 2 to 7 million underground tanks as the source of 45% of all groundwater contamination, with 95% of the leaking tanks containing motor fuel. The rest contain heating oil, industrial solvents, and liquid wastes, among others.

The EPA estimates that more than half of all underground storage tanks installed before 1993 have developed leaks. Approximately 40% of the 210,000 retail service stations in the U.S. have had accidental releases of gasoline and diesel to the subsurface. Current legislation requires all new tanks that are installed after 1993 to have a leak detection system. Starting in 1998, tanks containing petroleum products or hazardous chemicals are required to have overfill and spill prevention devices, and double walls or concrete vaults. This goes a long way to prevent the problem from getting worse, but existing subsurface contamination must still be dealt with.

5.2   GENERAL CHARACTERISTICS OF PETROLEUM

Petroleum liquids are complex mixtures of hundreds of different hydrocarbons, with minor amounts of nitrogen, oxygen, sulfur, and some metals. Nearly all petroleum compounds are nonpolar and not very soluble in water. The behavior of these compounds in a groundwater environment depends on the physical and chemical nature of the particular hydrocarbon blend as well as the particular soil environment. For example, the migration potential and partitioning coefficients of each compound depend on the composition of the petroleum mixture in which it is found, the properties of the pure compound, and the characteristics of the surrounding soil. Furthermore, the properties of petroleum contaminants change as the petroleum ages and weathers.

Many nonfuel organic pollutants, such as chlorinated hydrocarbons and pesticides, are more soluble in petroleum than in water. Therefore, if an oil spill occurs where organic contamination already exists, the older pollutants tend to concentrate from soil surfaces and pore-space water into the fresh oil phase. An oil spill into an already contaminated soil can mobilize other pollutants that have been immobilized there by sorption and capillarity. As freshly spilled oil moves downward through the soil, immobilized pollutants can dissolve into the moving liquid oil and be carried along with it. Analysis of spilled petroleum products will often detect other organic compounds that were previously sorbed to the soil.

TYPES OF PETROLEUM PRODUCTS

The first step in refining crude oil into petroleum products is usually through fractional distillation which is a process that separates the oil components according to their boiling points. The resulting products are groups of mixtures, or fractions, each of which have boiling points within a specified range. All but the lightest fractions can contain up to hundreds of different hydrocarbon compounds. The fractions are often classified into the general groups described in Table 5.1. In addition, several pure *petrochemicals* may be produced, such as butane, hexane, benzene, toluene, and xylene, for use as solvents, for production of plastics and fibers, and for reblending into fuel mixtures. Refined petroleum products are further modified by catalytic cracking, blending, and reformulation processes to enhance desirable properties.

Rule of Thumb

The larger the hydrocarbon compound and the more carbon atoms it contains, the higher are its boiling point and viscosity, and the lower its volatility.

GASOLINES

Gasolines are among the lightest liquid fractions of petroleum and consist mainly of aliphatic and aromatic hydrocarbons in the carbon number range C4–C12. *Aliphatic hydrocarbons* consist of

TABLE 5.1
Principal Petroleum Fractions from Fractional Distillation

Boiling Range (°C)	Dominant Composition Range[a]	Fraction	Uses
−160 to +30	C1–C4	Gases	LPN, methane, gaseous fuels, feedstock for plastics
30–60	C5–C7	Petroleum Ether	Solvents, gasoline additives
90–130	C6–C9	Ligroin, Naphtha	Solvents
40–200	C4–C12	Gasoline	Motor fuel
60–200	C7–C12	Mineral Spirits	Solvents
150–300	C10–C16	Kerosene	Jet fuel, diesel fuel, lighter fuel oils
300–350	C16–C18	Fuel Oil	Diesel oil, heating oil, cracking stock
>350	C18–C24	Lubricating Stock	Lubricating oil, mineral oil, cracking stock
Solid Residue	C25–C40	Paraffin Wax	Candles, toiletries, wax paper
Solid Residue	>C40	Residuum	Roofing tar, road asphalt, waterproofing

[a] The notation used here gives the number range of carbon atoms in the fraction compounds. For example, C5–C7 means hydrocarbon compounds containing between 5 and 7 carbon atoms. As this Table indicates, as the number of carbon atoms in a hydrocarbon molecule increases, so do its boiling temperature and its viscosity. The volatility decreases as the number of carbon atoms in a compound increases.

- *Alkanes:* Are saturated hydrocarbons (all carbons are connected by single bonds) having linear, branched, or cyclic carbon-chain structures such as pentane, octane, decane, isobutane, or cyclohexane.
- *Alkenes:* Are unsaturated hydrocarbons having one or more double bonds between carbon atoms. They also may have linear, branched, or cyclic carbon-chain structures.
- *Alkynes:* Are unsaturated hydrocarbons having one or more triple bonds between carbon atoms. They also may have linear, branched, or cyclic carbon-chain structures.

Aromatic hydrocarbons (also called *arenes*) are hydrocarbons based on the benzene ring as a structural unit. They include monocyclic hydrocarbons such as benzene, toluene, ethylbenzene, and xylene (the *BTEX* group, see Figure 5.1), and polycyclic hydrocarbons such as naphthalene and anthracene.

Rules of Thumb for Gasoline Properties

1. Gasoline mixtures are volatile, somewhat soluble, and mobile in the groundwater environment.
2. Gasolines contain a much higher percentage of the BTEX group of aromatic hydrocarbons (benzene, toluene, ethylbenzene, and the xylene isomers) than do other fuels, such as diesel. They contain lower concentrations of heavier aromatics like naphthalene and anthracene than do diesel and heating fuels. Therefore, the presence of BTEX is often a useful indicator of gasoline contamination.
3. Oxygenated compounds such as alcohols (methanol and ethanol) and ethers (methyltertiary-butyl ether, MTBE) are normally added as octane boosters and oxygenators. MTBE is the most commonly used of these. Modern gasolines (since 1980) may contain around 15% MTBE by volume.

MIDDLE DISTILLATES

Middle distillates cover a broad range of hydrocarbons in the range of C6–C25. They include diesel fuel, kerosene, jet fuels, and lighter fuel oils. Typical middle distillate products are blends of up to 500 different compounds. They tend to be denser, more viscous, less volatile, less water soluble, and less mobile than gasoline. They contain low percentages of the lighter weight aromatic BTEX group, which may not be detectable in older releases due to degradation or transport.

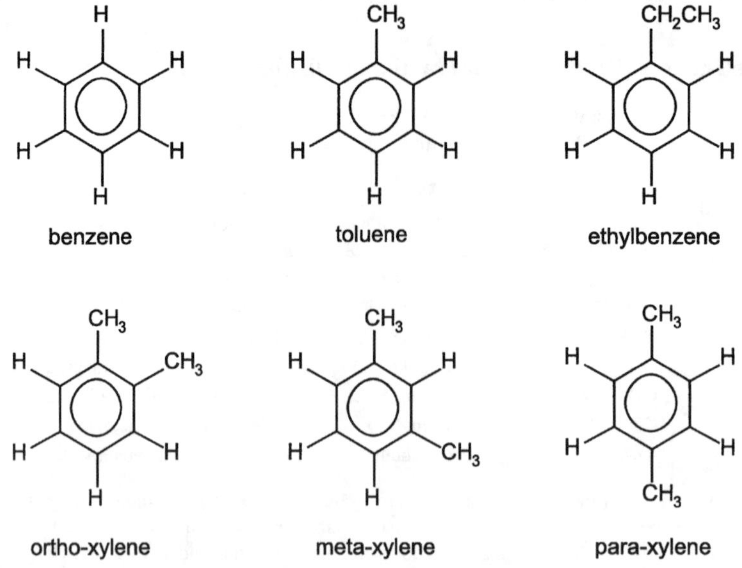

FIGURE 5.1   The BTEX group of aromatic hydrocarbons.

HEAVIER FUEL OILS AND LUBRICATING OILS

These are composed of heavier molecular weight compounds than the middle distillates, encompassing the approximate range of C15–C40. They are more viscous, less soluble in water, and less mobile in the subsurface than the middle distillates.

Figure 5.2 relates the carbon number of a petroleum compound to its properties, uses, and the instrumental methods used for its analysis. The following are notes for Figure 5.2:

- EPA 418.1 = infrared spectroscopy. It is used as a low cost screening method for total petroleum hydrocarbons (TPH).
- EPA 8015 = gas chromatography (GC) with a flame ionization detector.
- EPA 8020 = GC with photo ionization detector. It is used for total BTEX analysis.
- EPA 8260 = GC with mass spectrometer detector (GC/MS). It is used for volatile organic compounds.
- EPA 8270 = GC/MS. It is used for extractable organic compounds.
- ECD = electron capture GC detector.
- ELCD = electrolytic conductivity GC detector.
- FID = flame ionization GC detector.
- PID = photo ionization GC detector.

5.3   BEHAVIOR OF PETROLEUM HYDROCARBONS IN THE SUBSURFACE

Because of their low water solubilities, most of the compounds classified as petroleum hydrocarbons are generally considered as *nonaqueous phase liquids* (NAPL). If mixed into water, NAPLs separate into a distinct liquid phase with a well-defined boundary between the NAPL and the water, like oil and water or milk and cream.

NAPLs are further subdivided into *light nonaqueous phase liquids* (LNAPL) and *dense nonaqueous phase liquids* (DNAPL). LNAPLs are liquid hydrocarbon compounds or mixtures that are less dense than water, such as gasoline and diesel fuels and their individual components. DNAPLs

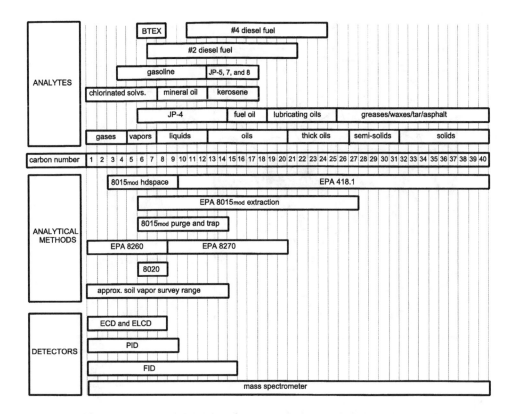

FIGURE 5.2 Hydrocarbon ranges, corresponding uses, and analytical methods.

are liquid hydrocarbon compounds or mixtures that are more dense than water, such as creosote, PCBs, coal tars, and most chlorinated solvents (chloroform, methylene dichloride, etc.).

The distinction between LNAPLs and DNAPLs is important because of their different behavior in the subsurface. LNAPL spills travel downward through soils only to the water table, where they remain "floating" on the water table surface. DNAPLs sink through the water-saturated zone to impermeable bedrock, where they collect in bottom pools. Obviously, remediation methods are different for LNAPLs and DNAPLs.

SOIL ZONES AND PORE SPACE

As illustrated in Figure 5.3, the subsurface soil may be divided into a water-*unsaturated* zone, from the soil surface down to just above the water table (also called the *vadose zone*), and a water-*saturated* zone, from the water table down to bedrock. Capillary action extends the saturated zone somewhat above the water table with a region of transition between the unsaturated and saturated zones. The capillary fringe can vary from a fraction of an inch in coarse-grained sediments to several feet in fine-grained sediments such as clay.

Each zone contains soil particles with pore spaces between them. In normally permeable soils, most of the pore spaces are continuous, allowing movement of water and liquid contaminants through them. In the absence of contaminants, pore spaces in the unsaturated zone contain air with some water adsorbed to the soil particles. Pore spaces in the saturated zone contain mainly water.

When contaminants enter the subsurface region as spilled liquid petroleum (free product),

- Volatile compounds vaporize from the free product mixture into the atmosphere and into air in the soil pore spaces.

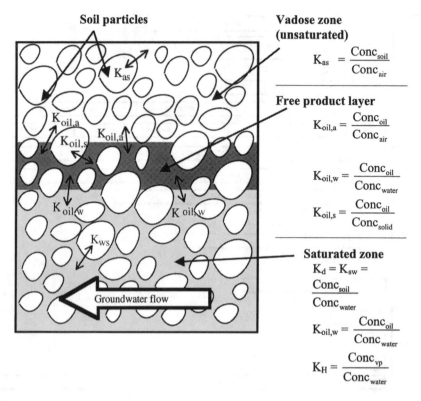

FIGURE 5.3   Soil zones and partitioning behavior of free product pollutant. All the Ks are partition coefficients. They quantitatively describe how the pollutant distributes itself among water, soil, air, and free product.

- Many compounds in the free product partially dissolve into water contained on soil particle surfaces, into water percolating down from the ground surface, and into groundwater in the saturated zone.
- A small fraction of the free product is taken up by microbiota.
- The remaining free product adsorbs to soil particles and, where free product is abundant, fills the pore spaces.

PARTITIONING OF LIGHT NONAQUEOUS PHASE LIQUIDS (LNAPLS) IN THE SUBSURFACE

Before a petroleum release occurs, the voids of vadose zone earth materials are filled with air and water. After a release, some voids contain immobile petroleum held by capillary forces and sorbed to soil surfaces. There may also be liquid petroleum moving downward through the pore interstices under gravity. If LNAPL reaches the water table, its buoyancy will prevent further downward movement and it will spread out horizontally over the water table to form a layer of free product, "floating" above the saturated zone. The individual components of the petroleum become partitioned into air, water and solid phases that come in contact with the free product.[3]

OIL MOBILITY THROUGH SOILS

Oil pollutants moving through soil, dissolved in water, or migrating as liquid free product leave a trail of contamination sorbed on soil particles and trapped in soil pore spaces. This trapped contamination is not easily removed by water flushing or air sparging. In the subsurface environment, a significant portion of oil contamination must be regarded as "permanent," with a lifetime of well over 25 years, unless deliberate efforts are made to mobilize, degrade, or remove it.[2,4] Immobilized oil contaminants in the subsurface act as a long-term source of groundwater contamination, as the more soluble components continue to diffuse to the oil-water interface and dissolve into the water.

This is nothing new to oil field workers. Liquid petroleum fields are found in rock formations of 10% to 30% porosity. Up to half of the pore space contains water. Primary recovery of oil, which relies on pumping out the portion of oil that is mobile and will accumulate in a well, collects only 15% to 30% of the oil in the formation. Secondary recovery techniques force water under pressure into the oil-bearing rocks to drive out more oil. Primary and secondary techniques together extract somewhat less than 50% of the oil from a formation. Tertiary recovery techniques use pressurized carbon dioxide to lower oil viscosity along with detergents to solubilize the oil. Even with using tertiary techniques, producers expect 40% of the oil to remain immobile and unrecoverable.

PROCESSES OF SUBSURFACE MIGRATION

After part of the spilled petroleum has partitioned from the free product into other phases, hydrocarbons (HCs) are present in solid, liquid, dissolved, and vapor phases.

1) *Solid phase HCs* are sorbed on soil surfaces or diffused into micropores and mineral grain lattices. They are immobile and degrade very slowly.
2) *Liquid phase HCs* exist in the subsurface as
   • Immobile residual liquids held by capillary forces and as a thin layer sorbed to sediments in the unsaturated zone and capillary fringe.
   • Free mobile liquids in the unsaturated zone above the capillary fringe.
   • Immobile residual liquids trapped below the water table in the saturated zone.
3) *Dissolved phase HCs* are found in
   • Water infiltrating through the unsaturated zone.
   • The residual films of groundwater sorbed to sediments in the capillary fringe and elsewhere in the HC plume.
   • The groundwater of the saturated zone.
4) *Vapor phase HCs* are found
   • Mostly in void spaces of the unsaturated zone not occupied by water or liquid HCs. Here, they are mobile.
   • As small bubbles trapped in the HC plume and in the water-bearing zone below the plume. Here, they are immobile.
   • Dissolved in the groundwater of the saturated zone, where they move with the groundwater.

BEHAVIOR OF LNAPL IN SOILS AND GROUNDWATER

LNAPL movement in the subsurface is a continual process of partitioning different components among different phases that are present in the subsurface matrix. Spilled LNAPL at or near the soil surface penetrates and thoroughly saturates the soil because there is little trapped water or air to block its movement. Under the influence of gravity, the LNAPL sinks vertically downward, leaving behind in the soil a trail of residual LNAPL trapped by sorption and capillary forces. Capillary forces cause the LNAPL to spread horizontally as well as vertically downward, creating an inverted funnel-shaped zone of soil contamination.

As liquid free product moves downward through soil, a significant portion becomes immobilized by sorption to soil particle surfaces and capillary entrapment in soil pore space. This continually reduces the amount of mobile contaminant. If the liquid free product is not replenished by a continuing leak, it eventually is completely depleted by entrapment in the soil and becomes essentially immobilized. However, even when immobilized, the trapped free product continues to lose mass into the dissolved and vapor phases during biodegradation.

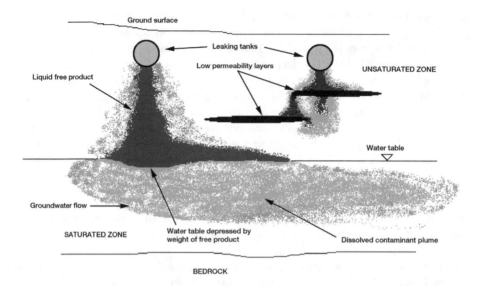

FIGURE 5.4 LNAPL fuel leaking from underground storage tanks migrates downward under gravity. Enough fuel free product has leaked from the left tank to reach the saturated zone and spread out above the water table, moving in the direction of groundwater flow. The smaller spill from the right tank is insufficient to reach the water table and has become immobilized within the unsaturated zone by sorption and capillary forces. The more soluble components of the free product are present in the dissolved plume, which extends beyond the free product plume into the saturated zone. There also is a vapor plume in the unsaturated zone of the most volatile components. The vapor plume extends in all directions independent of gravity. It may enter underground cavities such as sewers and basements, and may escape through the ground surface into the atmosphere.

When all the spilled oil has entered the subsurface, the LNAPL "front" continues downward, leaving behind an ever-widening "inverted funnel" of contaminated soil containing residual "immobile" LNAPL sorbed to soil surfaces and trapped in pore spaces. The term "immobile" is used loosely and really means that, although some mobility may still occur, it will be very slow compared to the remediation time frame of interest.

A spill may or may not reach the water table. If the groundwater table is deep enough or if the amount of spilled LNAPL is small enough, the mobile LNAPL can be completely depleted by entrapment in the soil before it reaches the groundwater. If the spill is large enough or the groundwater table is shallow, mobile LNAPL, commonly called free product, will contact the groundwater. The weight of the free product depresses the water table locally below the free product column (see Figure 5.4).

Free product will continue to spread laterally as a layer over the water table, leaving a trail of residual LNAPL entrapped in the soil, until it spreads out to a saturation level so low that it all becomes immobile. The lateral spreading of the free product is influenced by a viscous "frictional" interaction at the water-LNAPL interface, which tends to move the free product preferentially in the direction of groundwater movement, along the hydraulic gradient. The relative downgradient velocities of water and free product depend on their relative viscosities and the soil conductivities for the different liquids.

When the water table rises and falls, the "floating" free product is moved vertically, "smearing" LNAPL into a region thicker than the free product thickness. Still more residual LNAPL becomes immobilized in this "smear zone." The end result is that the smear zone of entrapped LNAPL extends above and below the average level of the water table.[1]

Summary of **LNAPL** Behavior

The following summarizes the behavior of spilled LNAPL:

1. Spilled liquid hydrocarbons move downward through the unsaturated zone under gravity.
2. A large fraction sorbs to the subsoil surfaces as trapped residual free product.
3. Some horizontal spreading occurs in the unsaturated zone because of attractive forces to mineral surfaces and capillary attractions.
4. Free product tends to accumulate and spread horizontally above layers of low permeability (low hydraulic conductivity).
5. At the water-bearing region of the capillary fringe, the free liquid phase floats on the water and begins to move laterally.
6. If the spill is small enough, LNAPL may not reach the water table. However, a portion that dissolves in downward percolating water will be carried to the water table and will contaminate it.
7. The vapor phase spreads widely in the unsaturated zone and can escape to the atmosphere and accumulate in cellars, sewers, and other underground air spaces.

"Weathering" of Subsurface Contaminants

With time, the composition of immobilized oil changes in the following ways:

- Less viscous components move downgradient through the soils.
- Volatile components are lost into the atmosphere.
- Soluble components are lost into the groundwater.
- Biodegradable components are lost to bacterial activity.

However, the total mass of immobilized oil decreases slowly because the loss processes are usually slow unless they are artificially enhanced as part of a remediation program. The natural rate of depletion becomes progressively slower with time, as the remaining contaminants are increasingly rich in those components that resist the loss mechanisms. The remaining oil becomes more and more firmly fixed in the subsurface soil, continually releasing its more soluble components in slowly decreasing concentrations to the groundwater.

Rules of Thumb

1. Less than 1% of the total mass of a gasoline spill will dissolve into water in the vadose and saturated zones.
2. Since more than 99% of a fuel spill remains as adsorbed or free product, it is impossible to clean up groundwater fuel contamination simply by "pump-and-treat" without eliminating the source residual and free product remaining in the soil.

5.4 PETROLEUM MOBILITY AND SOLUBILITY

The environmental impact of a contaminant release is determined mainly by the mobility and water solubility of the different components of the contaminant. The most important parameters determining the mobility of LNAPL free product are

- Average soil pore size which determines soil capillarity.
- Percent of soil pore space (soil porosity).

TABLE 5.2
Densities and Viscosities of Selected Fluids

Fluid	Density (g/mL)			Viscosity (centipoise)		
	0°C	15°C	25°C	0°C	15°C	25°C
Water	1.000	0.998	0.996	1.8	1.14	0.9
Automobile gasoline	0.76	0.73	0.68	0.8	0.62	—
Automotive diesel fuel	0.84	0.83	—	3.9	2.7	—
Kerosene	0.84	0.84	0.83	3.4	2.3	2.2
No. 5 jet fuel	0.84	—	—	—	—	—
No. 2 fuel oil	0.87	0.87	0.84	7.7	—	4.0
No. 4 fuel oil	0.91	0.90	0.90	—	47	23
No. 5 fuel oil	0.93	0.92	0.92	—	215	122
No. 6 fuel oil or Bunker C	0.99	0.97	0.96	7.4×10^7	—	3200

- Density and viscosity of the moving liquid contaminant. Density = mass per unit volume. Most petroleum hydrocarbons have a density less than the density of water, which is 1 g/mL. Viscosity measures resistance of fluid to flow. Gasoline is less viscous than water and can flow through smaller pores and fissures more easily than water. The heavier petroleum fractions, such as diesel fuel and fuel oils, are more viscous than water and flow less easily. Values of density and viscosity for several fuel products are listed in Table 5.2.
- Capillary attraction for the liquid contaminant to soil particles.
- Soil zone in which free product is present, as in whether the pore space contains air, water, or contaminant.
- Magnitude of pressure and concentration gradients acting on the liquid free product.

LNAPL solubility in water is variable and depends on the chemical mixture. Literature data for solubility of pure compounds can be misleading because the solubility of a specific compound decreases when it is part of a blend, as shown in Table 5.3.

Rule of Thumb

The aqueous solubility of a particular compound in a multi-component NAPL can be approximated by multiplying the mole fraction of the compound in the NAPL mixture by the aqueous solubility of the pure compound.

$$\text{Solubility of component } \mathbf{i} \text{ in a NAPL mixture} = S_i = X_i S^0_i \qquad (5.1)$$

where: S_i = solubility of component $\mathbf{i}$ in the mixture
X_i = mole fraction of component $\mathbf{i}$ in the mixture
S^0_i = solubility of pure component $\mathbf{i}$

5.5   FORMATION OF PETROLEUM CONTAMINATION PLUMES

In the subsurface soil environment, petroleum compounds are present in four phases and four plumes. The four phases are

1. Liquid petroleum free product
2. Petroleum compounds adsorbed to soil particles

3. Dissolved petroleum components
4. Vaporized petroleum components

TABLE 5.3
Solubility Variability of Gasoline Components from Different Mixtures

Concentration Dissolved in Water (mg/L)

Compound	Regular Leaded	Regular Unleaded	Super Unleaded	Pure Compound
Benzene	30.5	28.1	67.0	1740–1860
Toluene	31.4	31.1	107.0	500–627
Ethylbenzene	4.0	2.4	7.4	131–208
1,2-dichloroethane	1.3	—	—	8524
Methyl-*t*-butyl ether (MTBE)	43.7	35.1	966.0	48,000
t-butyl alcohol	22.3	15.9	933.0	miscible
m-xylene	13.9	10.9	11.5	134–196
o-, *p*-xylene	6.1	4.8	5.7	157–213
1,2-dibromoethane	0.58	—	—	4300

Each phase behaves differently and poses different remediation problems. The liquid free product originates directly from the contamination source and initially has the same composition. The adsorbed, dissolved, and vapor phases are extracted from the liquid free product as it contacts water, soil, and air in the soil pore space. Each phase moves independently in its own distinct contaminant plume.

In general, the vapor contaminant plume moves the most rapidly. The dissolved plume moves more slowly at groundwater velocity or less, depending on its retardation factor. The free product plume moves slower than the dissolved plume. The adsorbed plume may be immobilized, or in the saturated zone part of it can be sorbed to mobile colloids and move at approximately the groundwater velocity.

Rules of Thumb

1. With respect to the total mass of fuel contaminant in the subsurface soil environment, the floating free product and immobilized oil (trapped by capillary forces and adsorbed to soil particles) are generally more than 99% of the total mass.
2. When free product is present, the dissolved phase in the groundwater is generally less than 1% of the total mass. The dissolved plume is just the tip of the contamination iceberg.

Example 5.1: Comparing Dissolved and LNAPL Free Product Masses

A leaking underground storage tank (UST) released 1000 gallons of gasoline (density about 0.7 g/mL) to the subsurface. After 1 year the resulting dissolved-phase plume is about 1000 ft long, 100 ft wide, and 10 ft deep. The average concentration of hydrocarbons in the plume is 0.002 mg/L (estimated by measuring and adding the total volatile hydrocarbon [TVH] and total extractable hydrocarbon [TEH] concentrations). The porosity of the aquifer is 0.30. If no hydrocarbon is lost due to volatilization or biodegradation, how much of the original release is in the dissolved phase and how much is in the LNAPL phase?

Answer:

Total mass released = (1000 gal)(3.78 L/gal)(1000 mL/L)(0.7 g/mL)(1 kg/1000 mg) = 2646 kg.

Volume of contaminated groundwater = (1000 ft)(100 ft)(10 ft)(0.30)(28.3 L/ft^3) = 8.49 × 10^6 L.

Mass of dissolved hydrocarbons = (8.49 × 10^6 L)(0.002 mg/L)(1 kg/1000 mg) = *17.0 kg.*

Mass of LNAPL free product = 2646 − 17.0 = *2629 kg.*

Percent of total mass that is dissolved = (17 kg/2646 kg) × 100 = *0.64%.*

Percent of total mass that is LNAPL = (2629/2646) × 100 = *99.36%.*

DISSOLVED CONTAMINANT PLUME

Water solubility is the most important chemical property for assessing the impact of a contaminant on the environment. Dissolved contaminants arise when the free product comes in contact with water. The water may be in the form of moisture retained in the soil, precipitation percolating downward through the soil, groundwater flowing through contaminated soil or groundwater lying under a layer of free product. Both crude and refined petroleum products contain hundreds of different components with different water solubilities, ranging from slightly soluble to insoluble.

Rules of Thumb

1. In general, the lightweight aromatics such as the BTEX group (benzene, toluene, ethylbenzene, and xylene) are the most soluble components of fuel mixtures. If MTBE additive is present, it is the most soluble component by far.
2. The overall water solubility of commercial gasoline without additives ranges between 50 mg/L and 150 mg/L, depending on its exact composition. When free product gasoline is present, the dissolved portion generally accounts for less than 1% of the total contaminant mass present in the subsurface.
3. The overall solubility of fresh No. 2 diesel fuel in water is around 0.4–8.0 mg/L, again depending on its composition. When free product diesel fuel is present, the dissolved portion generally accounts for less than 0.1% of the total contaminant mass present in the subsurface.
4. Nevertheless, dissolved contaminants can greatly exceed concentration levels where water is regarded as seriously polluted.

The composition of the free product and dissolved fractions is very different, as indicated in Figure 5.5, because the more soluble compounds become concentrated in the water-soluble fraction.

Dissolved contaminants become a part of the water system and move with the groundwater but they usually move at a lower velocity because of their retardation by sorption processes. Sorption to soil and desorption back into the dissolved phase is a continual process that retards the movement of the dissolved phase. The amount of retardation depends mainly on the organic content of the soil. Retardation is greater in soils with more organic matter. Because their water solubilities are low, dissolved fuel contaminants continue to partition between the dissolved phase and soil particle surfaces, especially in soils with a high organic content.

Rule of Thumb

Typical retardation factors for BTEX in sandy soil range from 2.4 for dissolved benzene (groundwater moves 2.4 times faster than benzene) to 6.2 for dissolved xylene.

VAPOR CONTAMINANT PLUME

Vapor phase contaminants arise from the volatile components of the free product escaping into adjacent air. Lower mass hydrocarbon components commonly associated with the gasoline fraction

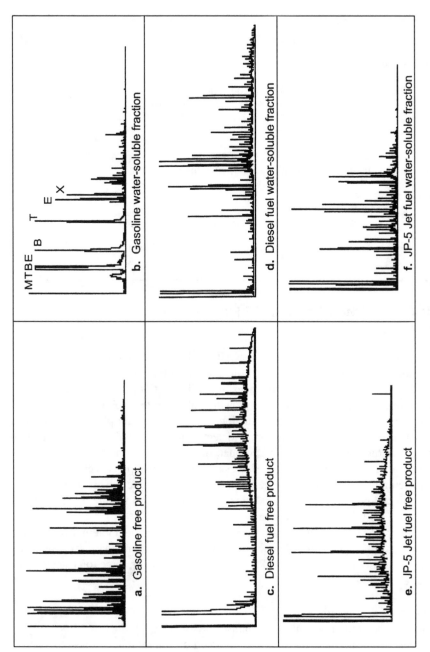

FIGURE 5.5 GC/FID chromatograms of gasoline, diesel, and JP5 fuels and their respective water-soluble fractions. Time of elution, which corresponds roughly to the number of carbons in the eluted compound, increases from left to right. Thus, peaks corresponding to heavier compounds appear farther to the right in each figure. The composition of free product and dissolved fractions are very different in each case because the more soluble compounds become concentrated in the water-soluble fraction. The water-soluble fractions are composed mainly of 1-, 2-, and 3-ring aromatic hydrocarbons. Using calibration standards for the water-soluble fractions improves the accuracy of identifying water sample contaminants.

are the most volatile. Vapor movement is not influenced by groundwater motion and only weakly by gravity. It follows the most conductive pathways through the subsurface, from regions of higher to lower pressure. Much of the vapor remains trapped in soil near its origin, slowly escaping to the surface atmosphere. A small portion of vapor phase contaminants may dissolve into soil-water, but it is generally insignificant.

Rules of Thumb

1. A measurable vapor concentration will be produced if either

Henry's constant ($K_H = C_a/C_w$) > 0.0005 atm m^3 mol^{-1}
(this produces significant partitioning from water to air),

or

Vapor pressure > 1.0 torr at 20°C
(this results in significant diffusion upward through the vadose zone).

2. Characteristic vapor pressures for gasoline:

Fresh gasoline: 260 torr (0.34 atm).

Weathered gasoline (2–5 years old): 15 to 40 torr (0.02 to 0.05 atm).

5.6 ESTIMATING THE AMOUNT OF FREE PRODUCT IN THE SUBSURFACE

The first steps in the remediation of a site where an LNAPL spill has occurred is to try to limit the movement of contaminant plumes and to remove from the subsurface as much free product as possible. As long as free product is present, it continues to partition into the sorbed, dissolved, and vapor contaminant plumes, continually feeding their growth. Only after the mobile free product has been removed from above the water table can remediation of the contaminant plumes be effective.

LNAPL free product in the subsurface is generally detected and measured by its accumulation in wells. To design a program for removing free product, one must obtain a reliable estimate of the volume of free product that must be removed. However, the relationship between the thickness of free product that accumulates in a well and the thickness of free product distributed above the water table is easily misinterpreted. This relationship is influenced by soil texture, fluctuations of the water table level, and the thickness of the free product layer.

Figure 5.6 illustrates some of the factors that affect free product accumulation in a well. In the subsurface away from a well, liquids are influenced by capillary attractions that draw them into small pore spaces and interstices. Where no LNAPL free product is present, three forces determine the aquifer water table elevation:

1. Gravity pulls water downward.
2. Water pressure in the aquifer acts upward against gravity.
3. Capillary forces at the interface between the saturated and unsaturated zones also act upward against gravity.

Within a well, there are no upward acting capillary forces affecting the liquid levels. Only the balance between gravity and water pressure in the aquifer determines the water level in a well.

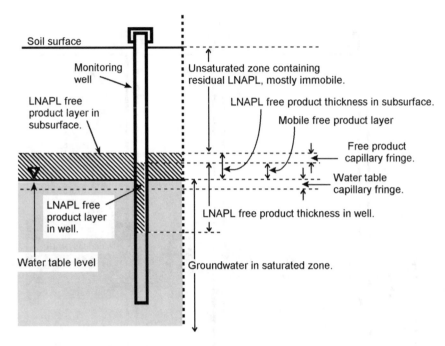

FIGURE 5.6   Thickness of LNAPL accumulated in a well compared to thickness in adjacent subsurface.

If LNAPL is present, it also develops a capillary fringe at its interface with the unsaturated zone. Capillary fringes occur at the upper boundaries of both the water table and the free product layer. In the capillary fringes, liquids are drawn upward against gravity and are largely immobile, especially in horizontal directions. The thickness of the capillary fringes depends on the soil texture. In coarse soils and sands with few capillary-size pore spaces and interstices, capillary fringe layers may be only a few millimeters thick; in fine soils and sands, they may be several meters thick.

Where LNAPL free product floats in contact with the water table, the water level is lowered by the weight of LNAPL (see Figure 5.7). Where the LNAPL free product layer is thin, it lies largely above the water capillary fringe because LNAPL cannot easily displace water from this region. Where the LNAPL layer is thick, its greater weight makes it penetrate farther into, or even through, the water capillary fringe, moving the free product-water interface even lower.

When an appropriately screened well passes through an LNAPL free product layer into the saturated zone, water and free product flow into the well from the surrounding subsurface. Liquid movement into the well occurs from the water and free product regions below their respective capillary fringes, where liquid mobility exists. LNAPL from the mobile zone around the well flows into the wells, and the additional weight of LNAPL lowers the water level in the well to below the normal water table in the aquifer. LNAPL flows into the well until the top level of LNAPL in the well is the same as the top of the mobile zone in the surrounding soil.

Within a well, where no capillary forces exist, the weight of LNAPL lowers the LNAPL-water interface farther than in the surrounding subsurface. LNAPL will continue to flow into the well, lowering the water table, until the upward pressure of aquifer water balances the weight of LNAPL. The end result is that LNAPL accumulates in a well to a greater thickness than in the surrounding subsurface, where capillary forces buoy up both the water level and free product layer. The upper level of LNAPL in the well is lower than the upper level in the surrounding subsurface by the thickness of the LNAPL capillary fringe. The LNAPL-water interface in the well is lower than in the subsurface by an amount that depends on the soil texture and the thickness of the subsurface mobile layer of LNAPL. This behavior is illustrated in Figures 5.7 and 5.8.

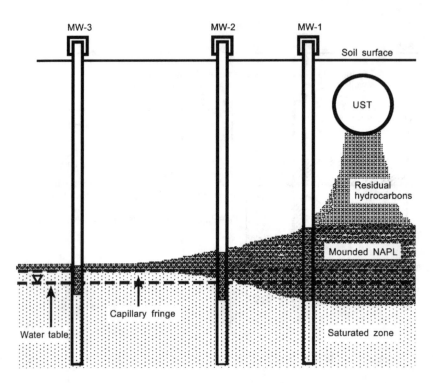

FIGURE 5.7   Comparison of LNAPL thickness in wells with thickness in adjacent subsurface.

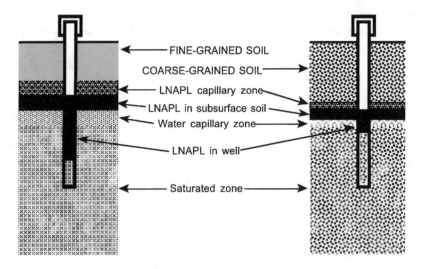

FIGURE 5.8   Effect of soil texture on LNAPL thickness in a well.

Equation 5.2 may be used to calculate the water table level in the adjacent subsurface from well measurements where LNAPL is present. Use of Equation 5.2 is necessary for evaluating and plotting groundwater elevations when LNAPL is present in the wells.

$$\text{WTE} = \text{WE}_{well} + (\text{LNAPL density} \times \text{LNAPL thickness in well}) \qquad (5.2)$$

where: WTE   = water table elevation in subsurface adjacent to the well.
         WE_{well} = water elevation at the water/LNAPL interface in the well.

EFFECT OF LNAPL SUBSURFACE LAYER THICKNESS ON WELL THICKNESS

Referring to Figure 5.7, in the subsurface near wells MW-2 and MW-3, the weight of LNAPL free product is not sufficient to force water downward through the capillary fringe. Near well MW-1, where the LNAPL is thicker because it has formed a dome, the weight of LNAPL is great enough to force water downward through the capillary fringe and below the original water table. Within the wells, there are no capillary forces acting upward on the water. In wells MW-2 and MW-3, the LNAPL thickness in the wells is greater than the LNAPL thickness in the surrounding soils because the absence of capillary forces in the wells reduces the net upward forces acting on the water, and the water level is lower. In well MW-1, where LNAPL has pressed down through the capillary fringe, upward forces on the water are due only to aquifer pressure and are the same in the well and in the surrounding soil. Thus, the LNAPL thickness and the water level in well MW-1 are the same as in the surrounding subsurface.

EFFECT OF SOIL TEXTURE

Soil texture determines the magnitude of the capillary forces that exert upward forces on subsurface water and LNAPL. Capillary forces are much larger in fine-grained soil than in coarse-grained soil. Consequently, the difference between LNAPL thickness in a well and LNAPL thickness in the adjacent subsurface is greater in fine-grained soil. The effect of soil texture is illustrated in Figure 5.8.

Details for calculating the recoverable volume of LNAPL free product in the subsurface from well measurements have been published.[6,7,8] Computer programs are also available for this as well as other related calculations.

Rules of Thumb

1. Measured LNAPL thickness in a well typically exceeds the corresponding LNAPL thickness in the surrounding subsurface by a factor of 2 to 10, because LNAPL above the water table flows into the well and depresses the well water level.
2. The LNAPL thickness ratio, h_{well}/h_s, generally increases with decreasing soil particle size, increasing capillary fringe thickness, and increasing LNAPL density.
3. A crude estimate, ignoring the soil properties, of LNAPL thickness in the adjacent subsurface can be made from

$$h_{well}/h_s \approx (\text{LNAPL density})/(\text{water density} - \text{LNAPL density}). \qquad (5.3)$$

4. Since fuel LNAPL (gasoline and diesel) generally has a density of 0.7–0.8 g/cm³, Equation 5.3 gives $h_{well}/h_s \approx 0.7/0.3$ to $0.8/0.2$, or 2.3 to 4.0.

EFFECT OF WATER TABLE FLUCTUATIONS ON LNAPL IN SUBSURFACE AND WELLS

Water table fluctuations promote vertical spreading of LNAPL in the subsurface and influence the thickness of LNAPL that collects in monitoring wells.

When the groundwater table rises

- Some floating LNAPL free product is driven up into the unsaturated zone. Sorbed residual LNAPLs in the formerly unsaturated zone can be remobilized by dissolving into the free product, causing further lateral spreading.
- Some free product remains trapped in pore spaces below the water table within the saturated zone.
- The mobile free product layer above the water table becomes thinner.

When the groundwater table falls

- LNAPL free product floating on the water table moves downward as the water table drops, leaving behind an immobilized fraction of free product as sorbed residual LNAPL that is retained in the newly unsaturated zone above the lowered water table.
- The mobile free product layer above the water table may become thicker because free product formerly trapped below the water table is free to migrate downward to the new water table.

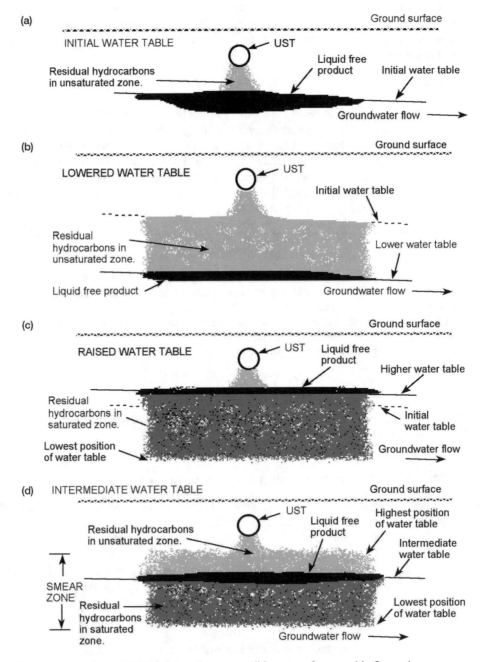

FIGURE 5.9 Spreading of LNAPL into a "smear zone" because of water table fluctuations.

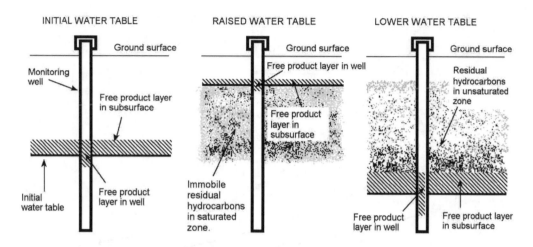

INITIAL WATER TABLE   RAISED WATER TABLE   LOWER WATER TABLE

FIGURE 5.10   Effect of fluctuating water table on LNAPL accumulation in a well.

The rising and falling of the water table leaves behind a "smear zone" of contamination that lies partially in the saturated zone and partially in the unsaturated zone. This behavior is illustrated in Figure 5.9.

Effect of Water Table Fluctuations on Well Measurements

LNAPL spills are often first detected by the appearance of a free product layer above the water in downgradient wells. If the well free product layer diminishes during a remediation program, it is tempting to believe that the cleanup effort is working successfully. An increase in the well free product thickness often initiates a search for new LNAPL sources. However, unless fluctuations in groundwater depth are taken into account, basing such conclusions on changes in the free product layer thickness in wells can lead to serious errors.

When groundwater rises, the thickness of the free product layer in wells generally decreases because a portion of the free product becomes trapped below the water table and becomes immobile, thinning the mobile free product layer. When the water table falls, free product formerly trapped in the saturated zone becomes mobile again and can accumulate as a free product layer over the water table where it is free to flow into wells. This behavior is illustrated in Figure 5.10.

5.7   ESTIMATING THE AMOUNT OF RESIDUAL LNAPL IMMOBILIZED IN THE SUBSURFACE

Residual LNAPL in the subsurface is the portion that will not flow into a well. It is the part of an LNAPL spill that cannot be removed by pumping to the surface. Residual LNAPL must be remediated by biodegradation, soil flushing, or excavation. Residual LNAPL is retained in the unsaturated zone by adsorption and capillary forces. Therefore, small soil particles and large surface area both increase the amount of residual LNAPL retained. The soil retention factor (volume of LNAPL per volume of soil) depends mainly on the soil pore size distribution, soil wettability, LNAPL viscosity, and LNAPL density.

Usually more LNAPL is immobilized in the saturated zone than in the unsaturated zone because part of the residual LNAPL in the unsaturated zone eventually drains down to the water table. In the saturated zone, water is the wetting fluid and LNAPL the nonwetting fluid. LNAPL becomes trapped in larger pores of the saturated zone by immobile water.

In the unsaturated zone, LNAPL is the wetting fluid and tends to spread into smaller pores and drain from the larger pores. Figure 5.11 shows soil retention factors for several kinds of LNAPL

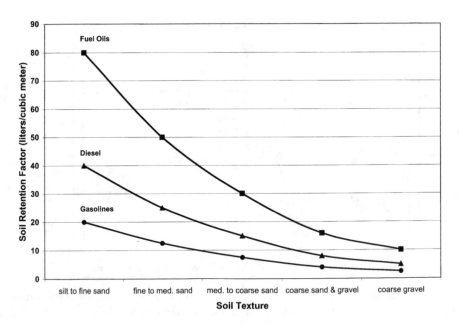

FIGURE 5.11 Soil retention factors for LNAPL fuels in different soils, plotted from data in Reference 9. Calculations assume a soil bulk density of 1.85 g/cm³ and LNAPL densities of 0.7, 0.8, and 0.9 g/cm³ for gasolines, diesel fuel, and fuel oils, respectively.

in soils of different textures. The retention of LNAPL in soils above the water table usually ranges between about 80 L per cubic meter of soil, for fuel oil in silt, to 2.5 L per cubic meter, for gasoline in coarse gravel. LNAPL in the unsaturated zone can often be remediated without excavation by some combination of soil washing, volatilization, or bioremediation.

Example 5.2: Using Soil Retention Factors

One thousand gallons of fuel oil were spilled on a soil consisting mostly of medium to coarse sand. How much soil is required to immobilize 1000 gallons? If the spill area was confined by a berm to 100 ft², how deep into the soil will the oil penetrate? Could it endanger a shallow aquifer 35 ft below the surface?

Answer: From Figure 5.11, the soil retention factor is about 30 L/m³ for fuel oil. The volume of soil needed to contain the entire spill is

$$V_{soil} = \left(\frac{1000 \text{ gal}}{30 \text{ L/m}^3} \right) \left(\frac{3.785 \text{ L}}{1 \text{ gal}} \right) \left(\frac{35.3 \text{ ft}^3}{1 \text{ m}^3} \right) = 4454 \text{ ft}^3.$$

Assume the oil plume travels downward without spreading, so that its cross-section is 100 ft². Then a volume of 4454 ft³ will extend downward by

$$\text{Depth of oil penetration until it all is retained and immobilized} = \frac{4454 \text{ ft}^3}{100 \text{ ft}^2} = 44.5 \text{ ft.}$$

Oil is likely to reach the aquifer at 35 ft.

TABLE 5.4
Relative Importance of Different Subsurface Loci in Sandy Soils for Retention of Gasoline Contamination

	Loci of Subsurface LNAPL Retention	Average Gasoline Retention in Sandy Soils (mg/cm³)	Percent of Total Retention in Sandy Soils
1.	Gasoline vapors in soil pores in the unsaturated zone.	0.095	<0.1%
2.	Liquid gasoline sorbed to dry soil particles in the unsaturated zone. Locus 2 is especially important in the soil volume immediately below a spill, but not downgradient of the spill.	36	9.88
3.	Gasoline dissolved in water on wet soil particles in the unsaturated zone.	0.0010	<0.1%
4.	Liquid gasoline sorbed to wet soil particles in the saturated and unsaturated zones.	0.076	<0.1%
5.	Liquid gasoline in soil pore spaces within the saturated zone. Locus 5 contaminants may generally be regarded as immobile.	38	10.4
6.	Liquid gasoline in soil pore spaces in the unsaturated zone. Contaminants enter locus 6 mainly from free product floating on the groundwater table when the table rises and then falls.	110	30.2
7.	LNAPL gasoline free product floating on top of the groundwater table. The most important loss mechanism from locus 7 occurs when a fluctuating water table moves contaminant into loci 5 and 6, where some of it remains trapped.	180	49.4
8.	Gasoline dissolved in groundwater.	0.020	<0.1%
9.	Gasoline sorbed to colloidal particles in water in the saturated and unsaturated zones.	0.00013	<0.1%
10.	Liquid gasoline diffused into mineral grains in the saturated and unsaturated zones.	0.000060	<0.1%
11.	Gasoline sorbed onto or into microbiota in the saturated and unsaturated zones.	0.010	<0.1%
12.	Gasoline dissolved into the mobile pore water of the unsaturated zone.	0.030	<0.1%
13.	Liquid gasoline in rock fractures in the saturated and unsaturated zones.	0.21	0.17

Source: Calculated using data from Reference 10.

SUBSURFACE PARTITIONING LOCI OF LNAPL FUELS

As part of an ongoing effort to provide an authoritative, defensible engineering basis for predicting contaminant behavior in soils, the EPA has identified 13 soil loci among which petroleum contaminants become partitioned.[10] Contaminants may move within a given locus under the influence of pressure and concentration gradients or from one locus to another (for example, when soluble components in the free product layer dissolve into the groundwater).

In Table 5.4, the partitioning behavior of gasoline in sandy soils has been calculated, using the methods and data recommended by EPA.[10] The table shows how much gasoline LNAPL is expected to be retained in the 13 different partitioning loci.

It is clear from Table 5.4 that when free product is present in locus 7 (gasoline LNAPL floating on top of the groundwater table), this condition is the controlling factor for the distribution of contaminants in other zones. With free product present above the water table, loci 2, 5, 6, and 7 are by far the most important in terms of mass and account for 99.9% of the total soil and groundwater contamination.

The mass of gasoline in the vapor and dissolved states (loci 1, 3, 8, and 12) is insignificant compared to that in the free product above the water table and in the smear zone created by a rising and falling water table (loci 2, 5, 6, and 7). Of course for remediation purposes, all loci are important and those with relatively small amounts of contaminant may be the most difficult to remediate to a regulated level.

Example 5.3: Calculation of Contaminant Plume Volume Required to Immobilize 1 Million Gallons of Gasoline

For a point source of contamination, we can assume that only loci 5 and 6, the smear zone caused by a fluctuating water table, are effective for immobilizing contaminant. Locus 2 lies primarily under the area of the initial spill or leak and probably is small compared to the total free product plume volume. LNAPL in locus 7 may be regarded as mobile. As LNAPL in locus 7 flows downgradient and is subjected to vertical movement caused by a fluctuating water table, it continually loses mass into loci 5 and 6.

Assume an average sandy soil with a hydraulic gradient of 0.009 ft/ft. Using the methods followed in U.S. EPA,[10] the estimated flow velocity for LNAPL gasoline in locus 7 is about 1.3 ft/day.* This may be compared with a groundwater velocity of about 10 ft/day for the same conditions. Further assume that the seasonal groundwater table fluctuations average around ± 1.5 ft, giving a free product smear zone 3 ft in thickness. Taking the smear zone to be of uniform thickness over the plume volume and assuming that, on average, half of the smear zone is in the saturated zone and half in the unsaturated zone, we can say loci 5 and 6 are each 1.5 ft thick everywhere over the free product plume.

Locus 5 retention

Take the density of aged gasoline LNAPL to be 0.74 g/cm³, or 2800 g/gal. From Table 5.4, locus 5 retains 38 mg of gasoline per cm³ of soil, or about 16,500 gal/acre-ft. For a locus 5 thickness of 1.5 ft, 24,750 gallons of gasoline LNAPL will be immobilized per acre of contamination plume cross-sectional area.

Locus 6 retention

Similarly, retention in locus 6 will be about 73,000 gal/acre. The total amount of fuel retained in loci 5 and 6 is

(24,750 + 73,000) gal/acre = 97,750 gal/acre, or close to 100,000 gal/acre.

For a 1-million-gallon gasoline spill in sandy soil, where the water table fluctuates ± 1.5 ft annually, the soil contaminant plume could become immobilized after it had spread into 30 acre-ft, or a horizontal cross-section of about 10 acres.

5.8   DNAPL FREE PRODUCT PLUME

Dense nonaqueous phase liquids, DNAPL, are nonaqueous phase liquids that are denser than water. If sufficient DNAPL is present, the free product sinks downward through the water table to the bottom of an aquifer. Only an impermeable obstruction, such as bedrock, stops the downward movement of DNAPL mobile free product. Some examples of DNAPL are

- Halogenated solvents (mostly chlorinated or brominated), either pure compounds or mixtures of solvents
- Wood preservatives

---

* Because of its greater viscosity, diesel fuel would move about one-third as fast.

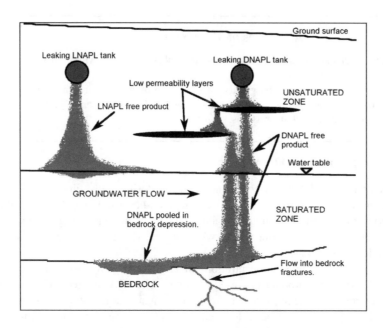

FIGURE 5.12 Comparison of DNAPL and LNAPL movement in the subsurface, after spills.

- Coal tars
- Creosote
- Polychlorinated biphenyls (PCBs)
- Many pesticides

Free product from DNAPL spills moves downward through the vadose zone (leaving a trail of residual soil-sorbed DNAPL) until it reaches the capillary zone of the water table. To continue moving downward, the DNAPL must displace water held by capillary forces. Consequently, downward movement slows while DNAPL piles up and spreads laterally. If sufficient weight of DNAPL accumulates, it presses downward through the capillary zone and continues down through the saturated zone (see Figure 5.12).

Where there is a decrease in soil permeability — whether in the unsaturated or saturated zone — DNAPL behaves similarly, the downward movement slows, and more lateral spreading occurs. This leads to the formation of pools and fingers. Eventually, impermeable bedrock is reached where DNAPL collects in pools. If the bedrock is slanted, DNAPL will migrate down the physical slope, even if the direction is opposite to the groundwater movement.

Many DNAPL compounds, especially chlorinated hydrocarbons, are less viscous than water. They can penetrate small fractures and micropores and become inaccessible to *in situ* remediation. Chlorinated hydrocarbons interact with some clays, causing them to shrink and crack. Clay-lined ponds are not reliable containment for liquid wastes containing chlorinated hydrocarbons.

Aerobic biodegradation techniques that work well with many LNAPL hydrocarbons have not been very successful with chlorinated DNAPL. Although there have been recent reports of successful biodegradation procedures involving anaerobic steps and carefully selected microbes, biodegradation and weathering are slow and DNAPL may persist for long time periods.

Solubilities are generally low; therefore, DNAPL will continue to dissolve slowly into the groundwater without significant diminution. Even with a moderate DNAPL release, dissolution can continue for hundreds of years under natural conditions before all the DNAPL has dissolved or degraded. Once in the subsurface, it is difficult or impossible to recover all of the trapped residual DNAPL. DNAPL that remains trapped in the soil/aquifer matrix acts as a continuing source of groundwater contamination.

Rules of Thumb for DNAPL

1. Chlorinated hydrocarbons are generally more dense than water (DNAPL). They sink to the bottom of the water table.
2. DNAPL movement is affected by gravity far more than by groundwater movement. It moves with the slope of the bedrock below the aquifer, independent of the direction of groundwater movement, and forms pools in bedrock depressions.

TESTING FOR THE PRESENCE OF DNAPL

It is very difficult to locate DNAPL free product with monitoring wells. First of all, DNAPL remains in the bottom of the well and may go unnoticed. Second, because DNAPL free product generally collects in pools at the bottom of an aquifer and in locations unrelated to groundwater movement, there often are no obvious guidelines as to where a well should be placed or how it should be screened to collect free product.

For these reasons, dissolved concentrations of DNAPL-related chemicals are often the only evidence that DNAPL free product is present at a site. The EPA has recommended an empirical approach for determining whether DNAPL free product is near a monitoring well where dissolved DNAPL-related compounds have been detected.[11] In order to use this approach, it is necessary to measure concentrations of DNAPL-related compounds dissolved in groundwater, to know the composition of the suspected DNAPL, and to calculate the *effective solubility* of the measured DNAPL components.

The effective solubility is the theoretical solubility in water of a single component of a DNAPL mixture. It may be approximated by multiplying the component's mole fraction* in the mixture by its pure phase solubility.

$$S_{eff}(a) = X_a S_{pure}(a), \qquad (5.4)$$

where: $S_{eff}(a)$ = effective solubility of component a in a DNAPL mixture, in mg/L.
X_a = mole fraction of compound a in the mixture.
$S_{pure}(a)$ = pure-phase solubility of compound a, in mg/L.

Conditions That Indicate the Presence of DNAPL

If any of the following conditions exist, there is a high probability that DNAPL free product is in the vicinity of the sampling location.

- Groundwater concentrations of DNAPL-related chemicals are >1% of either their pure phase solubility (S_{pure}) or their effective solubility (S_{eff}).
- Soil concentrations of DNAPL-related chemicals are >10,000 mg/kg (1% of soil mass).
- Groundwater concentrations of DNAPL-related chemicals increase with depth or appear in anomalous upgradient/cross-gradient locations with respect to groundwater flow.
- Groundwater concentrations of DNAPL-related chemicals calculated from water/soil partitioning relationships are greater than pure phase solubility or effective solubility.

---

* The mole fraction of compound a in a mixture of several compounds is written X_a.

$$X_a = \frac{\text{moles of } a}{\text{total moles of all compounds}}.$$

For a mixture with 1 mole of CCl_4 and 3 moles of $CHCl_3$, $X_{CCl4} = 1/4 = 0.25$ and $X_{CHCl3} = 3/4 = 0.75$. Note that the sum of all mole fractions must equal unity. The mole fraction of any pure substance equals unity.

Calculation Method for Assessing Residual DNAPL

1. Calculate $S_{eff}(a)$ as above.
2. Find K_{oc}, the organic carbon-water partition coefficient in Table 4.6, from published literature, or estimate it from $\log K_{oc} = \log K_{ow} - 0.21$.
3. Determine f_{oc}, the fraction of organic carbon (oc) in the soil by lab analysis. Values for f_{oc} typically range from 0.03 to 0.00017 (mg oc)/(mg soil). Convert values reported in percent (mg oc/100 mg soil) to (mg oc)/(mg soil).
4. Determine or estimate the dry bulk density of the soil (d_b). Typical values range from 1.8 to 2.1 g/cm^3 (kg/L). Determine or estimate the water-filled porosity (p_w) of the soil.
5. Determine K_d, the soil-water partition coefficient from $K_d = K_{oc} \times f_{oc}$, as in Equation 4.16.
6. Using $C_{soil}(a)$, the measured concentration, in mg/kg, of DNAPL-related compound a in saturated soil, calculate the theoretical concentration of a in pore water, $C_w(a)$ from

$$C_w(a) = \{C_{soil}(a) \times d_b\}/\{(K_d \times d_b) + p_w\}.$$

7. Compare $C_w(a)$ with $S_{eff}(a)$ from Step 1:
   - If $C_w(a) > S_{eff}(a)$ there is a possible presence of DNAPL near the sampling location.
   - If $C_w(a) < S_{eff}(a)$ there is a possible absence of DNAPL near the sampling location.

5.9   CHEMICAL FINGERPRINTING

Environmental professionals often need to do more than locate and clean up pollutants. They may need to help identify the sources of the contamination. Someone always has to pay for a remediation effort. The high costs that are often involved mean that any prudent potentially responsible party (PRP) will want convincing proof that he or she must accept responsibility for the pollution. Polluted sites often have a history of having different owners and uses, any or all of which may have contributed to the present site contamination. Off-site sources may also be responsible for all or part of the current pollution problems. Allocating responsibility for the expenses involved in remediation often becomes a legal issue. The methods used in building a legally defensible case that assigns responsibility for soil and groundwater pollutants have become part of a branch of environmental investigations known as *environmental forensics.*

One of the most common applications of environmental forensics is identifying the sources of fuel contamination. The use of petroleum fuels has been increasing for the past 100 years and fuel contaminants are pervasive wherever they have been used. There are many potential sources of fuel pollutants, from numerous small gasoline stations to large refinery operations, roadway accidents to criminal acts of disposal, pipeline breaks to tank corrosion. It is no wonder that the origin of any particular site contamination may often be legitimately questioned. Chemical fingerprinting of fuels is one tool in the toolbox of chemical forensics.

In chemical fingerprinting, the chemical characteristics of fuel contaminants are used to help distinguish among different possible sources of the pollutants. For example, knowing that only gasoline was ever used on a certain site, but that the groundwater hydrocarbon contaminants found there are from diesel fuel, clearly points to an off-site source.

FIRST STEPS IN CHEMICAL FINGERPRINTING OF FUEL HYDROCARBONS

Successful chemical fingerprinting of hydrocarbon contamination frequently consists of the sequential application of several investigative steps:

- Identify the fuel type of the contaminants. Are they gasoline, diesel fuel, or fuel oils? Each of these classes has unique chemical characteristics.

- If possible, distinguish between the ages of different contaminated samples. Weathering processes often introduce predictable changes in the chemical makeup of fuels. If samples have weathered differently because of different exposure times, they will have different chemical profiles. Certain compounds in fuel samples are more readily biodegraded, solubilized, or volatilized than others. This causes changes in the overall fuel composition as time passes. The exact nature of the changes is always site specific, but sometimes the ages of different samples can be clearly demonstrated to be significantly different. In some cases, the presence of certain degradation products may be a useful indicator of age. Also, fuel compositions have changed historically, as more efficient refinery production practices were developed and as clean air legislation and automotive engine development mandated changes. For example, the elimination of lead-based octane boosters in gasolines began in the early 1980s. The presence of lead may indicate a vintage rather than recent spill, especially if accompanied by dichloroethane and/or dibromoethane, which were lead scavenger gasoline additives.
- Look for unique chemical compounds that can serve as "markers" which may be present in contamination from one source, but not from another. An example might be the presence of MTBE or ethylene dichloride in some fuel-contaminated samples but not in others. Different production practices at different refineries sometimes produce subtle but distinctive differences in their fuel products, which can help to distinguish between possible sources. This approach generally requires extensive knowledge about the chemical composition of contamination at the site and, sometimes, knowledge about the chemicals previously and currently used at the site.

Chemical fingerprinting is sometimes fairly simple and sometimes very complicated, even impossible. A few approaches are described here to indicate the possibilities. Complicated cases will require the help of experienced forensic chemists.

IDENTIFYING FUEL TYPES

The most useful analytical tool for identifying different fuel types is the gas chromatograph (GC), especially with a mass spectrometer detector. Such instruments are referred to as GC/MS. When fuels are analyzed in a gas chromatograph, retention times of different hydrocarbon compounds closely correspond to their boiling points; the higher the boiling point, the longer the retention time. Thus, lower boiling point gasoline components elute from the gas chromatograph column earlier than higher boiling point diesel components.

Physical and chemical characteristics of different fuels were described in Section 5.2. Table 5.1 and Figure 5.2 show that carbon number ranges characteristic of different fuel types correspond to different boiling point ranges. In general, the more carbons in a petroleum molecule, the higher is its boiling point (see the discussion of London forces in Chapter 2). In a gas chromatogram of fuel hydrocarbons, the longer the retention time of a peak, the higher the boiling point of the corresponding compound and the more carbons in the molecule. Although the presence of structural differences, such as the presence of carbon side chains, introduces some variability, this general principle remains useful. Gas chromatograms of different types of fuel are different from one another and are useful for identifying fuel types.

In Figure 5.13, note that the gasoline signature contains more lightweight components (peaks farther to the left, indicating fewer carbon atoms) than do diesel fuel or lubricating oil. When fuel hydrocarbons are weathered by volatilization, dissolution, and biodegradation, the lighter components on the left side of the fresh gasoline and diesel fuel signatures are lost first from the free product. Thus, weathered diesel does not resemble gasoline because it lacks the lightweight components that are characteristic of gasoline. Likewise, weathered gasoline does not resemble diesel fuel because it lacks the heavier components that are characteristic of diesel fuel.

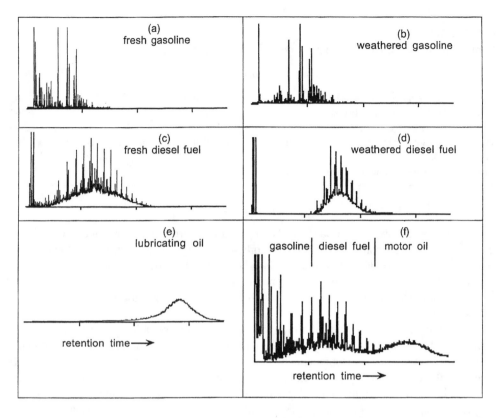

FIGURE 5.13 Gas chromatograms showing the differences in chromatographic signatures between different types of fresh and weathered petroleum hydrocarbon free product. Figure (f) at lower right is a GC of free product containing a mixture of gasoline, diesel fuel, and motor oil. Humps, where the chromatogram rises above the baseline, are due to hundreds of different hydrocarbon compounds, which are not chromatographically resolved.

A simple and common indicator of the age of gasoline spills is the ratio of benzene to other components of the fuel. Benzene is more soluble, volatile, and degradable than other common gasoline constituents and is depleted relative to other compounds in samples with older contamination.

Rules of Thumb for Estimating the Age of Gasoline Contamination

Current gasolines contain about 2% to 5% of benzene. Older gasolines may have as little as 1%. EPA promotes the reduction of benzene to 1% or less for health reasons. Because benzene is the most soluble of the BTEX group

1. A sample from water in contact with fresh gasoline will have a benzene-to-toluene ratio of about 0.6 to 1.5.
2. A gasoline-contaminated water sample containing much less benzene than toluene (benzene/toluene < 0.1) indicates older gasoline contamination, perhaps 6 months to 2 years old.
3. A water sample where the concentration ratio of benzene-to-total BTEX is less than 0.001 may be 5 to 10 years old, or even older.
4. Another clue to the age of gasoline contamination is based on tetraethyl lead, organic manganese compounds, ethylene dibromide (EDB), and ethylene dichloride (EDC), which were common additives to gasoline before 1980. Therefore, the presence of these compounds is supporting evidence for gasoline contamination originating before 1980. EDB and EDC, however, are present in some agricultural chemicals and must be used as an indicator with caution.

TABLE 5.5
Approximate Composition of Fresh Diesel Fuel (C-6 to C-24)

Class of Compound	Percent
N-alkanes (degrade fastest; smaller alkanes degrade faster than larger alkanes).	40%
Iso- and cyclo-alkanes (degrade slower).	36%
Isoprenoids (degrade very slowly).	3–4%
Aromatics (most soluble; mainly parent and alkylated benzenes, naphthalenes, hydrindenes, phenanthrenes, and fluorenes).	20%
Polar (water soluble sulfur, nitrogen, and oxygen compounds).	1%

Age-Dating Diesel Oils

Normal alkanes (linear hydrocarbons) in the approximate carbon number range C-6 to C-24 are the most abundant components of fresh diesel oils — although many other types of hydrocarbons are also present (see Table 5.5). The table also shows that the different types of hydrocarbons in diesel oils biodegrade at different rates. Isoprenoids are branched, unsaturated hydrocarbons that biodegrade more slowly than linear alkanes with similar masses because their chemical structure inhibits biodegradation.

The solubilities and volatilities of oil range n-alkanes and isoprenoids are quite low and similar for the two chemical classes. Therefore, they have similar rates of weathering by nonbiological processes. However, because isoprenoids biodegrade much more slowly than n-alkanes, the abundance ratio of n-alkanes to isoprenoids changes over time where biodegradation is occurring. Because the passive biodegradation rates of diesel in soils are fairly uniform at similar sites,[5] the corresponding changes in composition can be used for estimating the age of the diesel contamination.

In 1993, Christenson and Larsen[5] found, as others had, that by comparing gas chromatograph peak-height ratios of pristane (C-19 isoprenoid) with n-heptadecane (linear C-17 alkane) an estimate can be made of the degree of biodegradation that the sample has undergone. The ratios of phytane (C-20 isoprenoid) and n-octadecane (linear C-18 alkane) can be compared in a similar way. Using the relative extents of biodegradation, the relative ages of fuel contamination in different samples from similar sites can be estimated.

When the composition of fresh diesel fuel is compared with weathered fuels that have biodegraded in the subsurface environment, certain composition changes are apparent. In particular, n-alkanes dominate the composition of fresh fuels and isoprenoids dominate the composition of highly degraded fuels. The C-17/pristane peak-height ratio falls from about 2 (for fresh diesel) to 0 (after about 20 years). Figure 5.14 compares gas chromatograms of fresh and biodegraded No. 2 diesel oil. Changes in the relative peak heights of n-alkanes and isoprenoids are readily apparent.

Christenson and Larsen[5] also found a linear relation between the age of diesel oils and the C-17/pristane peak-height ratio. From their data, it is possible to determine the age of a diesel spill to within about 2 years for diesel that is between 5 and 20 years old, that was created by a single sudden spill event, and that has not been "weathered" significantly except by biodegradation. A linear best-fit to their data yields the equation

$$\text{Age of diesel fuel (yr)} = -8.3(\text{C-17/pristane ratio}) + 19.5. \qquad (5.5)$$

Example 5.4: Fingerprinting Fuel Contaminants at an Industrial Site

The following GC/MS data, taken from soils at an industrial site with contamination by gasoline, diesel, and heavy oil hydrocarbons, indicate that at least two different diesel spill events occurred, separated by approximately 6 to 10 years. The mass spectrometer detector was sometimes operated in the single-ion mode, which increases sensitivity. Single-ion monitoring consists of leaving the mass spectrometer tuned to a fixed mass number as the gas chromatogram peaks elute, rather than continually scanning the entire mass range for each GC peak. The difference between single-ion and full-range monitoring is shown in Figures 5.15a and 5.15b.

In order to estimate the age of the diesel fuel shown in Figure 5.15, the GC/MS spectra in the pristane/phytane region (15–17 min retention time) must be expanded. Expanded single-ion spectra are shown in Figure 5.16 for samples from Boreholes A and B, which are separated by about 200 yd.

Pristane, at 16.06 min in Figure 5.16a, biodegrades much more slowly than the C-17 *n*-alkane at 16.00 min. The same is true for phytane, at 16.92 min, and the C-18 *n*-alkane at 16.83 min. The retention times for these peaks are slightly different in Figure 5.16b.

In the Borehole A sample, the C-17/pristane peak-height ratio is 0.33 and the C-18/phytane ratio is 0.26, indicating significant biodegradation. Although, site-specific conditions will determine how much time this amount of degradation would require, the diesel in this sample is at least 15–20 years old. Equation 5.4 indicates an age of about 19 years.

In the Borehole B sample, the C-17/pristane ratio is 1.2, which indicates a more recent diesel spill. Equation 5.4 suggests that the diesel fuel from Borehole B is about 9 years old.

Rules of Thumb

1. Fresh diesel contains more *n*-alkanes than isoprenoids, such as pristane or phytane. Therefore, if relative depletion of the *n*-alkanes is observed, it indicates that biodegradation has occurred.
2. In a fresh diesel fuel, the C-17/pristane and C-18/phytane ratios are close to 2. Any ratio less than about 1.5 indicates that biodegradation has occurred.

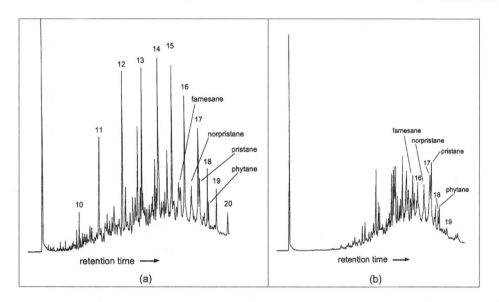

FIGURE 5.14

(a) A gas chromatogram of typical fresh No. 2 diesel oil. The numbered peaks are linear *n*-alkanes, where the number represents the carbon number (e.g., C-17) of the alkane. The *n*-alkanes are the most abundant compounds in fresh diesel fuels and dominate the composition. The peaks of several isoprenoids are labeled. The peak-height ratio of C-17/pristane is about 2:1. (b) A gas chromatogram of biodegraded No. 2 diesel oil. Isoprenoids are more abundant than *n*-alkanes. The peak-height ratio of C-17/pristane is about 0.8, indicating an age of about 13 years according to Equation 5.5.

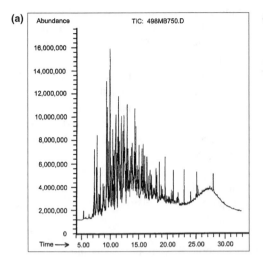
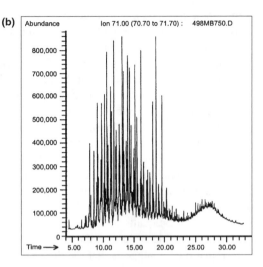

FIGURE 5.15   Full-range and single-ion GC/MS spectra of a contaminated soil sample from Borehole A. (a) GC/MS full-range spectrum of fuel contamination extracted from soil at Borehole A. The distribution of compounds indicates that the soil contains gasoline, diesel, and heavy oil hydrocarbons. (b) Single-ion spectrum at mass 71 of the same sample. Mass 71 is a fragment ion common to many hydrocarbon compounds of C-7 and larger. It is useful for diesel fingerprinting because the mass 71 ion is produced abundantly in the fragmentation of pristane and phytane.

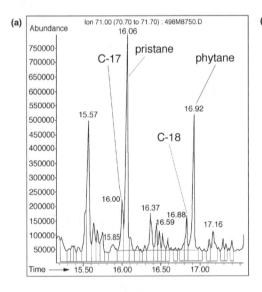
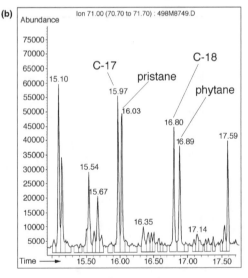

FIGURE 5.16   Expansion of mass 71 single-ion spectra in the pristane/phytane region of contaminated soil from Boreholes A and B. (a) Expansion of the mass 71 spectrum in the 15–17 min retention-time region for the Borehole A sample. The C-17/pristane peak-height ratio is about 0.26 indicating significant biodegradation. (b) At another location on the site, Borehole B, the C-17/pristane peak-height ratio is about 1.2 indicating diesel contamination that is younger and much less biodegraded.

SIMULATED DISTILLATION CURVES AND CARBON NUMBER DISTRIBUTION CURVES

A simple approach to chemical fingerprinting is to generate simulated distillation curves (SDCs) and carbon number distribution curves (CNDCs). These plots allow a relative comparison of the concentrations of volatile and semivolatile compounds that are present in a sample, without requiring a detailed analysis of each compound. The curves can be generated automatically from computerized gas chromatogram data without identifying the individual peaks.

A simulated distillation curve shows the percentage of the sample that would be volatilized at various temperatures. Since the volatility of a petroleum compound is closely related to the number of carbon atoms in that compound (compounds with fewer carbon atoms are lighter in weight and boil at lower temperatures than compounds with more carbon atoms), one can estimate the general chemical makeup of petroleum contaminants from an SDC. A carbon number distribution curve gives similar information but is often easier to interpret.

SDC and CNDC curves allow a visual comparison of the mass distribution of chemical compounds that are present in the analytical samples, based on their boiling points and masses. The shapes of these curves are distinctly different for different types of hydrocarbon mixtures. Gasoline, for example, contains relatively high concentrations of lightweight hydrocarbons such as benzene, while diesel fuel normally has very low concentrations of these lightweight compounds, but higher concentrations of heavier compounds. Weathering of organic compounds produces predictable changes in the shapes of SDCs and CNDCs.

When CNDCs and SDCs have distinctive shapes, they can be used as a "fingerprint" for determining whether contamination at one location is different from or similar to contamination at another location. Sometimes this preliminary analysis is sufficient for identification purposes, as in Example 5.5. In other cases, it serves to guide further study.

Example 5.5: Fingerprinting Diesel Contamination With Simulated Distillation and Carbon Number Curves

Mr. A, the owner of a newly purchased mountain home, frequently, but not always, detected strong fuel odors in his basement shortly after diesel fuel was delivered to a neighbor's underground storage tank (UST). Mr. B, the neighbor, was cooperative and allowed Mr. A to take samples from the UST for analysis and comparison with soil samples from around Mr. A's new home.

In Figure 5.17a, the simulated distillation curve for No. 2 diesel fuel from Mr. B's UST lies somewhat to the left of the laboratory standard for diesel, showing that it contains a higher percentage of lower boiling (lighter weight) compounds. This could indicate that the UST fuel was contaminated slightly with gasoline (perhaps from a tanker truck used to carry both types of fuel), or that it was specially formulated for use in cold weather. In Figure 5.17b, the carbon number distribution curve for the UST fuel has a distinctive fingerprint that includes two prominent peaks at C-11 and C-13.

Figures 5.18a and 5.18b show simulated distillation and carbon number distribution curves for a soil sample collected from a free product seep adjacent to the foundation of Mr. A's home. Not only is the SDC for the foundation seep sample very similar to the SDC for the UST sample, but also its CNDC contains the two distinctive peaks at C-11 and C-13 that were also seen in the UST sample.

Mr. B acknowledged that this data strongly implicated his UST to be the source of contamination at Mr. A's home, even if it did not explain why the leak appeared to be erratic. When his tank was leak tested, it was found that the filler pipe had cracked where it was fastened to the tank below the soil's surface. It leaked only when the tank was overfilled and diesel rose in the fill pipe above the crack.

REFERENCES

1. A Guide to the Assessment and Remediation of Underground Petroleum Releases, American Petroleum Institute, Washington D.C., Publ. No. 1628, 2nd edition, 1989.
2. Bredehoeft, J., Much contaminated groundwater can't be cleaned up, *Groundwater*, 30, pp. 834, 1992.
3. Freeze, R.A. and Cherry, J.A., *Groundwater*, Prentice-Hall, Englewood Cliffs, NJ, 1979, chap. 9.
4. MacDonald, J.A. and Kavanaugh, M.C., Restoring contaminated groundwater: an achievable goal?, *Environ. Sci. and Technol.*, 13(4), pp. 142–149, 1994.

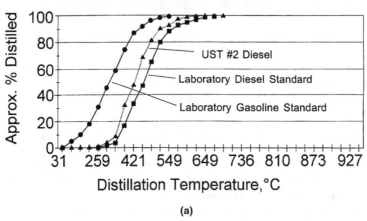

(a)

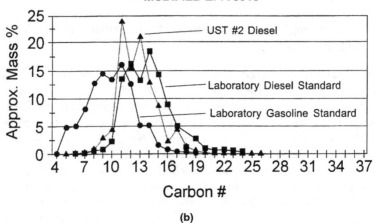

(b)

FIGURE 5.17a,b   Simulated distillation curves and carbon number curves comparing No. 2 diesel fuel from an underground storage tank with laboratory standards of diesel fuel and gasoline.

5. Christensen L.B. and Larsen T.H., Method for determining the age of diesel oil spills in the soil, *Groundwater Monitoring and Remediation*, pp. 142–149, 1993.
6. Farr, A.M., Houghtalen R.J., and McWhorter D.B., Volume estimation of light nonaqueous phase liquids in porous media, *Groundwater*, 28(1), pp. 48–56, 1990.
7. Lenhard, R.J. and J.C. Parker, Estimation of free hydrocarbon volume from fluid levels in monitoring wells, *Groundwater*, 28(1), pp. 57–67, 1990.
8. Parker, J.C., Waddill D.W., and Johnson J.A., *UST Corrective Action Technologies: Engineering Design of Free Product Recovery Systems,* EPA/600/SR-96/031, Order No. PB96-153556, National Technical Information Service, Springfield, VA.
9. Mercer J. and Cohen R., A review of immiscible fluids in the subsurface: properties, models, characterization and remediation, *J. Contamin. Hydrol.*, 6, pp. 107–163, 1990.
10. U.S. EPA, *Assessing UST Corrective Action Technology, A Scientific Evaluation of the Mobility and Degradability of Organic Contaminants in Subsurface Environments*, EPA/600/2-91/053, September 1991.
11. U.S. EPA, *Estimating Potential for Occurrence of DNAPL at Superfund Sites,* Publication 9355.4-07FS, Office of Emergency and Remedial Response, Washington, D.C., NTIS PB92-963338, January 1992.

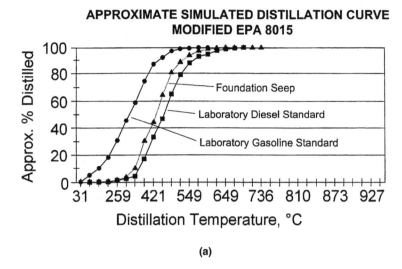

(a)

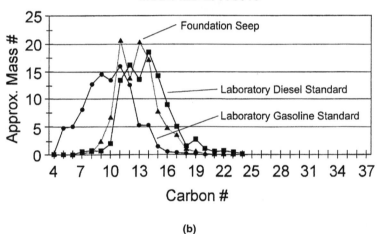

(b)

FIGURE 5.18a,b Simulated distillation curves and carbon number curves comparing free product from a seep adjacent to a house foundation with laboratory standards of diesel fuel and gasoline.

6 Selected Topics in Environmental Chemistry

CONTENTS

6.1   ACID MINE DRAINAGE

The main cause of acid mine drainage is oxidation of iron pyrite. Iron pyrite, FeS_2, is the most widespread of all sulfide minerals and is found in many ore bodies. During mining operations, particularly coal mining, iron pyrite in the ore is exposed to air and water, causing it to be oxidized to sulfuric acid and ferrous ion:

$$FeS_2 + \tfrac{7}{2}\, O_2 + H_2O \leftrightarrow Fe^{2+} + 2\, SO_4^{2-} + 2\, H^+. \tag{6.1}$$

There is almost always enough moisture in mine wastes and mine workings to allow Equation 6.1 to occur releasing acidity and dissolved ferrous ion into the water. Next, dissolved ferrous ion (Fe^{2+}) is oxidized slowly by dissolved oxygen to ferric ion (Fe^{3+}), consuming some acidity:

$$Fe^{2+} + \tfrac{1}{4}\, O_2 + H^+ \leftrightarrow Fe^{3+} + \tfrac{1}{2}\, H_2O. \tag{6.2}$$

Above pH 4 and in the absence of iron-oxidizing bacteria, Equation 6.2 is the rate-limiting step in the reaction sequence. However, below pH 4 and in the presence of iron-oxidizing bacteria, the rate of Equation 6.2 is greatly accelerated by a million-fold or more. Ferric ion formed in Equation 6.2 can further oxidize pyrite, as in Equation 6.3, where ferric ion is reduced back to Fe^{2+}, releasing much more acidity.

$$FeS_2(s) + 14\, Fe^{3+} + 18\, H_2O \leftrightarrow 15\, Fe^{2+} + 2\, SO_4^{2-} + 16\, H^+. \tag{6.3}$$

By Equation 6.3, eight times more acidity is generated when ferric ion oxidizes pyrite than when dissolved oxygen serves as the oxidant (16 equivalents of H^+ compared to 2 equivalents, per mole of FeS_2). In the pH range from 2 to 7, pyrite oxidation by Fe^{3+} (Equation 6.3) is kinetically favored over abiotic oxidation by oxygen (Equation 6.2). In addition, Equation 6.3 returns soluble Fe^{2+} to the reaction cycle via Reaction 2.

Overall, 4 equivalents of acid are formed for each mole of FeS_2 oxidized in the cyclic reaction sequence of Equations 6.2 and 6.3. If bacterially mediated oxidation is occurring in Equation 6.2, the reaction cycle can be accelerated by over a millionfold.

Ferric ion also hydrolyzes (reacts with water), releasing more acid to the water and forming insoluble ferric hydroxide, which can coat streambeds with the yellow-orange deposits known as *yellow boy*:

$$Fe^{3+} + 3\, H_2O \leftrightarrow Fe(OH)_3(s) + 3\, H^+. \tag{6.4}$$

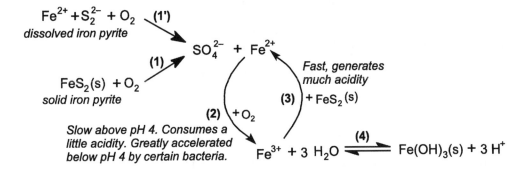

FIGURE 6.1   Reaction scheme for generation of acid mine drainage by pyrite oxidation.

$Fe(OH)_3$ precipitates serve as a reservoir for dissolved Fe^{3+}. If the generation of Fe^{3+} by Equation 6.3 is stopped because of lack of oxygen, then Fe^{3+} is supplied by dissolution of solid $Fe(OH)_3$ and is available to react via Equation 6.3.

SUMMARY OF ACID FORMATION IN MINE DRAINAGE

The steps of the reaction are summarized below and illustrated in Figure 6.1.

- *Step 1:* Iron pyrite, dissolved or solid, is oxidized by dissolved oxygen (Equation 6.1), producing Fe^{3+}, SO_4^{2-}, and lowering the pH.
- *Step 2:* Fe^{2+} formed in Step 1 is oxidized slowly by dissolved oxygen to Fe^{3+} (Equation 6.2). This is the rate-limiting step in the reaction sequence in the absence of iron-oxidizing bacteria. The abiotic rate decreases with lower pH. However, iron-oxidizing bacteria can greatly accelerate this step when the pH falls below 4.
- *Step 3:* Fe^{3+} from Step 2 is reduced rapidly back to Fe^{2+} by pyrite (Equation 6.3), generating much acidity. Ferrous ion, Fe^{2+}, generated in Step 3 re-enters the reaction cycle via step 2.
- *Step 4:* A portion of ferric ion, Fe^{3+}, reacts with water to form ferric hydroxide precipitate, $Fe(OH)_3(s)$, releasing more acidity (Equation 6.4). When the Fe^{3+} concentration diminishes, $Fe(OH)_3$ precipitate can dissolve, acting as a reservoir for replenishing Fe^{3+} and maintaining the acid producing cycle.

As pH is lowered, Step 1 becomes less important and the abiotic rate of Step 2 decreases. However, Step 2 can be greatly accelerated by certain bacteria such as Metallogenium, Ferrobacillus, Thiobacillus, and Leptospirillum, which derive energy from the oxidation of Fe^{2+} to Fe^{3+}. Below pH 4, these bacteria catalyze Step 2, speeding up the overall reaction rate by a factor as large as 1 million, and can lower the pH to 2 or less. Furthermore, these bacteria can tolerate high concentrations of dissolved metals (e.g., 40,000 mg/L Zn and Fe; 15,000 mg/L Cu) before experiencing toxic effects. They thrive in mine drainage waters as long as a minimal amount of oxygen is present. Once bacterial acceleration occurs, it is hard to reverse.

Rule of Thumb

Oxidation of iron pyrite is the most acidic of all common weathering reactions. The production of acid mine drainage can be a rapid, self-propagating, cyclic process that is accelerated by low pH and the presence of iron-oxidizing bacteria. The process will continue as long as oxygen, pyrite, and water are present.

NONIRON METAL SULFIDES DO NOT GENERATE ACIDITY

The oxidation by dissolved O_2 of noniron metal sulfides *does not* generate significant amounts of acidity. The metals are released as dissolved cations but acidity is not produced. For example

$$CuS(s) + 2\ O_2 \leftrightarrow Cu^{2+}(aq) + SO_4^{2-}(aq).$$

$$ZnS(s) + 2\ O_2 \leftrightarrow Zn^{2+}(aq) + SO_4^{2-}(aq).$$

$$PbS(s) + 2\ O_2 \leftrightarrow Pb^{2+}(aq) + SO_4^{2-}(aq).$$

Two possible reasons for the lack of acid formation when noniron sulfides are oxidized are

1. The oxidation state of sulfur is different in iron pyrite than in other sulfides, occurring as S_2^{2-} in iron pyrite and as S^{2-} in other sulfides. The respective oxidation reactions of these two sulfur forms indicate that acid is produced only with S_2^{2-}:

   (iron pyrite) $S_2^{2-} + \frac{7}{2}\ O_2 + H_2O \leftrightarrow 2\ SO_4^{2-} + 2\ H^+.$

   (other sulfides) $S^{2-} + 2\ O_2 \leftrightarrow SO_4^{2-}.$

2. Cu^{2+}, Zn^{2+}, Pb^{2+}, etc., do not hydrolyze as extensively as does Fe^{3+}, so noniron sulfides do not react significantly by reactions equivalent to Equation 6.4:

   $$Fe^{3+} + 3\ H_2O \leftrightarrow Fe(OH)_3(s) + 3\ H^+.$$

Ferric ion, Fe^{3+}, can oxidize other metal sulfides, such as ZnS (sphalerite), CuS (covellite), PbS (galena), and $CuFeS_2$ (chalcopyrite), in a similar fashion to its oxidation of FeS_2, releasing metal cations into the water without generating acidity.

ACID-BASE POTENTIAL OF SOIL

The acid-base potential (ABP) is a measure of how effectively the alkalinity (neutralization potential) in a solid sample can neutralize the acid-producing potential resulting from the presence of pyrite of the sample. The acid-base potential is equal to the equivalents of calcium carbonate ($CaCO_3$) in excess of the amount needed to neutralize the acid that could potentially be produced from oxidation of pyritic sulfur.

Example 6.1: Determining the Acid-Base Potential (ABP)

The ABP is calculated by

$$ABP = (alkalinity) - (31.25)(wt\ \%\ pyritic\ sulfur) \qquad (6.5)$$

where ABP is given in tons of $CaCO_3$-equivalents per 1000 tons of solid material.

Any rock or earth material with an ABP of −5.0 represents a soil with a net potential deficiency of 5.0 tons $CaCO_3$/1000 tons material, and is defined as a potentially toxic material.[27]

Rules of Thumb

1. If the ABP is positive, leachate from the sample is likely to be basic.
2. If the ABP is negative, leachate is likely to be acidic.
3. If the ABP is −5, or more negative, the earth material may be defined as a potentially toxic material.

6.2 AGRICULTURAL WATER QUALITY

Most water-quality related problems in irrigated agriculture fall into four general types:

1. *High salinity*: Dissolved salts (TDS) in the water may reduce water availability to the plants affecting the crop yield. The effect is caused by a lowering of the osmotic pressure that the plants can exert for absorbing water across their root membranes. Salinity problems can often be mitigated by proper irrigation practices.
2. *Low water infiltration rate*: Relatively high sodium or low calcium and magnesium water content in irrigation water may reduce the water permeability of the soil to the extent that sufficient water cannot flow through the root zone at an adequate rate for optimal plant growth. The effect takes place when an excess of sodium ions adsorbed on clay particles causes the soils to swell, thereby reducing pore size and water permeability. The sodium absorption ratio (SAR) measures the excess of sodium over calcium and magnesium, and provides a guide to potential soil permeability problems (see the discussion of sodium absorption ratio later in this chapter).
3. *Specific ion toxicity*: Certain ions can accumulate in the leaves of sensitive crops in concentrations high enough to cause crop damage and reduce yields. Ion toxicity arises mainly from sodium, chloride, and boron. Many other trace elements are also toxic to plants in low concentrations; however they usually are present in groundwater in such low concentrations that they seldom are a problem. Concentrations of concern for specific ion toxicity are lower for sprinkler irrigation than for surface irrigation because toxic ions can be absorbed directly into the plant through leaves wetted by the sprinkler water. Direct leaf absorption speeds the rate of accumulation of toxic ions.
4. *Excessive nutrients*: Nitrogen ion concentrations can be too high resulting in excessive vegetative growth, weak supporting stalks, delayed plant maturity, and poor crop quality.

Measuring the following set of parameters will allow an adequate evaluation of agricultural water quality:

The importance of these parameters is indicated in Tables 6.1 and 6.2.

TDS	calcium	chloride	bicarbonate
SAR	magnesium	selenium	nitrate + nitrite
sodium	boron	copper	pH

Tables 6.1 and 6.2 list quality parameters of potential concern in water that will be used for agricultural irrigation purposes. Most of the parameters listed as trace elements need to be monitored only for certain sensitive crops.

Table 6.1 lists parameters and maximum levels that will cause no crop growing restrictions for sensitive plants. Table 6.2 gives additional information concerning degrees of restriction for different parameter levels and the influence of the form of irrigation (sprinkler or surface watering).

6.3 BREAKPOINT CHLORINATION FOR REMOVING AMMONIA

Chlorination can be used to remove dissolved ammonia and ammonium ion from wastewater by the chemical reactions

$$NH_3 + Cl_2 \rightarrow NH_2Cl + Cl^- + H^+. \tag{6.6a}$$

$$NH_4^+ + Cl_2 \rightarrow NH_2Cl + Cl^- + 2\ H^+. \tag{6.6b}$$

Ammonia is converted stoichiometrically to monochloramine (NH_2Cl) at a 1 to 1 molar ratio or a 5 to 1 ratio by weight of Cl_2 to NH_3-N. $NHCl_2$ (dichloramine), and NCl_3 (nitrogen trichloride or

TABLE 6.1
Suggested Maximum Parameter Levels in Water Used for Crop Irrigation[a]

General Problem	Parameter	Units	Suggested Maximum Value
Salinity	total dissolved solids (TDS)	mg/L	450
	specific conductivity	mS/cm	700
Water Infiltration	SAR		3–9[b]
Specific Ion Toxicity	sodium (Na)	mg/L	70
	chloride (Cl)	mg/L	100
	boron (B)	mg/L	1–3[c]
	Trace Elements[d]		
	aluminum (Al)	mg/L	5.0
	arsenic (As)	mg/L	0.1
	beryllium (Be)	mg/L	0.1
	cadmium (Cd)	mg/L	0.01
	cobalt (Co)	mg/L	0.05
	chromium (Cr)	mg/L	0.10
	copper (Cu)	mg/L	0.20
	fluoride (F)	mg/L	1.0
	iron (Fe)	mg/L	5.0
	lithium (Li)	mg/L	2.5
	manganese (Mn)	mg/L	0.20
	molybdenum (Mo)	mg/L	0.01
	nickel (Ni)	mg/L	0.20
	lead (Pb)	mg/L	5.0
	selenium (Se)	mg/L	0.02
	vanadium (V)	mg/L	0.10
	zinc (Zn)	mg/L	2.0

[a] Based on data from "Water Quality for Agriculture," FAO Irrigation and Drainage Paper No. 29, Rev. 1, Food and Agriculture Organization of the United Nations, 1986, and Colorado water quality standards for agricultural uses.
[b] Depends on salinity. At given SAR, infiltration rate increases as water salinity increases.
[c] Depends on sensitivity of crop.
[d] Suggested maximum value is for a water application rate consistent with good agricultural practice (about 10,000 m³/year). Toxicity and suggested maximum value depend strongly on the crop. Trace elements normally are not monitored unless a problem is expected. Several trace elements are essential nutrients in low concentrations.

trichloramine) may also be formed, depending on small excesses of chlorine and pH. Further addition of chlorine leads to conversion of chloramines to nitrogen gas. The reaction for conversion of monochloramine is

$$2\ NH_2Cl + Cl_2 \rightarrow N_2(g) + 4\ H^+ + 4\ Cl^-. \tag{6.7}$$

The overall reaction for complete nitrification of ammonia by chlorine oxidation is

$$2\ NH_3 + 3\ Cl_2 \rightarrow N_2(g) + 6\ H^+ + 6\ Cl^-. \tag{6.8}$$

Equation 6.8 is theoretically complete at a molar ratio of 3 to 2 and a weight ratio of 7.6 to 1 of Cl_2 to NH_3-N. This process is called *breakpoint chlorination*. The reaction is very fast and both ionized (NH_4^+) and unionized (NH_3) forms of ammonia are removed.

TABLE 6.2
Water Parameter Levels of Potential Concern for Crop Irrigation[a]

Crop Growing Restrictions

Restriction Cause	Parameter Value	Degree of Restriction
Chloride toxicity (surface irrigation)[a]	less than 142 mg/L	none
	between 142 and 355 mg/L	moderate
	greater than 355 mg/L	severe
Chloride toxicity (sprinkler irrigation)[c]	less than 107 mg/L	none
	greater than 107 mg/L	moderate
Sodium toxicity (surface irrigation)[b]	less than 69 mg/L	none
	between 69 and 207 mg/L	moderate
	greater than 207 mg/L	severe
Sodium toxicity (sprinkler irrigation)[c]	less than 69 mg/L	none
	greater than 69 mg/L	moderate
Sodium absorption ratio[d]	SAR less than 3	none
	SAR between 3 and 9	moderate
	SAR greater than 9	severe
Nitrate[e]	less than 5 mg/L	none
	between 5 and 12 mg/L	slight
	between 12 and 30 mg/L	moderate
	greater than 30 mg/L	severe

[a] Based on data from "Water Quality for Agriculture," FAO Irrigation and Drainage Paper No. 29, Rev. 1, Food and Agriculture Organization of the United Nations, 1986, and Colorado water quality standards for agricultural uses.
[b] With surface irrigation, sodium and chloride ions are absorbed with water through plant roots. They move with the transpiration stream and accumulate in the leaves where leaf burn and drying may result. Most tree crops and woody plants are sensitive to sodium and chloride toxicity. Most annual plants are not sensitive.
[c] With sprinkler irrigation, toxic sodium and chloride ions can be absorbed directly into the plant through leaves wetted by the sprinkler water. Direct leaf absorption speeds the rate of accumulation of toxic ions.
[d] SAR values greater than 3.0 may reduce soil permeability and restrict the availability of water to plant roots.
[e] NO_3 levels greater than 5 mg/L may cause excessive growth, weakening grain stalks and affecting production of sensitive crops (e.g., sugar beets, grapes, apricots, citrus, avocados, etc.). Grazing animals may be harmed by pasturing where NO_3 levels are high.

Rules of Thumb

1. The rate of ammonia removal is most rapid at pH = 8.3.
2. The rate decreases at higher and lower pH. Since the reactions lower the pH, additional alkalinity as lime might be needed if $[NH_3] > 15$ mg/L. Add alkalinity as $CaCO_3$ in a weight ratio of about 11 to 1 of $CaCO_3$ to NH_3-N.
3. Rate also decreases at temperatures below 30°C.
4. The chlorine "breakpoint," (see Figure 6.2) occurs theoretically at a Cl_2:NH_3-N weight ratio of 7.6.
5. In actual practice, ratios of 10:1 to 15:1 may be needed if oxidizable substances other than NH_3 are present (such as Fe^{2+}, Mn^{2+}, S^{2-}, and organics).

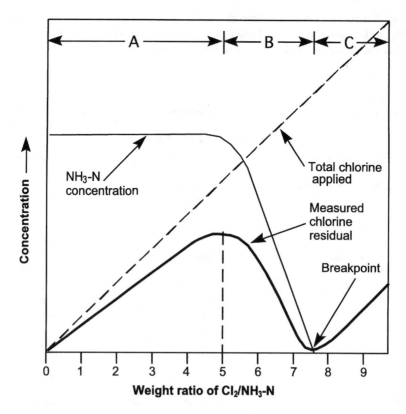

FIGURE 6.2 Breakpoint chlorination curves showing removal of ammonia from wastewater. *Region A*: Easily oxidizable substances such as Fe^{2+}, H_2S, and organic matter react. Ammonia reacts to form chloramines. Organics react to form chloro-organic compounds. *Region B*: Adding more chlorine oxidizes chloramines to N_2O and N_2. At the breakpoint, virtually all chloramines and a large part of chloro-organics have been oxidized. *Region C*: Further addition of chlorine results in a free residual of HOCl and OCl^-.

Example 6.2: Calculate the Chlorine Needed to Remove Ammonia

A waste treatment plant handles 1,500,000 L/day of sewage that contains an average of 50 mg/L of NH_3-N. How many grams of Cl_2(aq) must be present daily in the wastewater to remove all of the ammonia?

Answer: By equation 6.8, 3 moles of chlorine are needed for every 2 moles of ammonia nitrogen.

$$2 \ NH_3 + 3 \ Cl_2 \rightarrow N_2(g) + 6 \ H^+ + 6 \ Cl^-. \tag{6.9}$$

Molecular weights are

$$Cl_2 = 71 \text{ and } N = 14.$$

$$3 \text{ moles of } Cl_2 = 3 \times 71 = 213 \text{ g.}$$

$$2 \text{ moles of } N = 2 \times 14 = 28 \text{ g.}$$

Thus, the stoichiometric weight ratio is 213/28 = 7.6 g Cl_2 per gram of N (as ammonia).

One mole of NH_3 contains 14 g of N and 3 g of H. Thus, 50 mg/L of NH_3 contains 14/17 × 50 mg/L = 41.2 mg/L of N. In 1,500,000 L there will be

$$1{,}500{,}000 \text{ L} \times 41.2 \text{ mg/L} = 61{,}800{,}000 \text{ mg N, or } 61{,}800 \text{ g N/day.}$$

The theoretical amount of chlorine required is

$$\frac{7.6 \text{ g Cl}_2}{1 \text{ g N}} \times 61{,}800 \text{ g N} = 470 \text{ kg Cl}_2\text{/day, or about } 1036 \text{ lb/day.}$$

Depending on the quantity of other oxidizable substances in the wastewater, the plant operator should be prepared to use up to twice this amount of chlorine.

6.4 DE-ICING AND SANDING OF ROADS: CONTROLLING ENVIRONMENTAL EFFECTS

Road sanding and de-icing to enhance winter highway safety have the potential of contributing significant amounts of sediment and chemicals to the receiving waters of surface runoff. To minimize the impact on surface waters, it is often necessary to incorporate physical and operational controls that are designed to reduce the application of sand and de-icing chemicals and to manage surface flow from treated roads and stockpiled materials in a manner that retains sediment and infiltrates dissolved chemicals.

METHODS FOR MAINTAINING WINTER HIGHWAY SAFETY

Snow and ice on the roads reduce wheel traction and cause drivers to have less control of their vehicles. Highway departments currently use a site- and event-specific combination of three approaches for mitigating the effects of highway snow and ice:

1. Apply antiskid materials, such as sand or other gritty solids, to road surfaces to improve traction.
2. Apply de-icing chemicals that melt snow and ice by lowering the freezing point of water.
3. Plow roads to remove the snow and ice.

Although highway safety is the first concern in the use of snow control measures, environmental impact is also important. Many highway departments are evaluating the effectiveness of alternative chemicals and operating procedures for minimizing the environmental impact of sanding, de-icing and snow removal without compromising road safety.

ANTISKID MATERIALS

The most commonly used antiskid material is sand, usually derived either from rivers or crushed aggregate. Other abrasives such as volcanic cinders, coal ash, and mine tailings are sometimes used based on their local availability and cost. River sand is round and smooth and is somewhat less effective than crushed aggregate, which is rough and angular. However, river sand is cleaner and less contaminated than crushed aggregate. Between 3 and 30% by volume of de-icing chemicals are often mixed with sand for increased effectiveness. The amount of sand required is very site- and event-specific. For example, in the Denver, Colorado metro area, the average amount of sand applied per snow event is 800–1200 lb per lane mile of treated road — more sand generally is required in the western part than in the eastern part of the city.[20] In Glenwood Canyon, Colorado, where post-event sand removal is especially difficult, highway maintenance personnel have reduced the use of sand in recent years from 280 to 60 lb per lane mile by increasing the use of chemical de-icers.[23]

Environmental Concerns of Antiskid Materials

Air and water contamination are potential concerns with the use of sand and other antiskid grits. In Denver, fine particulates generated by traffic abrasion of road sand have been found to contribute around 45% of the atmospheric PM_{10} load (airborne particulate matter less than 10 μm in diameter) during winter. In 1997, EPA standards for PM_{10} were 50 mg/m^3, annual arithmetic mean, and 150 mg/m^3, 24-hour arithmetic mean. Efforts to attain compliance with these standards have compelled communities to increasingly use chemical de-icers in place of antiskid grits.[20]

Although airborne particulates from road sand are significant atmospheric polluters, they represent an insignificant fraction of the total mass of sand applied to the roads. Essentially all the sand applied for traction control becomes a potential washload that is eventually either flushed to receiving waters (including sewers, streams, and lakes), trapped in sediment control structures, or swept up and deposited in landfills.[9]

CHEMICAL DE-ICERS

A variety of water-soluble inorganic salts and organic compounds are used to melt snow and ice from the roads. The most commonly used road de-icer is sodium chloride because of its relatively low cost and high effectiveness. Other acceptable road de-icing agents are potassium chloride, calcium chloride, magnesium chloride, calcium magnesium acetate (CMA), potassium acetate, and sodium acetate.* These chemicals may be used in solid or liquid forms and are frequently combined with one another in various ratios. Different de-icer formulations have been rated for overall value based on their performance in melting, penetrating, and disbonding snow from the road surface, and based on their corrosivity, spalling of road surface, environmental impact, and cost.[20] Commercial formulations that use chloride salts usually include corrosion inhibitors which are generally regarded to be effective and worth the additional cost.

Chemical Principles of De-icing

Water containing dissolved substances always has a lower freezing point than pure water. Any soluble substance will have some de-icing properties. How far the freezing point of water is lowered by a solute depends only on the concentration, not the nature, of the dissolved particles. Given the same concentration of dissolved particles, the freezing point of water will be lowered the same amount by sodium chloride, calcium chloride, ethylene glycol, or any other solute. This behavior is called a *colligative* property. The solubility of each de-icing substance at the final solution temperature determines how many particles can go into solution. This is the ultimate limit on the lowest freezing point attainable: ice will melt as long as the outdoor temperature is above the lowest freezing point of the solute-water mixture. Pure sodium chloride theoretically can melt ice at temperatures as low as –6°F, but no lower. Calcium chloride is effective down to –67°F.

When a salt dissolves to form positive and negative ions, each ion counts as a dissolved particle. Ionic compounds such as sodium chloride (NaCl) and calcium chloride (CaCl$_2$) are efficient de-icers because they always dissociate into positive and negative ions upon dissolving forming more dissolved particles per mole than nonionizing solutes. One NaCl molecule dissolves to form two particles, Na$^+$ and Cl$^-$; one CaCl$_2$ molecule forms three particles, one Ca^{2+} and two Cl$^-$, whereas the organic molecule ethylene glycol (C$_2$H$_6$O$_2$) does not dissociate and dissolves as one particle. Three molecules of dissolved ethylene glycol are needed to lower the freezing point by the same amount as one molecule of calcium chloride. Another advantage of calcium and magnesium chlorides is that they dissolve exothermically, releasing a significant amount of heat that further

---

* Several de-icers, such as ethylene glycol, methanol, and urea, are used mainly for special purposes, such as airplane and runway de-icing, but are seldom used on the highways because of poor performance, high costs, toxicity, and/or difficulty of application.

helps to melt snow and ice. Conversely, sodium chloride does not release heat upon dissolving. The dissolution of sodium chloride is slightly endothermic and has a small cooling effect.

The difference in effectiveness for different de-icing chemicals is related primarily to their different solubilities at environmental temperatures, number of dissolved particles formed per pound of material, and exothermicity of dissolution. Organic de-icers, such as calcium magnesium acetate (CMA) and ethylene glycol, are said to be more effective than salts at breaking the bond between pavement and snow, allowing for easier plowing and snow removal. Organic de-icers also are believed to be stored in surface pores of pavement, helping in disbonding the snow and possibly prolonging the period of effective de-icing.

Corrosivity

The main advantage of organic de-icers, such as CMA, over inorganic chloride salts, such as sodium chloride, is their lower corrosivity. Corrosivity results from chemical and electrolytic reactions with solid materials. The chemical corrosivity of chloride salts arises mostly from the chemical reactivity of chloride ions and does not depend strongly on which salt is the source of the chloride ions. Electrolytic corrosivity affects metals, mainly iron alloys, and occurs when dissolved salt ions transfer electrons between zones of the metal surface with slightly different composition. The transfer of electrons allows atmospheric or dissolved oxygen to chemically react with the metal. Electrolytic corrosivity depends, in a complicated fashion, on the nature of the metal surface and the nature of the dissolved ions. However, electrolytic corrosivity for any surface will always increase as the total ion concentration (often measured as TDS or specific conductivity) increases.

When chloride is present, chemical and electrolytic corrosivity act synergistically to accelerate the overall corrosion rate. The addition of corrosion inhibitors to commercially formulated salt de-icers is reported to reduce salt corrosivity. The main reason for using chloride salts rather than nitrates, fluorides, or bromides, is the relatively low toxicity of chlorides to plants and aquatic life.

Environmental Concerns of Chemical De-icers

Corrosivity, not adverse environmental impact, has been the main problem associated with the use of chemical de-icers. While each of the common de-icers has potential environmental effects, studies show that none pose strong adverse threats.[19,41] Most de-icing residues are highly soluble and have low toxicity. They flush quickly through soils and waterways and rapidly become diluted to levels that cause no environmental problems. Under a worst-case scenario, undesirable effects are likely to be observed only near the points of application, where concentrations are the highest. Studies by the Michigan Department of Transportation show that the greatest impact has been to sensitive vegetation adjoining treated roadways. Stream and lake concentrations of chloride and other de-icing chemicals seldom reach levels that are detrimental to aquatic life. The state of Michigan has found little surface water and groundwater contamination directly attributable to de-icing practices. Much of the contamination that has been found is the result of spillage and poor storage practices.[19]

DE-ICER COMPONENTS AND THEIR POTENTIAL ENVIRONMENTAL EFFECTS

Chloride Ion

There usually are no stream standards for chloride ion which is generally regarded as a nondetrimental chemical component of state waters. Tests on fish showed no effect for concentrations of sodium chloride between 5000 to 30,000 mg/L, depending on species, exposure time, and water quality. Concentrations required to immobilize Daphnia in natural waters ranged between 2100 to 6143 mg/L.[18] It has been recommended that concentrations above 3000 mg/L be considered deleterious to both fish-food organisms and fish fry, and that a permissible limit of 2000 mg/L be established for fresh waters. However, these recommendations have not been acted upon at either the federal or state level. The U.S. EPA's secondary drinking water standard of 250 mg/L, based on average taste thresholds, was seldom found to be exceeded in a Michigan study of road de-icing impacts.[19]

Chloride in road splash can "burn" sensitive vegetation adjacent to treated roads by causing osmotic stress to the vegetation. Theoretically, chloride can form complexes that increase mobility of metals in soils, but very few such cases have been confirmed.[19] Spring thaw surges may temporarily create surface water chloride levels that are detrimental to aquatic biota. However, dilution quickly occurs and flowing streams have not been significantly impacted.[19] Spring thaw surges may temporarily cause high chloride levels in groundwater, although reports of excessive levels (>250 mg/L) are rare.[19]

Sodium Ion

Sodium is even less toxic to aquatic biota than chloride. There are no water quality standards for sodium ion. The main problems associated with sodium ion are its effects on agricultural soil permeability (see Sodium Absorption Ratio) and the necessity for restricted sodium intake by hypertensive people. Na^+ levels in groundwater can increase temporarily during spring thaws which may pose a health threat to people requiring low sodium intake.

Calcium, Magnesium, and Potassium Ions

Calcium, magnesium, and potassium ions are all plant, animal, and human nutrients, and there are no stream or drinking water standards for them. Calcium and magnesium improve soil aeration and permeability by decreasing the sodium absorption ratio. Calcium and magnesium also increase water hardness beneficially, reducing the toxic effects of dissolved heavy metals on aquatic life. Theoretically, these cations could increase heavy metal mobility in soils by exchange processes, but there is little documentation of such behavior.

Acetate

Acetate has no drinking water standards and has lower toxicity than sodium chloride. It biodegrades rapidly and does not accumulate in the environment. The only reported potential environmental problem with acetate is that, in large concentrations, it can deplete oxygen levels in surface waters by increasing BOD during biodegradation.

Impurities Present in De-icing Materials

De-icers contain trace amounts of heavy metals and sometimes phosphorus and nitrogen. These can be released with snow melt, especially during spring thaw. Because the heavy metal impurities become mostly associated with solids, they are best controlled by sediment containment. Phosphorus and nitrogen will be controlled by infiltration of snow melt into pervious areas, where they encourage vegetative growth.

6.5   DRINKING WATER TREATMENT

Clean drinking water is the most important public health factor. But well over 2 billion people worldwide do not have adequate supplies of safe drinking water. Worldwide, between 15 to 20 million babies die every year from water-borne diarrheal diseases such as typhoid fever, dysentery, and cholera. Contaminated water supplies and poor sanitation cause 80% of the diseases that afflict people in the poorest countries. The development of municipal water purification in the last century has allowed cities in the developed countries to be essentially free of water-carried diseases. Since the introduction of filtration and disinfection of drinking water in the U.S., water-borne diseases, such as cholera and typhoid, have been virtually eliminated.

However, in 1974, it was discovered that water disinfectants react with organic compounds that are naturally occurring in water and form unintended disinfection byproducts (DBPs) that may cause health risks.[3,7,24] Trihalomethane DBPs were regulated by the EPA in 1979.[10]

Since then, several DBPs (bromodichloromethane, bromoform, chloroform, dichloroacetic acid, and bromate) have been shown to be carcinogenic in laboratory animals at high doses. Some DBPs (bromodichloromethane, chlorite, and certain haloacetic acids) also can cause adverse reproductive or developmental effects in laboratory animals. In the belief that DBPs present a potential public health risk, the EPA published guidelines for minimizing their formation[38,39] and established standards in 1998 for drinking water concentrations of DBPs and disinfectant residuals (see Appendix A). The goal of EPA disinfectant and disinfection byproduct regulations is to balance the health risks of pathogen contamination, normally controlled by water disinfection, against DPB formation.

WATER SOURCES

Drinking water supplies come either from surface waters or groundwaters. In the U.S., groundwater sources or wells supply about 53% of all drinking water and surfacewater sources, such as reservoirs, rivers, and lakes, supply the remaining 47%. Groundwater comes from underground aquifers into which wells are drilled to recover the water. Wells range from tens to hundreds of meters deep. Generally, water in deep aquifers is replaced by percolation from the surface very slowly over hundreds to thousands of years. Water in the deep Ogallala aquifer in the Great Plains region of the U.S. is estimated to be thousands of years old and is called "fossil water." Replenishment of such aquifers occurs over thousands of years, and it is easy to withdraw water from them at a rate that greatly exceeds replacement. Such aquifers are essentially nonrenewable resources in our lifetime. The Ogallala aquifer has been depleted significantly over the past several decades, principally by agricultural irrigation.

Groundwater tends to be less contaminated than surface water. It is normally more protected from surface contamination and, because it moves more slowly, organic matter has time to be decomposed by soil bacteria. The soil itself acts as a filter so that less suspended matter is present.

Surface waters come from lakes, rivers, and reservoirs. It usually has more suspended materials than groundwater and requires more processing to make it safe to drink. Surface waters are used for purposes other than drinking and often become polluted by sewage, industrial, and recreational activities. On most rivers, the fraction of "new" water diminishes with distance from the head waters, as the water becomes more and more used. On the Rhine river in Europe, for example, communities near the mouth of the river receive as little as 40% "new" water in the river. All the other water has been previously discharged by an upstream city or originates as nonpoint source return flow from agricultural activities. Water treatment must make this quality of river water fit to drink. Filtration through sand was the first successful method of municipal water treatment, used in London in the middle 1800s. It led to an immediate decline in the amount of water-borne diseases.

WATER TREATMENT

Major changes are occurring in the water treatment field driven by increasingly tighter water quality standards, a steady increase in the number of regulated drinking water contaminants (from about 5 in 1940 to around 100 in 1999), and new regulations affecting disinfection and disinfection byproducts. Municipalities are constantly seeking to refine their water treatment and provide higher quality water by more economical means. A recent development in water treatment is the application of membrane filtration to drinking water treatment. Membrane filters have been refined to the point where, in certain cases, they are suitable as stand-alone treatment for small systems. More often, they are used in conjunction with other treatment methods to economically improve the overall quality of finished drinking water.

Basic Drinking Water Treatment

The purpose of water treatment is (1) to make water safe to drink by ensuring that it is free of pathogens and toxic substances, and (2) to make it a desirable drink by removing offensive turbidity, tastes, colors, and odors.

Conventional drinking water treatment addresses both of these goals. It consists of four steps:

1) Primary settling
2) Aeration
3) Coagulation and filtration
4) Disinfection

Not all four of the basic steps are needed in every treatment plant. Groundwaters, in particular, usually need much less treatment than surface waters. Groundwaters may need no settling, aeration, or coagulation. For clean groundwaters, only a little chlorine ($\approx$0.16 ppm) is added to protect the water while in the distribution system. The relatively new treatment technology of membrane filtration is increasingly being used in conjunction with the more traditional treatments and as a stand-alone treatment.

Primary Settling

Water, which has been coarsely screened to remove large particulate matter, is brought into a large holding basin to allow finer particulates to settle. Chemical coagulants may be added to form floc. Lime may be added at this point to help clarification if pH < 6.5. The floc settles by gravity, removing solids larger than about 25 microns.

Aeration

The clarified water is agitated with air. This promotes oxidation of any easily oxidizable substances — for example those which are strong reducing agents. Chlorine will be added later. If chlorine were added at this point and reducing agents were still in the water, they would reduce the chlorine and make it ineffective as a disinfectant.

Ferrous iron, Fe^{2+}, is a particularly troublesome reducing agent. It may arise from the water passing through iron pyrite (FeS_2) or iron carbonate ($FeCO_3$) minerals. With dissolved oxygen present, Fe^{2+} is oxidized to Fe^{3+}, which precipitates as ferric hydroxide, $Fe(OH)_3$, at any pH greater than 3.5. $Fe(OH)_3$ gives a metallic taste to the water and causes the ugly red-brown stain commonly found in sinks and toilets in iron-rich regions. The stain is easily removed with weak acid solutions, such as vinegar.

Coagulation and Filtration

The finest sediments, such as pollen, spores, bacteria, and colloidal minerals, do not settle out in the primary settling step. For the finished water to look clear and sparkling, these fine sediments must be removed. Hydrated aluminum sulfate, $Al_2(SO_4)_3 \cdot 18\ H_2O$, sometimes called alum or filter alum, applied with lime, $Ca(OH)_2$, is the most common filtering agent used for secondary settling.

$$Al_2(SO_4)_3 + Ca(OH)_2 \rightarrow Al(OH)_3(s) + CaSO_4. \tag{6.10}$$

At pH = 6–8, $Al(OH)_3(s)$ is formed as a light, fluffy, gelatinous flocculant having an extremely large surface area that attracts and traps small suspended particles, carrying them to the bottom of the tank as the precipitate slowly settles. In this pH range, $Al(OH)_3$ is near its minimum solubility and very little Al^{3+} is left in solution. Additonal filtration with sand beds or membranes may be used in a final polishing step before disinfection.

Disinfection

Killing bacteria and viruses is the most important part of water treatment. Proper disinfection provides a residual disinfectant level that persists throughout the distribution system. This not only kills organisms that pass through filtration and coagulation at the treatment plant, it prevents reinfection during the time the water is in the distribution system. In a large city, water may remain in the system for 5 days or more before it is used. Five days is plenty of time for any missed microorganisms to multiply. Leaks and breaks in water mains can permit recontamination, especially at the extremities of the system where the pressure is low. High pressure causes the flow at leaks to always be from the inside to the outside. But at low pressure, bacteria can seep in.

As a result of concerns about DBPs, the EPA and the water treatment industry are placing more emphasis on the use of disinfectants other than chlorine, which at present is the most commonly used water disinfectant. Another approach to reducing the probability of DBP formation is by removing DBP precursors (naturally occurring organic matter) from water before disinfection. However, use of alternative disinfectants has also been found to produce DBPs. Current regulations try to balance the risks between microbial pathogens and DBPs. DBPs include the following, not all of which pose health risks:

- Halogenated organic compounds, such as trihalomethanes (THMs), haloacetic acids, haloketones, and other halogenated compounds that are formed primarily when chlorine or ozone (in the presence of bromide ion) are used for disinfection.
- Organic oxidation byproducts, such as aldehydes, ketones, assimilable organic carbon (AOC), and biodegradable organic carbon (BDOC). The latter two DBPs result from large organic molecules being oxidized to smaller molecules, which are more available to microbes, plant, and aquatic life as a nutrient source. Oxidized organics are formed when strong oxidizing agents (ozone, permanganate, chlorine dioxide, or hydroxyl radical) are used.
- Inorganic compounds, such as chlorate, chlorite, and bromate ions. These are formed when chlorine dioxide and ozone disinfectants are used.

Disinfection Procedures

Most disinfectants are strong oxidizing agents that react with organic and inorganic oxidizable compounds in water. In some cases, the oxidant is produced as a reaction byproduct — hydroxyl radical is formed in this way. In addition to destroying pathogens, disinfectants are also used for removing disagreeable tastes, odors, and colors. They also can assist in the oxidation of dissolved iron and manganese, prevention of algal growth, improvement of coagulation and filtration efficiency, and control of nuisance water organisms such as Asiatic clams and zebra mussels.

The most commonly used water treatment disinfectant is chlorine. It was first used on a regular basis in Belgium in the early 1900s. Other disinfectants sometimes used are ozone, chlorine dioxide, and ultraviolet radiation. Of these, only chlorine and chlorine dioxide have residual disinfectant capability. With chlorine or chlorine dioxide, adding a small excess of disinfectant maintains protection of the drinking water throughout the distribution system. Normally, a residual chlorine or chlorine dioxide concentration of about 0.2 to 0.5 mg/L is sought. Disinfectants that do not provide residual protection are normally followed by a low dose of chlorine in order to preserve a disinfection capability throughout the distribution system.

Part of the disinfection procedure involves removing DBP precursors, mainly total organic carbon (TOC), by coagulation, water softening, or filtration. A high TOC concentration (greater than 2.0 mg/L) indicates a high potential for DBP formation. Typical required reduction percentages of TOC for conventional treatment plants are given in Table 6.3.

TABLE 6.3

Required Percentage Removal of Total Organic Carbon by Enhanced Coagulation[a] for Conventional Water Treatment Systems[b]

Source Water TOC (mg/L)	Source Water Alkalinity (mg/L as CaCO$_3$)		
	0 to 60	>60 to 120	>120[2]
>2.0 to 4.0	35.0%	25.0%	15.0%
>4.0 to 8.0	45.0%	35.0%	25.0%
>8.0	50.0%	40.0%	30.0%

[a] Enhanced coagulation is defined, in part, as the coagulant dose where an incremental addition of 10 mg/L of alum (or an equivalent amount of ferric salt) results in a TOC removal to below 0.3 mg/L.

[b] Applies to utilities using surface water and groundwater impacted by surface water.

DISINFECTION BYPRODUCTS AND DISINFECTION RESIDUALS

The principal precursor of organic DBPs is naturally occurring organic matter (NOM). NOM is usually measured as total organic carbon (TOC) or dissolved organic carbon (DOC). Typically, about 90% of TOC is in the form of DOC (DOC is defined as the part of TOC that passes through a 0.45 μm filter). Halogenated organic byproducts are formed in water when NOM reacts with free chlorine (Cl_2) or free bromine (Br_2). Free chlorine may be introduced when chlorine gas, chlorine dioxide, or chloramines are added for disinfection. Free bromine is a product of the oxidation by disinfectants of bromide ion already present in the source water.

Reactions of strong oxidants with NOM also form nonhalogenated DBPs, particularly when nonchlorine oxidants such as ozone and peroxone are used. Common nonhalogenated DBPs include aldehydes, ketones, organic acids, ammonia, and hydrogen peroxide.

Bromide ion (Br^-) may be present, especially where geothermal waters impact surface and groundwaters, and in coastal areas where saltwater incursion is occurring. Ozone or free chlorine oxidizes Br^- to form brominated DBPs such as: bromate ion, bromoform, cyanogen bromide, bromopicrin, and brominated acetic acid.

STRATEGIES FOR CONTROLLING DISINFECTION BYPRODUCTS

Once formed, DBPs are difficult to remove from a water supply. Therefore, DBP control is focused on preventing their formation. Chief control measures for DBPs are

- Lowering NOM concentrations in source water by coagulation, settling, filtering, and oxidation
- Using sorption on granulated activated carbon (GAC) to remove DOC
- Moving the disinfection step later in the treatment train, so that it comes after all processes that decrease NOM
- Limiting chlorine to providing residual disinfection, following primary disinfection with ozone, chlorine dioxide, chloramines, or ultraviolet radiation
- Protection of source water from bromide ion

Table 6.4 is a list of the cancer classifications assigned by the EPA for disinfectants and DBPs as of January 1999.

TABLE 6.4
EPA Cancer Classifications for Disinfectants and DBPs[38]

Compound	Cancer Classification[a]
Chloroform	B2
Bromodichloromethane	B2
Dibromochloromethame	C
Bromoform	B2
Monochloroacetic acid	—
Dichloroacetic acid	B2
Trichloroacetic acid	C
Dichloroacetonitrile	C
Bromochloroacetonitrile	—
Dibromoacetonitrile	C
Trichloroacetonitrile	—
1,1-Dichloropropanone	—
1,1,1-Trichloropropanone	—
2-Chlorophenol	D
2,4-Dichlorophenol	D
2,4,6-Trichlorophenol	B2
Chloropicrin	—
Chloral hydrate	C
Cyanogen chloride	—
Formaldehyde	B1
Chlorate	—
Chlorite	D
Bromate	B2
Ammonia	D
Hypochlorous acid	—
Hypochlorite	—
Monochloramine	—
Chlorine dioxide	D

[a] The EPA classifications for carcinogenic potential of chemicals are[37] A: Human carcinogen; sufficient evidence in epidemiologic studies to support causal association between exposure and cancer. B: Probable human carcinogen; limited evidence in epidemiologic studies (B1) and/or sufficient evidence from animal studies (B2). C: Possible human carcinogen; limited evidence from animal studies and inadequate or no data in humans. D: Not classifiable; inadequate or no animal and human evidence of carcinogenicity. E: No evidence of carcinogenicity for humans; no evidence of carcinogenicity in at least two adequate animal tests or in adequate epidemiologic and animal studies.

Note: Not all of the EPA cancer classifications are found among the listed disinfectants and DBPs. The EPA is in the process of revising these cancer guidelines.

CHLORINE DISINFECTION TREATMENT

At room temperature, chlorine is a corrosive and toxic yellow-green gas with a strong, irritating odor. It is stored and shipped as a liquefied gas. Chlorine is the most widely used water treatment disinfectant because of its many attractive features:

- It is effective against a wide range of pathogens commonly found in water, particularly bacteria and viruses.
- It leaves a residual that stabilizes water in distribution systems against reinfection.
- It is economical and easily measured and controlled.

- It has been used for a long time and represents a well-understood treatment technology. It maintains an excellent safety record despite the hazards of handling chlorine gas.
- Chlorine disinfection is available from sodium and calcium hypochlorite salts, as well as from chlorine gas. Hypochlorite solutions may be more economical and convenient than chlorine gas for small treatment systems.

In addition to disinfection, chlorination is used for

- Taste and odor control, including destruction of hydrogen sulfide.
- Color bleaching.
- Controlling algal growth.
- Precipitation of soluble iron and manganese.
- Sterilizing and maintaining wells, water mains, distribution pipelines, and filter systems.
- Improving some coagulation processes.

Problems with chlorine usage include

- Not effective against *Cryptosporidium* and limited effectiveness against *Giardia lamblia* protozoa.
- Reactions with NOM can result in the formation of undesirable DBPs.
- The hazards of handling chlorine gas require special equipment and safety programs.
- If site conditions require high chlorine doses, taste and odor problems may arise.

Chlorine dissolves in water by the following equilibrium reactions:

$$Cl_2(g) \leftrightarrow Cl_2(aq). \tag{6.11}$$

$$Cl_2(aq) + H_2O \leftrightarrow H^+(aq) + Cl^-(aq) + HOCl(aq). \tag{6.12}$$

$$HOCl(aq) \leftrightarrow H^+(aq) + OCl^-(aq). \tag{6.13}$$

At pH values below 7.5, hypochlorous acid (HOCl) is the dominant dissolved chlorine species. Above pH 7.5, chlorite anion (OCl⁻) is dominant (see Figure 6.3). The formation of H^+ means that chlorination reduces total alkalinity.

The active disinfection species, Cl_2, HOCl, and OCl⁻, are called the *total free available chlorine*. All these species are oxidizing agents, but chloride ion (Cl⁻) is not. HOCl is about 100 times more effective as a disinfectant than OCl⁻. Thus, the amount of chlorine required for a given level of disinfection depends on the pH. Higher doses are needed at a higher pH. At pH 8.5, 7.6 times as much chlorine must be used as at pH 7.0, for the same amount of disinfection. HOCl is more effective than OCl⁻ because, as a neutral molecule, it can penetrate cell membranes of microorganisms more easily than OCl⁻ can.

When chlorine gas is added to a water system, it dissolves according to Equations 6.11–6.13. All substances present in the water that are oxidizable by chlorine constitute the *chlorine demand*. Until oxidation of these substances is complete, all the added chlorine is consumed, and the net dissolved chlorine concentration remains zero as chlorine is added. When no chlorine-oxidizable matter is left, for example when the chlorine demand has been met, the dissolved chlorine concentration (chlorine residual) increases in direct proportion to the additional dose (see Figure 6.4).

If chlorine demand is zero, residual always equals the dose, and the plot is a straight line of slope = 1, passing through the zero. Chlorine is supplied as the bulk liquid under pressure, the boiling point of chlorine gas is –35°C at 1 atmosphere pressure. The total time of water in the

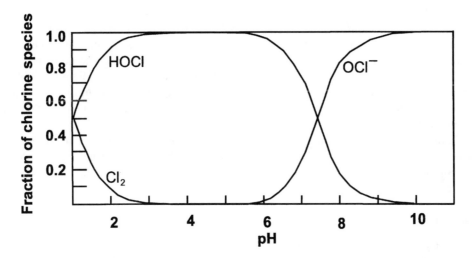

FIGURE 6.3 Distribution diagram for dissolved chlorine species. Free chlorine molecules, Cl_2, exist only below about pH = 2. At pH = 7.5, [HOCl] = [OCl⁻].

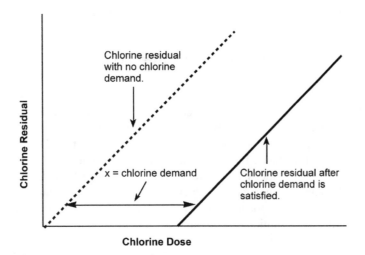

FIGURE 6.4 Relations among chlorine dose, chlorine demand, and chlorine residual.

chlorine disinfection tank is generally about 20–60 minutes. A typical concentration of residual chlorine in the finished water is 1 ppm or less.

Hypochlorite

In addition to chlorine gas, the active disinfecting species HOCl and OCl⁻ can be obtained from hypochlorite salts, chiefly sodium hypochlorite (NaOCl) and calcium hypochlorite ($Ca(OCl)_2$). The salts react in water according to Equations 6.14 and 6.15.

$$NaOCl + H_2O \rightarrow HOCl + Na^+ + OH^-. \tag{6.14}$$

$$Ca(OCl)_2 + 2\ H_2O \rightarrow 2\ HOCl + Ca^{2+} + 2\ OH^-. \tag{6.15}$$

Notice, that while adding chlorine gas to water lowers the pH, Equations 6.11–6.13, hypochlorite salts raise the pH.

Sodium hypochlorite salts are available as the dry salt or in aqueous solution. The solution is corrosive with a pH of about 12. One gallon of 12.5% sodium hypochlorite solution is the equivalent of about 1 lb of chlorine gas. Unfortunately, sodium hypochlorite presents storage problems. After one month of storage under the best of conditions (low temperature, dark, and no metal contact), a 12.5% solution will have degraded to about 10%. On-site generation of sodium hypochlorite is accomplished by passing low voltage electrical current through a sodium chloride solution. On-site generation allows smaller quantities to be stored and makes the use of more stable dilute solutions (0.8%) feasible.

Calcium hypochlorite is commonly available as the dry salt which contains about 65% available chlorine. 1.5 lbs of calcium hypochlorite are equivalent to about 1 lb of chlorine gas. Storage is less of a problem with calcium hypochlorite; normal storage conditions result in a 3–5% loss of its available chlorine per year.

Definitions

Chlorine dose: the amount of chlorine originally used.
Chlorine residual: the amount remaining at time of analysis.
Chlorine demand: the amount used up in oxidizing organic substances and pathogens in the water, for example the difference between the chlorine dose and the chlorine residual.
Free available chlorine: the total amount of HOCl and ClO$^-$ in solution. (Cl$_2$ is not present above pH = 2.)

DRAWBACKS TO USE OF CHLORINE: DISINFECTION BYPRODUCTS (DBPs)

Trihalomethanes (THMs)

The problem of greatest concern with the use of chlorine is the formation of chlorination byproducts, particularly trihalomethanes (CHCl$_3$, CHBrCl$_2$, CHBr$_2$Cl, CHBr$_3$, CHCl$_2$I, CHBrClI) and carbon tetrachloride (CCl$_4$) as possible carcinogens. It was once thought that THMs were formed by chlorination of dissolved methane. It is now known that they come from the reaction of HOCl, with acetyl groups in NOM, chiefly humic acids. Humic acids are breakdown products of plant materials like lignin. There is no evidence that chlorine itself is carcinogenic.

In addition to the general strategies for controlling DBPs listed earlier, another option is available with chlorine use. Addition of ammonia with chlorination forms chloramines (see Break-point Chlorination to Remove Ammonia). Chloramines are weaker oxidants than chlorine and are useful for providing a residual disinfectant capability with a lower potential for forming DBPs.

Chlorinated Phenols

If phenol or its derivatives from industrial activities are in the water, taste and color can be a problem. Phenols are easily chlorinated, forming compounds with very penetrating antiseptic odors. The most common chlorinated phenols arising from chlorine disinfection are shown in Table 6.5, with their odor thresholds, several of which are in the ppb (μg/L) range. At the ppm level, chlorinated phenols can make water completely unfit for drinking or cooking. If phenol is present in the intake water, treatment choices are to employ additional nonchlorine oxidation for removing phenol, to remove phenol with activated charcoal, or to use a different disinfectant. The activated charcoal treatment is expensive and few communities use it.

Example 6.3

Water has begun to seep into the basement of a home. The home's foundation is well above the water table and this problem had not been experienced before. The house is located about 50 ft

TABLE 6.5
Odor Thresholds of Phenol and Chlorinated Derivatives
from Drinking Water Disinfection With Chlorine

Phenol Compound	Chemical Structure	Odor Threshold in Water (ppb)
Phenol		>1000
2-chlorophenol		2
4-chlorophenol		250
2,4-chlorophenol		2
2,6-chlorophenol		3
2,4,6-chlorophenol		>1000

downgradient from a main water line and one possibility is that a leak has occurred in the pipeline. The water utility company tested water entering the basement for the presence of chlorine, thinking that if the water source was the pipeline, the chlorine residual should be detected. When no chlorine was found, the utility company concluded that they were not responsible for the seep. Was this conclusion justified?

Answer: No. Water would have to travel at least 50 ft through soil from the pipeline to the house. The chlorine residual should not exceed 4 mg/L (see Appendix A) and would almost certainly come in contact with enough oxidizable organic and inorganic matter in the soil to be depleted below detection. A better water source marker would be fluoride, assuming the water supply is fluoridated. Although fluoride might react with calcium and magnesium in the soil to form solid precipitates, it is more likely to be detectable at the house than is chlorine. However, neither test is conclusive. The simplest and best test would be to turn off the water in the pipeline long enough to observe any change in water flow into the house. This, however, might not be possible. Another approach would be to examine the water line for leaks, using a video camera probe or soil conductivity measuring equipment.

CHLORAMINES

Many utilities use chlorine for disinfection and chloramines for residual maintenance. Chloramines are formed in the reaction of ammonia with HOCl from chlorine — a process that is inexpensive and easy to control. The reactions are described in the section on breakpoint chlorination. Although the reaction of chlorine with ammonia can be used for the purpose of destroying ammonia, it also serves to generate chloramines, which are useful disinfectants that are more stable and longer lasting in a water distribution system than is free chlorine. Thus, chloramines are effective for controlling bacterial regrowth in water systems although they are not very effective against viruses and protozoa. The primary role of chloramines is their use as a secondary disinfectant to provide residual treatment — an application which has been practiced in the U.S. since about 1910. Being weaker oxidizers than chlorine, chloramines form far fewer disinfection byproducts. However, they are not useful for oxidizing iron and manganese. When chloramine disinfection is the goal, ammonia is added in the final chlorination step. Chloramines are always generated on site.

Optimal chloramine disinfection occurs when the chlorine:ammonia-nitrogen (Cl_2:N) ratio by weight is around 4, before the chlorination breakpoint occurs. Under these conditions, monochloramine (NH_2Cl) and dichloramine ($NHCl_2$) are the main reaction products and the effective disinfectant species. The normal dose of chloramines is between 1 and 4 mg/L. Residual concentrations are usually maintained between 0.5 and 1 mg/L. The maximum residual disinfection level (MRDL) mandated by the EPA is 4.0 mg/L.

CHLORINE DIOXIDE DISINFECTION TREATMENT

Chlorine dioxide (ClO_2) is a gas at temperatures above 12°C with high water solubility. Unlike chlorine, it reacts quite slowly with water, remaining mostly dissolved as a neutral molecule. It is a very good disinfectant, about twice as effective as HOCl from Cl_2 but also about twice as expensive. ClO_2 was first used as a municipal water disinfectant in Niagara Falls, NY in 1944. In 1977, about 100 municipalities in the U.S. and thousands in Europe were using it. The main drawback to its use is that it is unstable and cannot be stored. It must be made and used on site, whereas chlorine can be delivered in tank cars.

Much of its reactivity is due to being a free radical. ClO_2 cannot be compressed for storage because it is explosive when pressurized or when it is at concentrations above 10 percent by volume in air. It decomposes in storage and can decompose explosively in sunlight, when heated or agitated suddenly. So it is never shipped and is always prepared on site and used immediately. Typical dose rates are 0.1–1.0 ppm.

Sodium chlorite is used to make ClO_2 by one of three methods:

$$5\ NaClO_2 + 4\ HCl \leftrightarrow 4\ ClO_2(g) + 5\ NaCl + 2\ H_2O. \tag{6.16}$$

$$2\ NaClO_2 + Cl_2(g) \rightarrow 2\ ClO_2(g) + NaCl. \tag{6.17}$$

$$2 \text{ NaClO}_2 + \text{HOCl} \rightarrow 2 \text{ ClO}_2(g) + \text{NaCl} + \text{NaOH}. \tag{6.18}$$

Sodium chlorite is extremely reactive, especially in the dry form, and it must be handled with care to prevent potentially explosive conditions. If chlorine dioxide generator conditions are not carefully controlled (pH, feedstock ratios, low feedstock concentrations, etc.), the undesirable byproducts chlorite (ClO_2^-) and chlorate (ClO_3^-) may be formed.

Chlorine dioxide solutions below about 10 g/L will not have sufficiently high vapor pressures to create an explosive hazard under normal environmental conditions of temperature and pressure. For drinking water treatment, ClO_2 solutions are generally less than 4 g/L and treatment levels generally are between 0.07 to 2.0 mg/L.

Since ClO_2 is an oxidizer but not a chlorinating agent, it does not form trihalomethanes or chlorinated phenols. So it does not have taste or odor problems. Common applications for ClO_2 have been to control taste and odor problems associated with algae and decaying vegetation, to reduce the concentrations of phenolic compounds, and to oxidize iron and manganese to insoluble forms. Chlorine dioxide can maintain a residual disinfection concentration in distribution systems. The toxicity of ClO_2 restricts the maximum dose. At 50 ppm, ClO_2 can cause breakdown of red corpuscles with the release of hemoglobin. Therefore, the dose of ClO_2 is limited to 1 ppm.

OZONE DISINFECTION TREATMENT

Ozone (O_3) is a colorless, highly corrosive gas at room temperature, with a pungent odor that is easily detectable at concentrations as low as 0.02 ppmv — well below a hazardous level. It is one of the strongest chemical oxidizing agents available, second only to hydroxyl free radical (HO·), among disinfectants commonly used in water treatment. Ozone use for water disinfection started in 1893 in the Netherlands and in 1901 in Germany. Significant use in the U.S. did not occur until the 1980s. Ozone is one of the most potent disinfectants used in water treatment today. Ozone disinfection is effective against bacteria, viruses, and protozoan cysts, including *Cryptosporidium* and *Giardi lamblia*.

Ozone is made by passing a high voltage electric discharge of about 20,000 V through dry, pressurized air.

$$3 \text{ O}_2(g) + \text{energy} \rightarrow 2 \text{ O}_3(g). \tag{6.19}$$

Equation 6.19 is endothermic and requires a large input of electrical energy. Because ozone is unstable, it cannot be stored and shipped efficiently. Therefore, it must be generated at the point of application. The ozone gas is transferred to water through bubble diffusers, injectors, or turbine mixers. Once dissolved in water, ozone reacts with pathogens and oxidizable organic and inorganic compounds. Undissolved gas is released to the surroundings as off-gas and must be collected and destroyed by conversion back to oxygen before release to the atmosphere. Ozonator off-gas may contain as much as 3000 ppmv of ozone, well above a fatal level. Ozone is readily converted to oxygen by heating it to above 350°C or by passing it through a catalyst held above 100°C. OSHA currently requires released gases to contain no more than 0.1 ppmv of ozone for worker exposure. Typical dissolved ozone concentrations in water near an ozonator are around 1 mg/L.

The dissolved ozone gas decomposes spontaneously in water by a complex mechanism that includes the formation of hydroxyl free radical, which is the strongest oxidizing agent available for water treatment. Hydroxyl radical essentially reacts at every molecular collision with many organic compounds. The very high reaction rate of hydroxyl radicals limits their half life in water to the order of microseconds and their concentration to less than about 10^{-12} mol/L. Both ozone molecules and hydroxyl free radicals play prominent oxidant roles in water treatment by ozonation.

Ozone concentrations of about 4–6% are achieved in municipal and industrial ozonators. Ozone reacts quickly and completely in water, leaving no active residual concentration. Decomposition of ozone in water produces hydroxyl radical (a very reactive short-lived oxidant) and dissolved oxygen, which further aid in disinfection and diminishing BOD, COD, color, and odor problems. The air–ozone mixture is typically bubbled through water for a 10–15 minutes contact time. The main drawbacks to ozone use have been its high capital and operating costs and the fact that it leaves no residual disinfection concentration. Since it offers no residual protection, ozone can be used only as a primary disinfectant. It must be followed by a light dose of secondary disinfectant, such as chlorine, chloramine, or chlorine dioxide for a complete disinfection system.

Several ways to assist ozonation are adding hydrogen peroxide (H_2O_2), using ultraviolet radiation (UV), and/or raising the pH to around 10–11. Hydrogen peroxide decomposes to form the reactive hydroxyl radical, greatly increasing the hydroxyl radical concentration above that generated by simple ozone reaction with water. Reactions of hydroxyl radicals with organic matter cause structural changes that make organic matter still more susceptible to ozone attack. Adding hydrogen peroxide to ozonation is known as the Advanced Oxidation Process (AOP) or Peroxone process. UV radiation dissociates peroxide, forming hydroxyl radicals at a rapid rate. Raising the pH allows ozone to react with hydroxyl ions (OH^-, *not* the radical $HO\cdot$) to form additional hydrogen peroxide. In addition to increasing the effectiveness of ozone oxidation, peroxide and UV radiation are also effective as disinfectants. The use of these ozonation enhancers is known as the AOP process.

The equipment for ozonation is expensive, but the cost per gallon decreases with large scale operations. Generally, only large cities use ozone. ClO_2 is not as problem free as ozone, but it is cheaper to use for small systems.

In addition to disinfection, ozone is used for

- DBP precursor control
- Protection against *Cryptosporidium* and *Giardi*
- Taste and odor control, including destruction of hydrogen sulfide
- Color bleaching
- Precipitation of soluble iron and manganese
- Sterilizing and maintaining wells, water mains, distribution pipelines, and filter systems
- Improving some coagulation processes

Ozone DBPs

Although it does not form the chlorinated disinfection byproducts that are of concern with chlorine use, ozone can react to form its own set of oxidation byproducts. When bromide ion (Br^-) is present — where geothermal waters impact surface and groundwaters or in coastal areas where saltwater incursion is occurring — ozonation can produce bromate ion (BrO_3^-), a suspected carcinogen, as well as brominated THMs and other brominated disinfection byproducts. Controlling the formation of unwanted ozonation byproducts is accomplished by pretreatment to remove organic matter (activated carbon filters and membrane filtration) and scavenge BrO_3^- (pH lowering and hydrogen peroxide addition).

When bromide is present, the addition of ammonia with ozone forms bromamines — by reactions analogous to the formation of chloramines with ammonia and chlorine — and lessens the formation of bromate ion and organic DBPs.

POTASSIUM PERMANGANATE

Potassium permanganate salt ($KMnO_4$) dissolves to form the permanganate anion (MnO_4^-), a strong oxidant effective at oxidizing a wide variety of organic and inorganic substances. In the process, manganese is reduced to manganese dioxide (MnO_2), an insoluble solid that precipitates from solution. Permanganate imparts a pink to purple color to water and is, therefore, unsuitable as a residual disinfectant.

Although it is easy to transport, store, and apply, permanganate generally is too expensive for use as a primary or secondary disinfectant. It is used in drinking water treatment primarily as an alternative to chlorine for taste and odor control, iron and manganese oxidation, oxidation of DPB precursors, control of algae, and control of nuisance organisms, such as zebra mussels and the Asiatic clam. It contains no chlorine and does not contribute to the formation of THMs. When used to oxidize NOM early in a water treatment train that includes post-treatment chlorination, permanganate can reduce the formation of THMs.

PEROXONE (OZONE + HYDROGEN PEROXIDE)

The peroxone process is an advanced oxidation process (AOP). AOPs employ highly reactive hydroxyl radicals (OH·) as major oxidizing species. Hydroxyl radicals are produced when ozone decomposes spontaneously. Accelerating ozone decomposition by using, for example, ultraviolet radiation or adding hydrogen peroxide, elevates the hydroxyl radical concentration and increases the rate of contaminant oxidation. When hydrogen peroxide is used, the process is called peroxone.

Like ozonation, the peroxone process does not provide a lasting disinfectant residual. Oxidation is more complete and much faster with peroxone than with ozone. Peroxone is the treatment of choice for oxidizing many chlorinated hydrocarbons that are difficult to treat by any other oxidant. It is also used for inactivating pathogens and destroying pesticides, herbicides, and volatile organic compounds (VOCs). It can be more effective than ozone for removing taste- and odor-causing compounds such as geosmin and 2-methyliosborneol (MIB). However, it is less effective than ozone for oxidizing iron and manganese. Because hydroxyl radicals react readily with carbonate, it may be necessary to lower the alkalinity in water with a high carbonate level in order to maintain a useful level of radicals. Peroxone treatment produces similar DBPs as does ozonation. In general, it forms more bromate than ozone under similar water conditions and bromine concentrations.

ULTRAVIOLET (UV) DISINFECTION TREATMENT

Ultraviolet radiation at wavelengths below 300 nm is very damaging to life forms, including microorganisms. Low-pressure mercury lamps, known as germicidal lamps, have their maximum energy output at 254 nm. They are very efficient, with about 40% of their electrical input being converted to 254 nm radiation. Protein and DNA in microorganisms absorb radiation at 254 nm, leading to photochemical reactions that destroy the ability to reproduce. UV doses required to inactivate bacteria and viruses are relatively low, of the order of 20–40 mW·s/cm^2. Much higher doses, 200 mW·s/cm^2 or higher, are needed to inactivate *Cryptosporidium* and *Giardia lamblia*.

Color or high levels of suspended solids can interfere with transmission of UV through the treatment cell and UV absorption by iron species diminishes the UV energy absorbed by microorganisms. Such problems may necessitate higher UV dose rates or pretreatment filtration. To minimize these problems, UV reaction cells are designed to induce turbulent flow, have long water flow paths and short light paths (around 3 inches), and provide for cleaning of residues from the lamp housings. Wherever used, usually in small water treatment systems, UV irradiation is generally the last step in the water treatment process, just after final filtration and before entering the distribution system. UV systems are normally easy to operate and maintain although severe site conditions, such as high levels of dissolved iron or hardness, may require pretreatment.

UV does not introduce any chemicals into the water and causes little, if any, chemical change in water. Therefore, overdosing does not cause water quality problems. UV is used mostly for inactivating pathogens to regulated levels. Since it leaves no residual, it can serve only as a primary disinfectant and must by followed by some form of chemical secondary disinfection, generally chlorine or chloramine. UV water treatment is used more in Europe than in the U.S. Small-scale units are available for individuals who have wells with high microbial levels.

Characteristics of UV Treatment

- Short contact time of 1–10 seconds. Ozone and chlorine require 10–50 minutes, necessitating large reaction tanks. Ozonation can be run on a flow-through basis.
- Destroys most viruses and bacteria without chemical additives. The destruction of *Giardia lamblia*, however, requires prefiltration. Leaves no residual disinfection potential in the water so that, for water entering a distribution system, light chlorination is still needed to provide prolonged disinfection.
- Low overall installation costs. Ozone generators are expensive. Chlorine metering systems are not especially expensive, but large reaction tanks and safety systems are high cost items.
- Not influenced by pH or temperature. Chlorination and ozonation work best at lower pH (chlorine because it is in the HOCl form; ozone because it decomposes more rapidly at higher pH). Chlorination and ozonation both require longer contact time at lower temperatures.
- No toxic residues. It adds nothing to the water unless some organics are present that photoreact to form toxic compounds. The formation of THMs or other DBPs is minimal.

MEMBRANE FILTRATION WATER TREATMENT

Membrane filters are being used to treat groundwater, surface water, and reclaimed wastewater. Membrane filtration is a physical separation process that removes unwanted substances from water without utilizing chemical reactions that can lead to undesirable byproducts. The range of membrane filters available is shown in Figure 6.5, along with common substances that can be removed by filtering. Although membranes sometimes serve as a stand-alone treatment, they are more often combined with other treatment technologies. For example, currently available microfiltration (MF) and ultrafiltration (UF) membranes are not very effective in removing dissolved organic carbon, some synthetic organic compounds or THM precursors. Their performance is improved by adding powdered adsorbent material to the wastewater flow. Contaminants that might pass through the filters are adsorbed to the larger adsorbent particles and rejected by the filters.

Filter membranes are made of organic and inorganic materials. Organic membrane filters are made from several different organic polymer films, normally formed as a thin film supported on a woven or nonwoven fabric. Inorganic membrane filters are made from ceramics, glass, or carbon. They generally consist of porous supporting layer on which a thin microporous layer is chemically deposited. Inorganic membranes resist higher pressures, a wide pH range, and more extreme temperatures than do organic membranes. Their main disadvantages are greater weight and expense.

Filters can be fabricated to remove substances as small as dissolved ions. They are useful for removing total dissolved solids (TDS), nitrate sulfate, radium, iron, manganese, DBP precursors, bacteria, viruses, and other pathogens from water without adding chemicals. It must also be recognized that there will be imperfections in manufactured membrane filters through which contaminants may pass. This is of particular concern with pathogens. Therefore, filters must never be regarded as having 100% rejection for any size range. Furthermore, filters do not protect water from reinfection after it has entered a distribution system, so it is common to add chlorine or another residual disinfectant at the end of the treatment chain for this purpose. Because most organic matter has already been removed, end-of-treatment chlorination does not generate significant disinfection byproducts.

Unlike coarser filters operating in a "normal" mode, where all of the water passes through the filter surface, membrane filters operate in a *cross-flow* mode. In cross-flow filtration, the feed or influent stream is separated into two separate effluent streams. The pressurized feed water flows parallel to the membrane filter surface and some of the water diffuses through the filter. The remaining feed stream continues parallel to the membrane to exit the system without passing through the membrane surface. Filtered contaminants remain in the feed stream water, increasing in concentration until the feed stream exits the filter unit. The filtered water is called the *permeate* effluent, and the exiting feed stream water is called the *concentrate* effluent. Crossflow filtration provides a self-cleaning effect that allows continuous flushing away of contaminants, which, in "normal" filtration, would plug filters of small pore size very quickly.

Depending on the nominal size of the pores engineered into the membrane, cross-flow filters are used in filtering applications classified as reverse osmosis (RO), nanofiltration (NF), ultrafiltration (UF), and microfiltration (MF), listed in order of increasing pore size range. The pore sizes in these membranes are so small that significant pressure is required to force water through them; the smaller the pore size, the higher the required pressure. Figure 6.5 illustrates some of the uses for different membrane filter types.

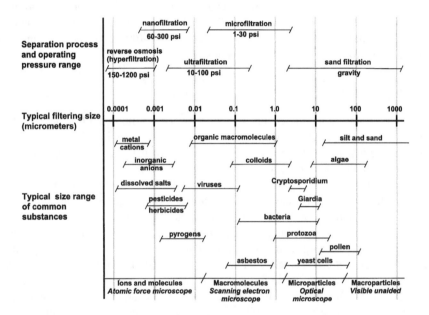

FIGURE 6.5   Comparison of filter processes and size ranges.

Reverse Osmosis (RO)

RO, sometimes called hyperfiltration, was the first cross-flow membrane separation process to be widely used for water treatment. It requires operating pressures of 150 to 1200 psi. It removes up to 99% of ions and most dissolved organic compounds. RO can meet most drinking water standards with a single-pass system. Although it might not be the most economical approach, using RO in multiple-pass systems allows the most stringent drinking water standards to be met. For example, rejection of 99.9% of viruses, bacteria, and pyrogens is achievable with a double-pass system.

Nanofiltration (NF)

NF membranes can separate organic compounds with molecular weights as small as 250 Da.* It will also separate most divalent ions and is effective for softening water (removing Ca^{2+} and Mg^{2+}). It allows greater water flow-through at lower operating pressure (60 to 300 psi) than RO.

---

* Da stands for dalton. 1 Da = 1 molecular weight unit. For example, the molecular weight of chloroform ($CHCl_3$) is 119. The mass of one chloroform molecule is 119 Da. Chloroform will not be separated by nanofiltration, which does not reject molecules smaller than 250 Da.

Ultrafiltration (UF)

UF membranes do not remove ions. They reject compounds greater than about 700 Da. The larger pore size permits lower operating pressures, in the range of 10 to 100 psi. UF is useful for separating larger organic compounds, such as colloids, bacteria, pyrogens, *Giardia lamblia*, and *Cryptosporidium*. Most ions and smaller molecules such as chloroform and sucrose will pass through the membranes.

Microfiltration (MF)

MF membranes reject contaminants in the 0.05 to 3.0 μm range and operate at pressures of 1 to 30 psi. They are available in polymer, metal, and ceramic materials. MF is sometimes used as a pretreatment for RO, increasing the efficiency and duty cycle of RO membranes significantly. MF will remove *Giardia lamblia* and *Cryptosporidium*, whose spores range between 3 and 18 μm in diameter, but will not remove viruses and most bacteria. MF-RO treatment trains are reported to be economical, easy to operate, and very reliable.

6.6   ION EXCHANGE

Ion exchange is the reversible interchange of ions between a solid and a liquid. Hydrated ions on a solid are exchanged, equivalent for equivalent, for hydrated ions in solution. *Cation exchange* involves the interchange of positive ions. *Anion exchange* involves the interchange of negative ions.

In the natural environment, most solid particles carry a surface charge that is either positive or negative. This is true for both organic and inorganic solids. As water passes through soil, dissolved ions can leave the water to become attached to oppositely charged sites on soil surfaces. This displaces ions of the same charge sign previously attached to the surface, so that they become dissolved and mobile in the water. In general, ions of higher charge and smaller hydrated diameter will displace ions of lower charge and larger hydrated diameter. Larger hydrated diameter correlates with smaller ionic diameter and smaller ionic charge. Thus, smaller ions of the same charge have the largest hydrated diameters, as do ions of approximately the same ionic diameter but with a smaller charge. Such ions (small ionic diameter and small charge) coordinate with more water molecules in their hydration sphere, resulting in a larger hydrated diameter. This is why sodium cation, Na^+, which is smaller than K^+, and has a smaller ionic charge than both Mg^{2+} and Ca^{2+}, causes greater swelling and loss of permeability of clayey soils than any of these other cations. See the discussion of sodium absorption ratio (SAR).

Rules of Thumb

1. Dissolved ions with higher binding strength tend to displace surface-bound ions of lower binding strength.
2. Binding strength increases with larger *nonhydrated ionic* diameter (smaller hydrated diameter) of the ions.
3. Binding strength increases with the charge on the ions.
4. There is also a concentration effect. Continual high concentrations of any ion eventually displace most other ions having the same charge sign.

The order of cation binding strengths to a negatively charged surface is

(strongest) $Cr^{3+} > Al^{3+} >> Ba^{2+} > Sr^{2+} > Ca^{2+} > Mg^{2+} >> Cs^+ > NH_4^+ > K^+ > Na^+ > H_3O^+ > Li^+$ (weakest).

For example, Cr^{3+} will displace Al^{3+} on a surface; Ca^{2+} will displace K^+; H_3O^+ will displace Li^+. There is no permanent change in the structure of the solid that serves as the ion-exchange material.

Why Do Solids in Nature Carry a Surface Charge?

Solid particle surfaces can acquire an electric charge in four ways. All four surface charge mechanisms can exist at the same time on mineral surfaces, and the latter two can exist at the same time on nonmineral (organic) surfaces.

1. *Lattice imperfections*: During the crystal growth of silica minerals, an Al^{3+} cation may enter a lattice location intended for Si^{4+}, or a Mg^{2+} may substitute for Al^{3+} (all are isoelectronic 3rd Period cations and, thus, are of similar sizes), resulting in a net negative charge on the crystal.
2. *Differential solubilities*: Ions at different locations on the surface of slightly soluble salt crystals may have different tendencies to dissolve into water, resulting in either a negative or positive charge imbalance.
3. *pH dependent chemical reactions at the particle surface*: Many solid surfaces (oxides, hydroxides, organics) contain ionizable functional groups, such as $-OH$, $-COOH$, or $-SH$. At high pH, these groups lose an H^+ (by: $H^+ + OH^- \rightarrow H_2O$), becoming charged as $-O^-$, $-COO^-$, or $-S^-$. At low pH, these groups gain an H^+, becoming $-OH_2^+$, $-COOH_2^+$, or $-SH_2^+$.
4. Adsorption of hydrophobic (low solubility) or surfactant ions: This can result in either positive or negative surfaces, and is not pH dependent.

Rules of Thumb for Permanent Surface Charge

1. Surface charge caused by lattice imperfections is permanent and is not pH dependent.
2. Permanent surface charge occurs on clays and most minerals.
3. The permanent surface charge on minerals and clays is generally negative.

Rules of Thumb for pH Dependent Surface Charge

1. At high pH, a negatively charged surface prevails.
2. At low pH, a positively charged surface prevails.
3. At some intermediate pH, the pH dependent surface charge is zero. This pH is called the point of zero charge (pzc).

Cation and Anion Exchange Capacity (CEC and AEC)

Cations are attracted to negative sites on a solid surface. Cation exchange capacity (CEC) is defined as the total number of negatively charged sites in a material at which reversible cation adsorption and desorption can occur. Operationally, it is measured by determining the total concentration (usually in meq/100 g of dry soil) of all exchangeable cations adsorbed. Thus, CEC is a measure of the reversible adsorptive capacity of a material for cations. At equilibrium, the total adsorbed cation charge equals the total negative charge on the solid. The portion of CEC not affected by pH changes is caused by adsorption to permanently charged sites. The portion of CEC that increases with pH is caused by pH dependent charged sites. Below about pH 5, H^+ ions are strongly bound to oxygen atoms at crystal edges, making these sites unavailable for cation adsorption. As pH increases above 5, H^+ ions are increasingly released into solution, making new sites available for cation adsorption.

Anion exchange capacity (AEC) arises mainly from protonation of hydroxyl groups on the surface of minerals and organic particles. It is mostly pH dependent.

$$(-OH + H^+ \rightarrow -OH_2^+).$$

Rules of Thumb (Refer to Figure 6.6)

1. pH dependent CEC does not change much as pH increases up to about pH 5.
2. Above pH 5, CEC increases rapidly with pH.
3. AEC increases as pH decreases. Gibbsite, kaolinite, goethite, and allophane clays exhibit small AECs.

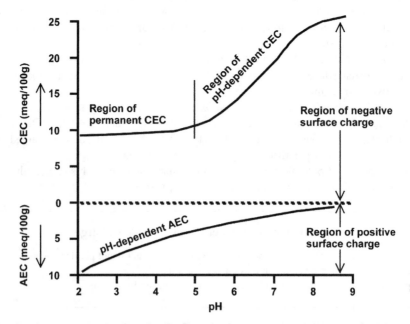

FIGURE 6.6 Ion exchange capacity dependence on pH. Cation exchange capacity (negative charge on the solid) is relatively constant up to about pH 5, due to permanent surface charge. It increases rapidly with pH above pH 5 due to pH dependent charge. Anion exchange capacity (positive charge on the solid) decreases with increasing pH due to pH dependent surface charge.

Exchangeable Bases: Percent Base Saturation

The primary exchangeable bases (exchangeable metal cations) are Na^+, Ca^{2+}, Mg^{2+}, and K^+. They usually occupy the majority of CEC sites in natural environments. The remaining CEC sites are occupied mainly by H^+. The surface concentration of H^+ is pH dependent.

Percent base saturation is defined as the percent of primary exchangeable base cations relative to the total adsorbed cation concentration, at, or near, pH 7 ($CEC_{pH=7}$):

$$\% \text{ base saturation} = (\text{primary exchangeable bases}/CEC_{pH=7}) \times 100.$$

If there is no pH dependent surface charge (i.e., only permanent surface charge), $CEC_{pH=7}$ will equal total exchangeable cations (including H^+) at any pH.

Example 6.4

Suppose a clay is measured to have the following cations (in meq/100 g):

$$Ca^{2+} = 16.2, Mg^{2+} = 4.4, K^+ = 0.1, Na^+ = 1.6, \text{ and } H^+ = 10.2.$$

What is its percent base saturation?

Answer: At the pH of the CEC measurement, the total CEC = 16.2 + 4.4 + 0.1 + 1.6 + 10.2 = 32.5 meq/100 g. The 22.3 meq/100 g of base exchange capacity due to Ca^{2+}, Mg^{2+}, K^+, and Na^+ represent 68.6% of the total CEC. The 10.2 meq/100 g of exchangeable H^+ is 31.4% of the total CEC. Assume that there is no pH dependent surface charge. Then,

$$CEC_{pH=7} = 32.5 \text{ meq/100 g, and percent base saturation} = 68.6.$$

Percent base saturation is related to the soil pH as follows:

- The higher the % base saturation, the higher the pH (more sites have been vacated by H^+ and occupied by metal cations).
- The lower the % base saturation, the lower the pH (more sites are occupied by H^+ and unavailable to metal cations).

Leaching of soils reduces base saturation but does not change CEC. Therefore, *soil leaching tends to increase soil acidity.*

Rules of Thumb

1. Soil pH is correlated with the percentage of base saturation.
2. The higher the base saturation, the higher the soil pH. *A base saturation of 90–100% indicates a soil pH around 7 or higher.*
3. Low base saturation (<90%) indicates acidic soils.
4. A 5% decrease in base saturation causes about 0.1 decrease in pH.

If a soil with a low base saturation is made into a slurry with water, the water pH will be acidic, around 4–5 or lower. Because a low base saturation means that many sites are occupied by H^+, acidic water results from exchangeable H^+ being released from the soil into solution. Similarly, passing acidic water through soil causes H^+ to exchange with soil-bound metal cations. The most weakly bound metals are displaced first and, if the pH is low enough, more strongly bound metals are displaced. At first, K^+ and Na^+ are removed, then Ca^{2+} and Mg^{2+} go into solution. Trivalent ions, such as Al^{3+} are the last to leave the soil and go into solution.

Example 6.5

Estimate the pH of the soil of Example 6.4.

Answer: The percent base saturation was measured to be 68.6%. At pH = 7, the % base saturation is about 100% (rule of thumb #2 above). Therefore, 68.6% represents a decrease of 100 − 68.6 = 31.4% from pH 7. A 5% decrease in base saturation causes about 0.1 decrease in pH (rule of thumb #4 above). Therefore, 31.4/5 ≈ 6 units of 5% decrease in base saturation, or 6 × 0.1 units decrease in pH. Therefore, 6 × 0.1 = 0.6 pH units decrease. Soil pH is around 7 − 0.6 = 6.4.

Rule of Thumb

The presence of an elevated concentration of Al^{3+} (>0.1 mg/L) is a characteristic of acid waters (pH <5).

CEC in Clays and Organic Matter

CEC in Clays

Clays adsorb cations from solution to the fixed total charge that represents their CEC. The total charge of adsorbed cations equals the total negative surface charge on the clay. A clay particle may be regarded as a system composed of (1) a relatively large, insoluble negatively charged particle and (2) a loosely held swarm of adsorbed exchangeable cations.

CEC depends on the clay crystalline layer structure. Clays have internal surfaces as well as external because of their layered structure. In some clays, the layers are held together more strongly than in others. Water can penetrate into weakly bound layers and force them apart, causing the clay to swell.

Different clays can differ markedly in their layer structures. Montmorillonite is a strongly swelling clay. Kaolinite does not swell significantly. Clays that swell in water have higher CECs than nonswelling clays. Swelling separates the layers and increases the surface area available for ion exchange. Surface charge on the interlayer surfaces attracts additional cations. Between the layers, adsorbed cations with large hydrated diameters (such as Na^+) exert separating forces that can greatly increase clay expansion. This decreases clay permeability to water (see SAR).

Rule of Thumb

A clay CEC > 10-15 meq/100 g indicates some degree of layer expansion by water swelling.

CEC in Organic Matter

One percent of humic organic material in a mineral soil contributes a CEC of around 2 meq/100 g of soil, which is about 4 times the CEC of an equal weight of clay.*

Rule of Thumb

A rough estimate of the CEC of a soil can be made as follows:

1. Estimate or determine the percentages by weight of silicate clay and organic matter.
2. Multiply the clay percentage by 0.5 to get the clay contribution.
3. Multiply the organic percentage by 2 to get the humus contribution.
4. Add the results to get the total CEC.

Example 6.6

Suppose a soil contained 2% organic matter and 16% clay. The soil CEC may be estimated to be around $(2 \times 2) + (16 \times 0.5) = 12$ meq/100 g.

Rates of Cation Exchange

The rates of exchange depend on the clay type. Kaolinite does not swell in water and there is no inner-layer access. Exchange reactions in kaolinites are almost instantaneous because they occur only on the outer surface. Illites swell slightly. Exchange reactions in illites can take several hours because a small part of the exchange occurs between inner crystal unit layers, to which cation

---

* Clay is much denser than humic matter. An equal weight of clay represents many fewer milliequivalents.

diffusion is slow. Montmorillonite expands considerably and most of the exposed surfaces are on the inner layers. Montmorillonites take still longer to reach ion-exchange equilibrium.

6.7 INDICATORS OF FECAL CONTAMINATION: COLIFORM AND STREPTOCOCCI BACTERIA

BACKGROUND

Detecting and preventing fecal contamination is of prime importance for all drinking water systems and recreation water managers. Fecal wastes may contain enteric pathogens (disease-causing organisms from the intestines of warm-blooded animals) such as viruses, bacteria, and protozoans (which include *Cryptosporidium* and *Giardia*). Fecal contaminated water is a common cause of gastrointestinal illness, including diarrhea, dysentery, ulcers, fatigue, and cramps. It also may carry pathogens that cause a host of other serious diseases, such as cholera, typhoid fever, hepatitis A, meningitis, and myocarditis.

So far, testing water directly for individual pathogenic organisms is impractical for several reasons, including

- There are so many different kinds of pathogens that a comprehensive analysis would be very expensive and time consuming. Time is of the essence for pathogen detection.
- Pathogens can be dangerous at small concentrations. They require large sample volumes for analysis which add to the time and cost of analysis.
- Reliable analytical methods for several important pathogens are difficult or not even available. Also, not all water-borne pathogenic microorganisms are known.
- A satisfactory alternative is available, namely the identification of "indicator" species that are easy to measure and are always present with enteric pathogens.

Hence, awareness of possible contamination by enteric pathogens is based on detecting the more easily identified "indicator" species whose presence indicates that fecal contamination may have occurred.

The five indicator species most commonly used today are total coliforms, fecal coliforms, *Escherichia coli* (*E. coli*), fecal streptococci, and enterococci. All are bacteria normally present in the intestines and feces of warm-blooded animals, including humans. All but *E. coli* consist of groups of bacterial species that are similar in shape, habitat, and behavior. *E. coli* is a single species within the fecal coliform group. These indicators themselves are usually not pathogens and do not pose a danger to humans or animals. However, if the indicators are present in water, the accompanying presence of enteric pathogens is a possibility.

All the indicator species are easier to measure than most pathogens but are harder to kill. Therefore, treatment that satisfactorily destroys the indicator species may be assumed to have also destroyed enteric pathogens that were present. For example, in waste water disinfection, it is assumed that a decrease in fecal coliforms to <200 fc/100 mL will have eliminated the great majority of pathogens.

Total coliforms and fecal coliforms are the "old reliable" indicators of fecal contamination, used since the 1920s to protect public health. However, both have limitations that stimulate regulators to continue seeking improved methods.

TOTAL COLIFORMS

Total coliform bacteria are widespread in nature. In addition to their animal intestine habitat, they occur naturally in plant material and soil. Therefore, their presence does not necessarily indicate fecal contamination. Total coliforms are not recommended as indicators of recreational water contamination, where they are usually present from soil and plant contact. Total coliforms are the standard test for contamination of finished drinking water, where contamination of a water supply or distribution system by fecal, plant, or soil sources is not acceptable. Federal drinking water standards are based on total coliform bacteria. The EPA maximum contaminant level (MCL) for drinking water is zero total coliforms per 100 mL of water for 95% of samples after treatment (chlorination, ozonation, UV). See Appendix A.

Fecal Coliforms

Fecal coliforms are a more fecal-specific subset of total coliform bacteria. However, even the fecal coliform group contains a genus, *Klebsiella,* with member species not necessarily fecal in origin. *Klebsiella* coliforms are found in large numbers in textile, pulp, and paper mill wastes. Fecal coliforms are widely used to monitor recreational waters and are the only indicator approved for classifying shellfish waters by the U.S. Food and Drug Administration's National Shellfish Sanitation Program. On the basis of statistical data, the EPA has recently begun recommending *E. coli* and enterococci as better indicators of health risk from water contact. However, many states still use fecal coliforms for this purpose, in part so that new data can be directly compared with historical data.

Natural surface waters almost always contain some background level of fecal coliforms, usually less than 15–20 fc/100 mL MPN (most probable number). In sewage entering a waste treatment plant, the fecal coliform count may be over 10 million fc/100 mL (fecal coliforms per 100 mL of sample). Satisfactory disinfection of secondary effluent from a waste treatment plant is defined by an average fecal coliform count of <200 fc/100 mL. Fecal coliforms are normally absent after wastewater percolates through 5 ft of soil.

State water standards for fecal coliform levels vary, but typical state standards are (geometric mean values)

- Class 1 primary contact recreational   200 fc/100 mL
- Class 1 secondary contact recreational   2000 fc/100 mL
- Domestic water supply (before treatment)   2000 fc/100 mL

E. coli

E. coli is a single species of fecal coliform bacteria that occurs only in fecal matter from humans and other warm-blooded animals. EPA studies[33] indicate that, in fresh water, *E. coli* correlates better with swimming-related illness than do fecal coliforms. Since 1986, the EPA has been recommending that states use *E. coli* as an indicator for fecal-contaminated freshwater recreation areas, instead of fecal coliform. States vary in their adoption of this recommendation.

Fecal Streptococci

The normal habitat of fecal streptococci bacteria is the gastrointestinal tract of warm-blooded animals. Because humans differ from animals in the relative amounts of coliforms and streptococci normally present in their intestines, it was believed that the ratio of fecal coliforms to fecal streptococci could be used to differentiate human fecal contamination from that of other warm-blooded animals. A ratio greater than 4 was considered indicative of human sources, and a ratio less than 0.7 suggested animal sources. However, this is no longer regarded as a reliable test because of variable survival half lives in water of different species of streptococci and the effects of wastewater disinfection and different analytical procedures on the measured coliform/streptococci ratio. Where contamination is recent and disinfectants are absent, the test may still be useful.

Enterococci

Enterococci are a subgroup of fecal streptococci that are differentiated by their ability to thrive in saline water over a wide temperature range. In this respect, they mimic many enteric pathogens better than other indicators do. The EPA recommends enterococci as the best indicator of health risk in salt recreational waters and as a useful indicator in fresh waters as well.

Rules of Thumb

1. To determine whether recreational water or wastewater meets state water quality requirements, find out which indicator (fecal coliform, *E. coli*, or enterococci) your state uses for recreational or wastewaters and measure that one.
2. For compliance with federal drinking water standards, measure total coliforms.
3. For a nonregulatory determination of the health risks from recreational water contact, measure *E. coli*, or enterococci.

6.8 MUNICIPAL WASTEWATER REUSE: THE MOVEMENT AND FATE OF MICROBIAL PATHOGENS

PATHOGENS IN TREATED WASTEWATER

Reuse of municipal wastewater requires special care to minimize hazardous exposure of humans and animals to water-borne pathogens. Wastewater that has received secondary treatment generally contains residual active viruses and other pathogens[25] which can persist to varying degrees after release to the environment. Removal of microorganisms is accomplished by filtration, adsorption, desiccation, radiation, predation, and exposure to other adverse conditions. Because of their large size, protozoa and helminths are removed primarily by filtration. Bacteria are removed by filtration and adsorption. Fecal coliforms are normally absent after wastewater percolates through 5 ft of soil. Virus removals are not as well documented,[32] but it is agreed that viruses are removed from subsurface water flow almost entirely by adsorption to soil particles. As seen below, removal is not the same as inactivation and is not permanent. Viruses adsorbed to soil particles can be rereleased still in an active state. Once viruses are mobilized in the environment, they can become inactivated by a variety of factors such as higher temperatures, pH, loss of moisture, exposure to sunlight, inorganic cations and anions, and the presence of antagonistic soil microflora.

Viral persistence in the environment can be prolonged due to low temperatures, adsorption to particulate surfaces, and moist conditions.[4] Under appropriate conditions, viruses can remain infective for several months in wastewater sludges and environmental waters.[14] In a study of virus occurrences in treated wastewaters in Arizona and Florida, average virus levels ranged from 13–130 pfu/100 L (plaque-forming units per 100 L).[25] In most of the viral-positive samples, viral effluent quality significantly exceeded the standard for unrestricted irrigation of 1 pfu/40 L. The authors performed a risk analysis which found that viral effluent quality in Arizona and Florida would potentially produce two infections per 1000 exposed people.

A review of the general literature shows that the persistence of pathogens associated with the use of municipal wastewater for aquifer recharge is very site specific. In particular, the behavior of viruses, which are more difficult to remove or inactivate than bacteria or parasites, depends strongly on environmental conditions and the types of viruses present. Some studies find that viruses are completely inactivated quickly after introduction to the soil/groundwater environment, while other studies find that they can persist for long periods of time (over a year)[22,32] and over long travel distances (hundreds of meters).[32]

There is no question that viruses have a potential for high mobility and persistence in groundwater. A National Research Council (NRC) Report[21] emphasizes that there are significant uncertainties associated with predicting the transport and fate of viruses in recharged aquifers, and that these uncertainties make it difficult to determine the levels of risk from any infectious agents still contained in the disinfected wastewater. The NRC report endorses groundwater recharge practices but cautions that current recharge technologies are "especially well-suited to nonpotable uses such as landscape irrigation" and that "potable reuse should be considered only when better quality sources are unavailable." This report states further that "water quality monitoring and operations management should be more stringent for recharge systems intended for potable reuse."

The state of California has been in the forefront of wastewater recycling applications because of chronic water shortages and the threat of saltwater incursions into freshwater aquifers. A 1987 survey reported that California had more than 200 wastewater reclamation plants and 854 water reuse areas that processed approximately 238 mgd of reclaimed water. California is projected to use about 738 mgd of reclaimed water by the year 2000. California's Title 22 regulation establishes the criteria for protecting public health which include extensive wastewater treatment, frequent water-quality monitoring, and strict use-area controls. The California water reuse regulations pay particular attention to enteric viruses (viruses that are shed in fecal matter) because of the possibility of contracting disease with relatively low doses of the viruses and the difficulty of routine examination of wastewater for their presence. The infectious dose for viruses is reported to be 1–10 viral units.[34] California requires essentially a virus-free effluent via a "full treatment" process for wastewater reuse applications with high potential for human exposure.[2]

TRANSPORT AND INACTIVATION OF VIRUSES IN SOILS AND GROUNDWATER

Viruses are the smallest wastewater pathogens, consisting of a nucleic acid genome enclosed in a protective protein coat called a capsid. A virus capsid contains many ionizable proteins that are subject to protonation and deprotonation reactions in water, depending on the pH and ionic strength of the water. At low pH, virus particles tend to carry a positive charge because of attached H^+ ions. As the pH rises, the positive charge on a virus particle decreases, then passes through zero at the isoelectric point, and becomes negative due to increasing numbers of attached OH^- ions. As a result, viruses can have the ion-exchange characteristics of either cations or anions, depending on the pH.

In groundwater and soils, viruses move as colloidal particles. Because of their small size, it is believed that viruses are not significantly removed from groundwater by mechanical filters that are coarser than reverse osmosis or nanofiltration. Viruses become attached to soil particles mainly by sorption forces arising from electrostatic interactions, London forces, hydrophobic forces, covalent bonding, and hydrogen bonding.[1,32] As a result, the extent to which viruses are sorbed to soils depends strongly on pH, temperature, ionic strength, and flow velocity of the water, as well as the mineral-organic composition and particle size distribution of the soil and the particular type of virus. Clay soils are more retentive than sandy soils and finely divided soils retard virus mobility more than coarser soils.[32]

The isoelectric point of enteric viruses is usually below pH 5, so that in most soils enteroviruses carry a negative charge, as do most soils. In general, virus adsorption to soil is enhanced at lower pH values (pH < 7), where soil and virus charges are opposite, and reduced at higher pH values (pH > 7).[40] If chemical conditions change or the flow velocity is increased, either locally by microscopic changes or overall by macroscopic changes, adsorbed virus particles can be detached from soil surfaces and returned to suspension in the flow. Waters with high TDS concentrations favor adsorption to soils because electrostatic repulsion is minimized in waters with high ionic strength. A rain event can dilute TDS levels near the surface and cause a burst of released viruses. The same "burst" effect can occur with a release of higher pH water which locally raises the water column pH from, 7.2 to 8 or 9.[32] For these reasons, virus adsorption to soils cannot be considered a process of absolute immobilization of the viruses from the water. Infective viruses are capable of release from soil particles after immobilization for long periods of time. Any environmental change that reduces their attraction to soil particles will result in their further movement with groundwater.

The presence of organic matter, such as humic and fulvic acids, in soils has been shown to inhibit the adsorption of viruses to soil surfaces by competing with viruses for adsorption sites.[8,12,17,26,28] In one study,[26] the presence of organic substances in an aqueous environment reduced the retention of viruses in a soil column from greater than 99% to less than 1.5%.

Adsorption to soil particles may prolong viral lifetimes in aqueous environments.[4,5,6,11,29,31,32] Viruses bound to solids are as infectious to humans and animals as the free viruses.[13,32] Virus survival in soil depends on the nature of the soil, temperature, pH, moisture, and the presence of antagonistic soil microflora.

In one study using f2 bacteriophage and poliovirus type 1, 60–90% of the viruses were inactivated at 20°C within 7 days after the initial release to the soil.[16] But, after the first 7 days, the inactivation rate slowed and polioviruses could still be detected at 91 days; f2 viruses survived longer than 175 days. At lower temperatures, up to 20% of the polioviruses survived longer than 175 days. Other studies indicate that virus lifetimes may range from 7 days to 6 months in soils, and from 2 days to more than 6 months in groundwater.[32] A proposed revision to the California Title 22 regulations would require that reclaimed water be held underground for at least 6 months prior to reuse to allow a high percentage of virus die-off.[22]

6.9   ODORS OF BIOLOGICAL ORIGIN IN WATER

Odors from anaerobic surface waters, groundwater, and domestic wastewater are usually from inorganic and organic gases generated by biological activity. Anaerobic decomposition of nitrogenous or sulfurous organic matter often produces gases that contain sulfur and/or nitrogen. Such gases are frequent causes of odors in water. The most common inorganic gases in water are carbon dioxide (CO_2), methane (CH_4), hydrogen (H_2), hydrogen sulfide (H_2S), ammonia (NH_3), carbon disulfide (CS_2), sulfur dioxide (SO_2), oxygen (O_2), and nitrogen (N_2). Of these inorganic gases, those with an odor always contain N or S in combination with H, C, and/or O, such as H_2S, NH_3, CS_2, and SO_2.

Hydrogen sulfide, from the anaerobic reduction of sulfate (SO_4^{2-}) by bacteria, usually is the most prevalent odor in natural waters and sewage. Sulfate, formed from the aerobic biodegradation of sulfur-containing proteins, is commonly present in domestic wastewater between 30 and 100 mg/L. Sulfate can arise in natural waters from sulfate minerals and aerobic decomposition of organic material. In addition to H_2S, other disagreeable odorous compounds may be formed by anaerobic decomposition of organics. The particular compounds that are formed depend on the types of bacteria and organic compounds present. Table 6.6 lists a number of common odiferous inorganic and organic compounds with their odor characteristics and odor threshold concentrations when dissolved in water. Sewage carrying industrial wastes may contain other volatile organic chemicals that can contribute additional odors.

ENVIRONMENTAL CHEMISTRY OF HYDROGEN SULFIDE

Under anaerobic aqueous conditions, in the presence of organic matter or sulfate-reducing bacteria, sulfate is reduced to sulfide ion (S^{2-}):

$$SO_4^{2-} + \text{organic matter/sulfate-reducing bacteria} \rightarrow S^{2-} + H_2O + CO_2. \qquad (6.20)$$

Sulfide ion is a strong base, reversibly reacting rapidly in water to form HS^- and gaseous hydrogen sulfide:

$$S^{2-} + 2\ H_2O \leftrightarrow OH^- + HS^- + H_2O \leftrightarrow H_2S(g) + 2\ OH^-. \qquad (6.21)$$

HS^- and S^{2-} are nonvolatile with no odor. H_2S is gaseous with a strong odor of rotten eggs. The equilibrium distribution between S^{2-}, HS^-, and H_2S depends mainly on the pH and somewhat on the temperature. In Figure 6.7, T = 30°C.

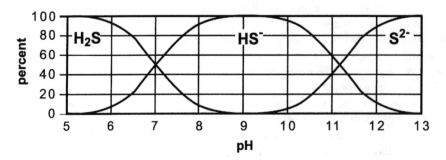

FIGURE 6.7   pH distribution of hydrogen sulfide species in water.

Rules of Thumb

In water, S^{2-} reacts according to Equation 6.21:

$$S^{2-} + 2\ H_2O \leftrightarrow OH^- + HS^- + H_2O \leftrightarrow H_2S + 2\ OH^-$$

1. Raising the pH shifts the equilibrium to the left, converting the malodorous gas H_2S into nonodorous and nonvolatile HS^- and S^{2-}.
2. Lowering the pH shifts the equilibrium to the right, creating more malodorous H_2S gas from the nonvolatile forms, HS^- and S^{2-}.
3. Lowering the temperature shifts the equilibrium to the right (more H_2S) at any pH.
4. Well water, groundwater, or stagnant surface water that smells of H_2S (rotten eggs) is usually a sign of sulfate reducing bacteria.
5. Water conditions promoting the formation of H_2S are

   - sulfate = >60 mg/L
   - oxidation-reduction potential = <200 mV;
   - pH = <6

CHEMICAL CONTROL OF ODORS

Depending on the odor-causing compound, chemical control of odors may be accomplished by a combination of pH control, eliminating the causes of reducing conditions, chemical oxidation and/or aeration, sorption to activated charcoal, air-stripping of volatile species, and chemical conversion (often microbially mediated, as in nitrification).

pH control

Hydrogen sulfide

For odors from H_2S, raising the pH (by adding NaOH or lime) shifts the equilibrium to the left for Equation 6.21:

$$S^{2-} + 2\ H_2O \leftrightarrow OH^- + HS^- + H_2O \leftrightarrow H_2S + 2\ OH^-. \tag{6.21}$$

This converts gaseous H_2S to the nonodorous ionic forms. However, the pH must be maintained above 9 for complete odor removal. Normally, odor control by removing the sulfur compounds is more practical.

TABLE 6.6
Odor Characteristics and Threshold Concentrations in Water

Substance	Formula	Odor Threshold Concentration in Water (mg/L)	Odor Characteristics
Allyl Mercaptan	$H_2C = CHCH_2SH$	0.00005	Very disagreeable, garlic-like
Ammonia	NH_3	0.037	Sharp, pungent
Benzyl Mercaptan	$C_6H_5CH_2SH$	0.00019	Unpleasant
Chlorine	Cl_2	0.010	Pungent, irritating
Chlorophenol	ClC_6H_4OH	0.00018	Medicinal
Crotyl Mercaptan	$CH_3CH = CHCH_2SH$	0.000029	Skunk-like
Diphenyl Sulfide	$(C_6H_5)_2S$	0.00005	Unpleasant
Ethyl Mercaptan	CH_3CH_2SH	0.00019	Decayed cabbage
Diethyl Sulfide (ethyl sulfide)	$(CH_3CH_2)_2S$	0.000025	Nauseating, ethereal
Hydrogen Sulfide	H_2S	0.0011	Rotten egg
Methyl Mercaptan	CH_2SH	0.0011	Decayed cabbage
Dimethyl Sulfide (methyl sulfide)	$(CH_3)_2S$	0.0011	Decayed vegetables
Pyridine	C_6H_5N	0.0037	Disagreeable, irritating
Skatole	C_9H_9N	0.0012	Fecal, nauseating
Sulfur Dioxide	SO_2	0.009	Pungent, irritating
Thiocresol	$CH_3C_6H_4SH$	0.001	Rancid, skunk-like
Thiophenol	C_6H_5SH	0.000062	Putrid, nauseating

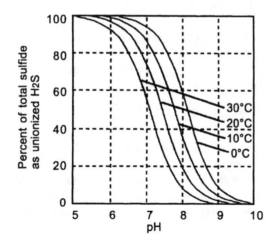

FIGURE 6.8 Fraction of hydrogen sulfide in unionized form (H_2S) as a function of temperature and pH.

Lowering the pH (by adding acid) shifts the equilibrium to the right, converting the ionic forms to gaseous H_2S (see Figure 6.8). At low pH, the gas can be removed from the water by air stripping in a manner similar to the air stripping of ammonia in Example 6.7. This of course does not destroy the H_2S; it moves it from the water to the air. Figure 6.8 gives the fraction of hydrogen sulfide that is in the volatile form of H_2S. Note that H_2S behaves the opposite of NH_3 (below). Stripping efficiency is increased with *decreasing* pH and lower temperatures.

Ammonia

pH control of odors from ammonia is opposite to that for hydrogen sulfide. To use air stripping to remove odor caused by ammonia, the pH must be raised (see Figure 6.9 and Example 6.7).

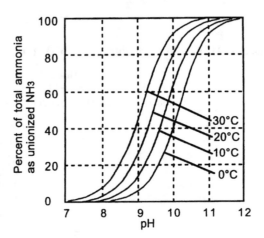

FIGURE 6.9   Fraction of ammonia in unionized form (NH_3) as a function of temperature and pH.

Example 6.7 Removing Ammonia by Air-stripping

A wastewater flow contains 30 g/L total ammonia nitrogen

$$\text{Total } NH_3\text{-N} = [NH_3] + [NH_4^+],$$

and has a 5 g/L discharge limit. The temperature varies from 0°C to 30°C and the pH is normally about 9. At what pH must the stripper be operated?

Calculation: pH must be adjusted so that enough ammonia is in the volatile form to meet the

discharge requirement. The ratio $\dfrac{[NH_3]}{[NH_3]+[NH_4^+]}$ must be at least as large as the fraction of NH_3-N

to be removed. Fraction of NH_3-N to be removed $= \dfrac{25}{30} = 0.83$.

Answer: Stripping efficiency is increased with *increasing* pH and *higher* temperatures. Figure 6.9 shows that to get 0.83 of the total NH_3-N in the form of volatile NH_3, the pH must be raised to about 10.8, for the worst case of 0°C. At higher temperatures, the removal efficiency will be higher. Lime (CaO) is the least expensive way to raise the pH, but it generates $CaCO_3$ sludge. NaOH is more expensive but it does not generate sludge. Note that NH_3 behaves the opposite of H_2S (above).

Oxidation

Add oxidizing agents such as Cl_2, NaOCl, H_2O_2, O_2, $KMnO_4$, or ClO_2. They oxidize hydrogen sulfide to odorless sulfate ion, SO_4^{2-}, and ammonia to odorless nitrogen compounds, including elemental nitrogen, N_2.

Rules of Thumb

1. The usual chlorine dose for odor control is 10 to 50 mg/L.
2. 8.9 mg of chlorine is required to oxidize 1 mg of hydrogen sulfide, H_2S.

Eliminate Reducing Conditions Caused by Decomposing Organic Matter

This often means mechanically cleaning out the organic slime and sludge in a well, sewer, drain, or wetland. Mechanical cleaning will aid the use of oxidizing agents and can also be accomplished sometimes by aerating the water. Increasing the dissolved oxygen level will shift conditions from reducing to oxidizing as the oxygen diffuses into the reducing zone of the water. Mixing currents in the water help this process. If the reducing zone is thick and the water stagnant and motionless, aeration control might be very slow.

Drying out wet soil with an H_2S odor also allows oxygen to diffuse into the organic matter, changing the decomposition process from anaerobic to aerobic.

Sorption to Activated Charcoal

Sorption to powdered or granular activated charcoal is a reliable "last resort" for removing bad tastes and odors. Powdered charcoal can be added as a slurry directly to a waste stream. Gases can be passed through a canister filled with granulated activated charcoal (GAC).

6.10   QUALITY ASSURANCE AND QUALITY CONTROL (QA/QC) IN ENVIRONMENTAL SAMPLING

Quality assurance and quality control (QA/QC) are the set of principles and practices which, if strictly followed in a sampling and analysis program, will produce data of known and defensible quality. The effectiveness of any monitoring effort depends on its QA/QC program.

A well designed QA/QC program reduces, as far as possible, the potential for undetected errors appearing in the final results. It has many components ranging from competency certification of personnel to replicate sampling. The overall goal is to minimize or correct all the errors that might occur, by introducing explicit steps into the sampling and analysis protocol for identifying, measuring, and controlling these errors.

QA/QC Has Different Field and Laboratory Components

Field QA/QC is intended to collect in the field and deliver to the laboratory a sample in which analyte properties have not changed significantly from their values in the environmental waters being tested. An obvious, but not trivial, part of field QA/QC is properly identifying all samples, so that measured analyte concentrations can be assigned with confidence to a known location and time.

Laboratory QA/QC is designed to produce a quantitative measure of certain properties of the sample with known limits of uncertainty. There are some overlapping QA/QC practices, such as including field duplicates and blind known samples with a field sample set to test a laboratory's performance.

This section deals only with field QA/QC protocols. Laboratory practices are generally out of the control of a sampling program manager, who ideally will have sufficient experience with the QA/QC standards of the laboratory being used to have confidence in the final data. However, laboratory performance can change, and it is prudent to remain alert to possible laboratory errors. Techniques for checking laboratory performance are discussed below.

Essential Components of Field QA/QC

Sample Collection

Proper sampling is a separate science in its own right. The choice of automated vs. hand sampling equipment, details of how to get a representative sample into the container, and decisions on appropriate grab, composite, or time/space-integrated samples, are not addressed here.* The importance of obtaining a representative sample for analysis is underscored when one recognizes that variability of the sampled media is the greatest source of variability in analytical results, far exceeding measurement uncertainties.

However, certain QA/QC practices are common to all sampling procedures:

- *Sample contamination must be avoided.* All parts of the sampling equipment and containers that come in contact with the sample must be scrupulously clean. Sample containers must be inert to the sample and preservatives.
- *Samples must be properly preserved.* All samples should be stored and transported at 4°C (ice temperature) or less. In addition, samples for certain analyses require chemical preservatives, e.g., acid addition to less than pH = 2 for metals and ammonia, base addition to greater than pH = 12 for cyanide.
- *Samples must be unambiguously identified.* If the sample container is not prelabeled, it must be securely labeled with a unique identification at the time of collection or removal from a sampler. The collector's name, date and time, station designation, form of preservation, and desired analyses should also be on the label. The sample identification must be entered on the report form without delay after labeling the sample container.
- *Samples must be packaged for transportation.* Generally, samples are transported in a cooler. The cooler lid must not accidentally open, container lids must not leak, and containers should be protected against breakage.
- *Chain-of-custody and field reporting documentation must be maintained.* Careful completion of chain-of-custody and sampling report forms are essential parts of field QA/QC. The chain-of-custody (COC) form helps to ensure sample integrity, from collection to data reporting. Entries on the COC form track the possession and handling of each sample from the time of collection, through analysis and final disposition, and demonstrate that all standard steps of sample control have been followed. The sample report form is needed for interpretation of the reported data and must be designed carefully to require an entry for all relevant information. The report form will always include the name or initials of the sampler and will have labeled areas to enter sampling location, sample identification, date, and sampling time for each sample. It might also include a space for other relevant information such as weather conditions, observations concerning the sampled waters (such as odor, color, flow, etc.), and sampling procedure (grab, time-integrated, etc.).

Most commercial laboratories provide coolers, ice packs, clean sample containers with preservatives appropriate for requested analyses, container labels, and chain-of-custody forms and seals as part of their analytical services.

The Field Sample Set

A field sample set consists of the environmental samples and several kinds of quality control samples. The project manager must determine what kinds of quality control samples should be collected, based on the purposes of the sampling program.

Quality control samples

Some field samples should be collected for quality control purposes. Sampling procedures can contribute to both systematic and random errors. Many environmental waters are inherently not uniformly mixed. Field blanks, spikes, and duplicates will help estimate errors caused by sampling bias (usually contamination and/or nonrepresentative samples) and calculate sampling precision.

---

* Detailed guidance for these procedures can be found in *Environmental Sampling and Analysis: A Practical Guide*,[15] and in reference works such as *Standard Methods*, 1995.

Field QC samples must be handled exactly the same way as the environmental samples, using identical sampling devices, sampling protocol, storage containers, preservation methods, storage times, and transportation methods. For a correct interpretation, quality control samples must be subject to the same holding time criteria as the environmental samples.

Blank sample requirements

There are several types of blank samples, each serving a distinct QC purpose. Wherever a possibility exists for a sample to become contaminated, a blank should be devised to detect and measure the contamination. The most commonly collected blanks include *field*, *trip*, and *equipment* blanks. Although it may be prudent to collect a full set of blanks, most of the blank samples usually do not need to be analyzed. Analysis costs can be reduced if the strategy of blank sampling is understood. First analyze only the field blanks, which are susceptible to the broadest range of contaminant sources. If these indicate no problems, the other more specific blanks can be discarded or stored. If a problem is suspected, other blanks can be analyzed for confirmation of the problem and discovering the source. Holding time limits must be observed with blanks as well as with environmental samples.

Field Blanks: At least one "clean" water sample should be exposed to the same sampling conditions as the environmental samples at each sampling site. The analytical laboratory can generally provide analyte-free distilled water for this purpose. Field blanks are transferred from one container to another, passed through automatic equipment, or otherwise exposed to the conditions at the sampling site. At a minimum, the field blank container is opened at the sampling site and exposed to the air for approximately the same time as the environmental samples. It is then capped, labeled, and sent to the laboratory with other samples. Field blanks measure incidental or accidental sample contamination throughout all the steps of transportation, sampling, analysis, and sample preparation at the laboratory. They help assure that artifacts are recognizable and are not mistaken as real data.

Trip Blanks: For each type of container and preservative, at least one container of analyte-free water should travel unopened from the laboratory to the sampling sites and back to the laboratory. Trip blanks serve to identify contamination from the container and preservative during transportation, handling, and storage.

Equipment Blanks: These are especially important with automated sampling equipment. Equipment blanks, sometimes referred to as *rinsate blanks,* document adequate decontamination of the sampling equipment. These blanks are collected by passing analyte-free water through the sampling equipment after decontamination and prior to resampling.

The analytical laboratory will run a similar set of blanks to document contamination and errors arising from handling and analysis procedures in the laboratory.

Field duplicates and spikes

Field duplicates are two separate environmental samples collected simultaneously at the identical source location and analyzed individually. Field duplicates are sensitive to the total sample variability, for example, variability from all sampling, storage, transportation, and analytical procedures. Where the goals of the sampling program warrant it, as in permit monitoring requirements, at least one field duplicate per day should be collected for each analyte. More than two replicates may be required in cases where it is difficult to obtain representative samples.

Field spikes are environmental samples to which known amounts of the analytes of interest are added. Ampoules containing carefully measured amounts of analytes can be purchased from chemical supply sources. Field spikes can identify storage, transportation, and matrix effects, such as loss of volatile compounds and analytical interferences caused by certain compounds that are present in the environmental source. In a spiked sample with no problems, the measured analyte concentration should be equal to the concentration present in the environmental sample plus the added spike concentration within the limits of the method's precision.

UNDERSTANDING LABORATORY REPORTED RESULTS

When a laboratory reports that a target compound was not detected, it does not mean that the compound was not present. It always means that the compound was not present in a concentration above a certain lowest reporting limit. There always is the possibility that the compound was present at a concentration below the reporting limit. The laboratory might even have identified the compound below the reporting limit but not reported it because the concentration could not be quantified within acceptable limits of error.

Reported results of analyte concentrations are never exact. There always is some margin of error. The only kind of measurement that can be exact is a tally of discrete objects, for example, dollars and cents or the number of cars in a parking lot. When measuring a quantity capable of continuous variation, such as mass, length, or the concentration of benzene in a sample, there always is some uncertainty. Like an irrational number, measurements capable of continuous variation can always be expressed to more significant figures, and any answer with a finite number of digits is always an approximation.

In addition, many repeated measurements of the same quantity, even under essentially identical conditions, will always yield results that are scattered randomly about some average value, due to uncontrollable variations in environmental, experimental, and operator behavior. The person who interprets experimental data must always bear in mind that no reported experimental value necessarily represents the "true" value. In other words, it is never possible to be completely certain of a result. Nevertheless, we still try to answer questions such as "Does this sample indicate a violation of a discharge limit?" or "Does the data indicate that my remediation activities are beginning to work?" Answers to such questions must always acknowledge that there is a range of error. The purpose of QA/QC controls is to assure that the reported value lies within a small enough range of error to be considered useful data.

Part of laboratory QA/QC involves determining how much scatter in the data is to be expected from different analytical procedures. The *precision* inherent in a particular procedure is a measure of how much deviation may be expected in repeated measurements of the same sample. The *standard deviation* is a common way of quantitatively expressing the precision, or reproducibility, of a measurement. A more thorough treatment of the standard deviation can be found in elementary statistics texts or in *Standard Methods*.[30] The procedure for using the standard deviation to expressing the reliability of measurements is described by the EPA in several documents (e.g., U.S. EPA, 1992b; 40 CFR 136, Appendix B). Here, it is only necessary to understand that, for many measurements of the same quantity, one standard deviation above and below the average value of the measurements will include about 68% of all the individual values. For example, suppose the concentration of benzene in a sample is measured repeatedly many times to yield an average value of 12.3 µg/L and a standard deviation of 2.7 µg/L. Then, 68% of all the individual measurements can be expected to lie within the range of 12.3 ± 2.7 µg/L, or between 9.6 and 15.0 µg/L. Two standard deviations will include about 95% of all the values.

The larger the standard deviation, the greater the scatter in the data, in other words, the poorer the precision, or reproducibility, of repeated measurements. The standard deviation and, hence, the measurement reproducibility, depend strongly on the matrix in which the analyte is being measured. Stream and groundwater samples can be measured more precisely than wastewater and soil samples. For this reason, it is not possible to make general statements about the reliability of measurements based only on the concentration values. The sample matrix is taken into account by determining the standard deviation.

In Figure 6.10, different types of measurement limits are defined in terms of the reported concentration expressed as the number of standard deviations above the instrumental zero. Only the method detection limit (MDL) is rigorously defined (40 CFR 136, Appendix B). A calculation of the MDL puts it about 3 standard deviations above the instrumental zero. The estimated

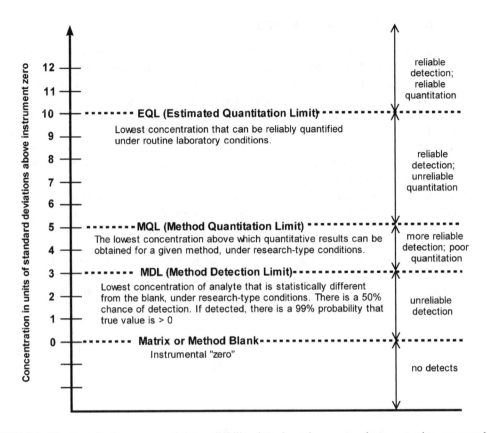

FIGURE 6.10   Statistical measures of data reliability, based on the measured concentration expressed in standard deviations above the instrumental zero. The standard deviation is determined by repeated measurements of a laboratory control sample. Other names for EQL are PQL (practical quantitation limit), LLQ (lower limit of quantitation), and LOQ (limit of quantitation).

quantitation limit (EQL) is an estimation of the lowest concentration above which reliable quantitative results can be obtained under routine laboratory conditions, and it is arrived at by evaluating the performance of different laboratories. If performance data are not available, the EQL can be estimated from the MDL. The EPA believes that setting the EQL at 5 to 10 times the MDL is generally a fair expectation for routine operation at most qualified government and commercial laboratories.[36] Many laboratories determine their own "routine performance" EQL and generally use 3 to 4 times the MDL as the lowest level for an EQL. The method quantitation limit (MQL) is an estimation of the lowest concentration above which reliable quantitative results can be obtained under research laboratory conditions. The MQL can be measured but is often estimated to be about 5 standard deviations above the instrumental zero. In addition, many laboratories include a reporting limit (RL) in their analytical reports, which represents the concentration above which that laboratory has confidence in its quantitative values. Depending on the laboratory, the RL for an analyte may be anywhere between its MDL and its EQL.

Rule of Thumb

When an "official" EQL has not been designated for a particular analysis, an unofficial EQL of 10 standard deviations above the instrument zero (or 3.3 times the MDL) is often used.

6.11   SODIUM ADSORPTION RATIO (SAR)

The sodium adsorption ratio (SAR) indicates the amount of sodium present in soils, relative to calcium and magnesium. If the sodium fraction is too large, soil permeability may be low, and the movement of water through the soil may be restricted. SAR is important to plant growth because its magnitude is an indication of the availability of soil pore water to plant roots.

The sodium ion has a large radius of hydration. This means that its hydration sphere of bound water molecules is larger than most cations. When dissolved sodium is adsorbed into clay-bearing soils, it causes them to disperse and swell. This reduces the soil pore size and causes the permeability to water to be greatly reduced. If the sodium ion is replaced by ion exchange with cations having a smaller radius of hydration, the soil dispersion is reduced.

In most natural waters, Ca^{2+}, Mg^{2+}, and Na^+ are by far the most abundant cations, so other cations can usually be neglected in their effect on soil dispersion. Also, Ca^{2+} and Mg^{2+}, being doubly charged, are more tightly adsorbed to clay surfaces than sodium and, therefore, are preferentially adsorbed. They also have smaller radii of hydration and cause less soil dispersion. The relative amounts of sodium, calcium, and magnesium adsorbed to soil are proportional to the amounts dissolved in groundwater. By measuring the concentrations of sodium, calcium, and magnesium in water used for irrigation, the SAR can be calculated using Equation 6.22 and the risk of low soil permeability evaluated.

$$SAR = \frac{\left[Na^+\right]}{\sqrt{\dfrac{\left[Ca^{2+}\right]+\left[Mg^{2+}\right]}{2}}}, \tag{6.22}$$

where

$[Na^+]$ = concentration of sodium in eq/L (water).
$[Ca^{2+}]$ = concentration of calcium in eq/L (water).
$[Mg^{2+}]$ = concentration of magnesium in eq/L (water).

What SAR Values Are Acceptable?

Acceptable SAR values for irrigation water depend on the particular water and soil characteristics and are often considered in conjunction with the specific conductivity of irrigation water. Since high conductivity means that the water contains high levels of dissolved solids, it generally will contain high levels of calcium and magnesium. In any case, it will contain many other cations that compete with sodium for soil adsorption sites. Thus, high conductivity limits the deleterious effects of sodium, as shown in Figure 6.11. It should be noted that water with specific conductivity greater than about 3000 μS/cm is of poor quality for irrigation regardless of the SAR value. When plant roots encounter water with a concentration of TDS that is too high, the osmotic pressure balance across the outer root membrane drives water from within the plant out into the soil, desiccating the plant.

Rules of Thumb

1. Water having SAR values less than 3 will not diminish soil permeability when specific conductivity is greater than about 200 μS/cm.
2. Water having SAR values between 3 and 6 and specific conductivity between 500 and 2000 μS/cm may require care with irrigation methods. Soil permeability may be diminished.
3. Waters having SAR values greater than 10 and specific conductivity less than 500 μS/cm will require considerable care with irrigation methods. Soil permeability may be too low to allow sufficient water to reach plant roots.

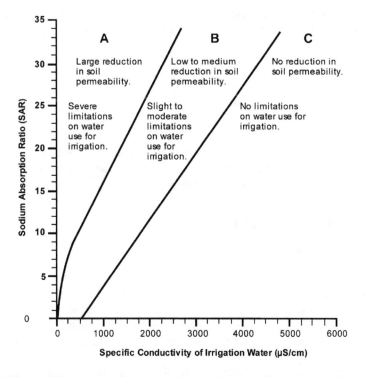

FIGURE 6.11 Effects of SAR and specific conductivity on soil permeability. Zone A exhibits large reductions in soil permeability to water and will severely restrict the amount of water available to plant roots. Zone B has low to moderate reductions in soil permeability and may require care in irrigation practices and choice of crops. Zone C water has little or no reduction in soil permeability.

Example 6.8: Determination of Sodium Adsorption Ratio

Determine the SAR for two different water supplies, A and B. Measured concentrations were

Supply A:	Ca^{2+} = 4.6 mg/L	Mg^{2+} = 1.1 mg/L	Na^+ = 144 mg/L	TDS = 255mg/L
Supply B:	Ca^{2+} = 108 mg/L	Mg^{2+} = 33 mg/L	Na^+ = 63 mg/L	TDS = 1050 mg/L

Solution:

1. Determine the Ca^{2+}, Mg^{2+}, and Na^+ concentrations in meq/L. (See discussion of equivalent weight in Chapter 1.)

		Concentration			
Constituent	Equivalent Weight (g/eq or mg/meq)	Water Supply A		Water Supply B	
		(mg/L)	(meq/L)	(mg/L)	(meq/L)
Ca^{2+}	20.0	4.6	0.23	108	5.40
Mg^{2+}	12.15	1.1	0.090	33	2.72
Na^+	23.0	255	11.1	63	2.74

2. Compute the SAR of the two waters.

a. For Supply A: $SAR = \dfrac{\left[Na^+\right]}{\sqrt{\dfrac{\left[Ca^{2+}\right]+\left[Mg^{2+}\right]}{2}}} = \dfrac{11.1}{\sqrt{\dfrac{0.23+0.090}{2}}} = 27.8.$

This is a very high SAR value, and the water would be unacceptable for irrigation. Note that the TDS value for water supply A is moderate (255 mg/L) and that Supply A water is excellent for domestic use.

b. For Supply B: $SAR = \dfrac{2.74}{\sqrt{\dfrac{5.40 + 2.72}{2}}} = 1.36.$

Water supply B has a low SAR value and would be an excellent source of irrigation water. The TDS value is high (1050 mg/L), and water supply B is an inferior source of domestic water compared with Supply A. Supply B will require more treatment to produce acceptable finished drinking water.

6.12   OIL AND GREASE (O&G)

OIL AND GREASE ANALYSIS

When you analyze for oil and grease, you are actually measuring a group of substances that have similar solubility characteristics in a designated solvent. "Oil and grease" is defined as any substance that is not volatilized during the analysis and that is recovered from an acidified sample by extracting it into a designated solvent. The extraction process is called liquid-liquid-extraction or LLE. After the sample is extracted, the extract may be measured either gravimetrically (drying and weighing) or by infrared spectroscopy.

The solvents that are used have the ability to dissolve not only oil and grease but also other organic substances. Some nonoil and grease materials commonly included in the determination of "oil and grease" are certain sulfur compounds, chlorophyll, and some organic dyes. There is no known solvent that will dissolve selectively only oil and grease. Some heavier residuals of petroleum (coal tar, used motor oil, etc.) may contain significant amounts of material that do not extract into the solvent. The method is entirely empirical and duplicate results with a high degree of precision can be obtained only by strict adherence to all details of the analytical procedure.

The designated solvent changed from petroleum ether and normal-hexane in 1965 to Freon-113 (trichlorotrifluoroethane) in 1976. While Freon-113 was still the prescribed solvent at present, an alternative solvent (80% n-hexane and 20% MTBE) was developed for gravimetric methods, to help phase out the use of freons. Then, on May 14, 1999, the EPA approved the new Method 1664 for analyzing oil and grease, which must replace all other methods. It differs mainly in details of the procedure and in the QA/QC requirements. Method 1664 uses pure n-hexane with either liquid-liquid-extraction (LLE) or solid phase extraction (SPE). Because n-hexane is less dense than water, LLE will not work well with all sample matrices.

Silica Gel Treatment

The "total oil and grease" designation includes oils and fats of biological origin (animal and vegetable fats and oils) as well as mineral oils (petroleum products). Silica gel has the ability to adsorb certain organic compounds known as "polar compounds." Since petroleum hydrocarbons are mostly nonpolar, and most hydrocarbons of biological origin are polar, silica gel is used to separate these different types of hydrocarbons. If a solution of nonpolar and polar hydrocarbons is mixed with silica gel, the polar hydrocarbons, such as fatty acids, are removed selectively from the solution. The materials *not* eliminated by silica gel adsorption are designated as "petroleum hydrocarbons" by this test. Normally, total oil and grease is measured first, and then petroleum hydrocarbons are determined after the animal and vegetable fats are removed by silica gel adsorption. The difference between "total oil and grease" and "petroleum hydrocarbons" is designated as biologically derived hydrocarbons.

REFERENCES

1. Aronheim, J.S., Virus Transport in Groundwater: Modeling of Bacteriophage PRD1 Transport Through One-Dimensional Columns and a Two-Dimensional Aquifer Tank, Master's thesis, Department of Civil, Environmental, and Architectural Engineering, University of Colorado, Boulder, 1992.
2. Asano, T., Leong L.Y.C., Rigby M.G., and Sakaji R.H., Evaluation of the California wastewater reclamation criteria using enteric virus monitoring data, *Water Sci. Technol.*, 26(7–8), pp. 1513–1524, 1992.
3. Bellar, T.A., Lichtenberg J.J., and Kroner R.C., The occurrence of organohalides in chlorinated drinking water, *J. AWWA*, 66(12), 703, December, 1974.
4. Bitton, G., Adsorption of viruses onto surfaces in soil and water, *Water Research*, 9, 473, 1975.
5. Bitton, G., Davidson J.M., and Farra S.R., On the value of soil columns for assessing the transport pattern of viruses through soils: a critical outlook, *Water, Air, and Soil Pollution*, 12, pp. 449–457, 1979.
6. Bitton, G., *Introduction to Environmental Virology,* John Wiley & Sons, New York, 1980.
7. Bull, R.J. and Kopfler F.C., Health effects of disinfectants and disinfection by-products, 90577, AWWA Research Foundation, Denver, 1991.
8. Burge, W.D. and Enkiri N.K., Adsorption kinetics of bacteriophage $\phi\chi$-174 on soil, *J. Environ. Qual.*, 7(4), pp. 536–541.
9. Campbell, D.H. and Stednick J.D., Transport of road derived sediment as a function of slope characteristics and time, Water Quality Laboratory, Department of Earth Resources, Colorado State University, submitted to USDA Forest Service, Contract 28-C2-214, Final Report, December, 1983.
10. *Federal Register*, 44:231:68624, November 29, 1979.
11. Foster, D.M., Emerson M.A., Buck C.F., Walsh D.S., and Sproul O.J., Ozone inactivation of cell- and fecal-associated viruses and bacteria, *J. Water Pollut. Control Fed.*, 52, 2174, 1980.
12. Gerba, C.P., Fate of wastewater bacteria and viruses in soil, *J. Irrig. Drain. Div.*, 101(IR3), pp. 157–173, 1975.
13. Hurst, C.J., Gerba C.P., and Cech I., Effects of environmental variables and soil characteristics on virus survival in soil, *Appl. Environ. Microbiol.*, 40(6), pp. 1067–1079,
14. Hurst, C.J., Fate of viruses during wastewater sludge treatment processes, *CRC Crit. Rev. Environ. Control*, 18(4), pp. 317–343, 1989.
15. Keith, L.H., *Environmental Sampling and Analysis: A Practical Guide,* Lewis Publishers, Inc., Ann Arbor, MI, 1992.
16. Lefler, E. and Kott Y., Enteric virus behavior in sand dunes, *Israel J. Technol.*, 12, pp. 298–304, 1974.
17. Lo, S.H. and Sproul O.J., Poliovirus adsorption from water onto silicate minerals, *Water Research*, 11, pp. 653–658, 1977.
18. McKee, J.E. and Wolf H.W., *Water Quality Criteria, State of California,* 2nd ed., Publication No. 3-A, The Resources Agency of California, State Water Quality Control Board, 1963.
19. Michigan Department of Transportation, *The Use of Selected De-icing Materials on Michigan Roads: Environmental and Economic Impacts,* prepared by Public Sector Consultants, Inc. for Michigan Department of Transportation, December, 1993.
20. Nien-Yin C., et al., *Final Report on Environmentally Sensitive Sanding and De-icing Practices,* ESSD Research Group, Department of Civil Engineering, University of Colorado at Denver, Transportation Institute and Colorado Department of Transportation CDOT-CTI-95-5, November 25, 1994.
21. National Research Council, *Groundwater Recharge: Using Waters of Impaired Quality,* National Research Council, Washington, D.C., 1994.
22. Pinholster, G., *Drinking Recycled Wastewater,* Environ. Sci. Technol., 29(4), pp. 174A–179A, 1995.
23. Rocky Mountain Construction, *Anti-icing: A Bold New Strategy,* Strategic Highway Research Program, p. S-14, September 30, 1995.
24. Rook, J., Formation of haloforms during chlorination of natural waters, *J. Soc. Water Treat. Exam.*, 23, 234, 1974.
25. Rose, J.B. and Gerba C.P., Assessing potential health risks from viruses and parasites in reclaimed water in Arizona and Florida, USA, *Water Sci. and Technol.*, 23, pp. 2091–2098, 1991.
26. Scheurman, P.R., Bitton G., Overman A.R., and Gifford G.E., *Transport of viruses through organic soils and sediments, J. Irrig. Drain. Div.*, Amer. Soc. of Civil Eng., 105, 1979.
27. Sobek, A.A., Schuller W.A., Freeman J.R., and Smith R.M., Field and Laboratory Methods Applicable to Overburdens and Minesoils, EPA-600/2-78-054, 1978.

28. Sobsey, M.D. and Hickey A.R., Effects of humic and fulvic acids on poliovirus concentration from water by microporous filtration, *Appl. Environ. Microbiol.*, 49(2), pp. 259–264, 1985.

29. Sobsey, M.D., Dean C.H., Knuckles M.E., and Wagner R.A., Interactions and adsorption of enteric viruses in soil materials, *Appl. Environ. Microbiol.*, 40, 92, 1980.

30. Eaton, A.D., Clesceri L.S., and Greenberg A.E., Eds., *Standard Methods for the Examination of Water and Wastewater*, 19th ed., American Public Health Association, American Water Works Association, and Water Environment Federation, 1995.

31. Stotsky, G., Schiffenbauer M., Lipson S.M., and Yu B.H., Surface interactions between viruses and clay minerals and microbes: mechanisms and implications, paper presented at the Int. Symp. Viruses and Wastewater Treatment, University of Surrey, Guilford, U.K., 1980.

32. U.S. EPA, *Process Design Manual for Land Treatment of Municipal Wastewater,* EPA 625/1-81-013, October, 1981.

33. U.S. EPA, *Bacteriological Ambient Water Quality Criteria for Marine and Fresh Recreational Waters,* EPA 440/5-84-002, NTIS PB-86-158-045, 1986.

34. U.S. EPA, *Manual: Guidelines for Water Reuse,* EPA 625/R-92-004, September, 1992.

35. U.S. EPA, *Test Methods for Evaluating Solid Waste,* Vol. 1A, Laboratory Manual Physical/Chemical Methods, SW 846, 3rd ed., July, 1992.

36. U.S. EPA, *Drinking Water Standard Setting: Question and Answer Primer,* EPA 811-K-94-001, November, 1994.

37. U.S. EPA, *Drinking Water Regulations and Health Advisories,* EPA 822/B-96-002, October, 1996.

38. U.S. EPA, *Small Systems Compliance Technology List for the Surface Water Treatment Rule*, EPA 815/R-97-002, August, 1997.

39. U.S. EPA, *Alternative Disinfectants and Oxidants Guidance Manual*, EPA 815/R-99-014, April, 1999.

40. Wallis, C., Henderson M., and Melnick J.L., 1972, Enterovirus concentration on cellulose membranes, *Appl. Microbiol.*, 23(3), pp. 476–480, 1972.

41. Watershed Research, *Rating De-icing Agents — Road Salt Stands Firm,* Technical Note 55, Vol. 1, No. 4, Summer 1995.

7 A Dictionary of Inorganic Water Quality Parameters and Pollutants*

CONTENTS

---

* For EPA drinking water standards, see Appendix A. For EPA aquatic life and human health criteria, see Appendix B. Information in Chapter 7 has been compiled from many different sources but particularly from the EPA World Wide Web pages.

7.1   INTRODUCTION

This section is a concise guide to useful information about frequently measured inorganic water quality parameters and pollutants, arranged alphabetically. Most of the parameters can have both natural and human origins. Several are more extensively described in Chapter 3. The information is focused chiefly on human health concerns and does not address effects on aquatic life. The EPA gives information about aquatic life effects in their *Gold Book*, EPA 440/5-86-001, *Quality Criteria for Water*. Appendix B in this book tabulates aquatic life water quality criteria.

Where CAS identification numbers have been assigned, they are included for each entry. CAS stands for Chemical Abstracts Service Registry, a division of the American Chemical Society that assigns unique identification numbers to each chemical compound and uses these numbers to facilitate literature and computer database searches for chemical information.

Although water quality components are listed alphabetically in the dictionary section, it is sometimes useful to classify them according to their typical abundance in natural waters. This is done in the listing below.

Water Quality Constituents: Classified by Abundance

Major chemical constituents — Those most often present in natural waters in concentrations greater than 1.0 mg/L. These are the cations calcium, magnesium, potassium, and sodium, and the anions bicarbonate/carbonate, chloride, fluoride, nitrate, and sulfate. Silicon is usually present as nonionic species and is reported by analytical laboratories as the equivalent concentration of silica (SiO_2). Several additional chemical parameters are sometimes included with the major constituents because of their importance in determining water quality and because some of them sometimes attain concentrations comparable to the parameters above. These are aluminum, boron, iron, manganese, nitrogen in forms other than nitrate (such as ammonia and nitrite), organic carbon, phosphate, and the dissolved gases oxygen, carbon dioxide, and hydrogen sulfide.

Minor chemical constituents — Those most often present in natural waters in concentrations less than 1.0 mg/L. These include the so-called trace elements and naturally occurring radioisotopes: antimony, arsenic, barium, beryllium, bromide, cadmium, cesium, chromium, cobalt, copper, iodide, lead, lithium, mercury, molybdenum, nickel, radium, radon, rubidium, selenium, silver, strontium, thorium, titanium, uranium, vanadium, and zinc.

Physical and chemical properties — Quantities that do not identify particular chemical species but are used as indicators of how water quality may affect water uses. These are acidity, alkalinity, hardness, hydrogen ion (measured as pH), biochemical oxygen demand (BOD), chemical oxygen demand (COD), color, corrosivity, gross alpha and beta emitters, odor, sodium adsorption ratio (SAR), Langelier Index, specific conductance (conductivity), specific gravity, temperature, total dissolved solids (TDS), total suspended solids (TSS), and turbidity.

7.2   ALPHABETICAL LISTING OF INORGANIC AND PHYSICAL WATER QUALITY PARAMETERS AND POLLUTANTS

Aluminum (Al), CAS # 7429-90-5

Background

Aluminum is the third most abundant element in the Earth's lithosphere (after oxygen and silicon) and its compounds are often found in natural waters. Aluminum is mobilized naturally in the environment by the weathering of rocks and minerals, particularly bauxite clays. It is a normal constituent of all soils and is found in low concentrations in all plant and animal tissues. Most naturally occurring aluminum compounds are of very low solubility between pH 6 to 9. Therefore, dissolved forms rarely occur in natural waters in concentrations greater than about 0.01 mg/L.

Concentrations in water that are greater than this usually indicate the presence of solid forms of aluminum, such as suspended solids and colloids.

The concentration of Al^{3+} in water is controlled by the solubility of aluminum hydroxide, $Al(OH)_3$, which increases by a factor of about 10^3 for every unit decrease in pH. Thus, the concentration of dissolved aluminum, Al^{3+}, is about 3×10^{-5} mg/L at pH 6, 0.03 mg/L at pH 5, and 30 mg/L at pH 4.

Also, at lower water pH (<pH 5) in the presence of clays and organic-rich soils, dissolved aluminum concentrations increase because of the release of Al^{3+} from the soil. Low pH means high H^+ concentrations. At pH < 5, the concentration of H^+ is high enough for H^+ to partially ion-exchange with other, more strongly bound, metals at ion-exchange sites on soil particles. Since Al^{3+} is bound more strongly than divalent and monovalent cations, it is among the last cations to be displaced by H^+ and requires a continued low pH to reach elevated dissolved levels.

Rule of Thumb

The presence of an elevated concentration of Al^{3+}, often exceeding the concentrations of Ca^{2+} and Mg^{2+}, is a common characteristic of acidic waters (pH < 5), including mine drainage and waters affected by acid rain.

Health Concerns

Naturally occurring aluminum has a very low toxicity to humans and animals. Only a few industrially important aluminum compounds, such as the fumigant aluminum phosphide, are considered acutely hazardous. Exposure to and ingestion of aluminum and its compounds is usually not harmful. Aluminum compounds are used in water treatment to remove color and turbidity, food packaging, medicines, soaps, dental cements, and drugstore items, such as antacids and antiperspirants, and are present in many foods because they are grown in soils containing aluminum.

Daily exposure to aluminum is inevitable due to its ubiquitous occurrence in nature and its many commercial uses. Estimated human consumption is about 88 mg per person of aluminum per day, mostly from food. Consumption of 2 L of water per day containing 1.5 mg/L of aluminum (well above the secondary drinking water standard, see below) only contributes 3.0 mg of aluminum per day, or less than 4% of the normal daily intake. Aluminum is not believed to be an essential nutrient. There is no human or animal evidence of carcinogenicity. There is some indication, however, that ingestion of large doses of aluminum in medicines may cause skeletal problems.

Drinking Water Standards

Maximum contaminant level goal: None.
Maximum contaminant level: None.
Secondary standard: 0.05 to 0.2 mg/L.

The EPA has no primary drinking water standard for aluminum. The EPA secondary drinking water standard (nonenforceable) is expressed as a range: 0.05 to 0.2 mg/L. The EPA recommends that 0.05 mg/L be met where possible but allows states to determine the required level on a case-by-case basis because water treatment technologies often use aluminum salts to remove color and turbidity and cannot always achieve the lower value. The EPA recommends that aluminum in drinking water does not exceed 0.2 mg/L because of taste and odor problems. In the presence of microorganisms, aluminum can react with iron, manganese, silica and organic material to form fine sediments that can appear at the consumer's tap. If dissolved aluminum exceeds 0.1 mg/L, levels of iron that are normally acceptable may produce discoloration and staining. No lifetime health advisory has been established. This is the concentration in drinking water that is not expected to cause any adverse effects over 70 years of exposure — with a margin of safety.

Other Comments

Because of its usually low concentrations, aluminum is normally of no concern in irrigation waters. However, a limit of 5.0 mg/L is recommended [National Academy of Sciences, 1982, *Drinking Water and Health*, Vol. 4, National Academy Press, Washington, D.C., p. 299] where irrigation waters are used regularly. Swimming pools treated with commercial grade aluminum sulfate compounds, known as alum, may cause eye irritation at concentrations greater than 0.1 mg/L.

AMMONIA/AMMONIUM ION (NH₃/NH₄⁺), CAS # 7664-41-7

Background (see Chapter 3 for a more detailed discussion).

In the biological decay of nitrogenous organic compounds, ammonia is the first nitrogenous product that does not contain carbon. It arises from the waste products and decay after the death of plants, animals, and other life forms. Thus, ammonia is present naturally in surface and groundwaters. It also is discharged in many waste streams, particularly from municipal waste treatment. Unpolluted waters have very low ammonia concentrations, generally less than 0.2 mg/L as N.

Continued oxidation of ammonia leads sequentially to nitrite and nitrate. Ammonia gas is very soluble in water, where it reacts as a base raising the pH and forming an ammonium cation and a hydroxyl anion

$$NH_3 + H_2O \leftrightarrow NH_4^+ + OH^-. \tag{3.16}$$

The unionized form (NH_3) is of greatest environmental concern because of its greater toxicity to aquatic life. However, because NH_4^+ readily converts to NH_3 by its pH dependent equilibrium (Equation 3.16), total ammonia is normally regulated in discharges. Concentrations of unionized (NH_3) or total ($NH_3 + NH_4^+$) ammonia are often reported in terms of the nitrogen content only, e.g., NH_3 = 10 mg/L-N, or $NH_3–N$ = 10 mg/L. This means that the sample contains unionized ammonia and the nitrogen portion of the unionized ammonia weighs 10 mg/L of sample. The weight of the hydrogen in the ammonia molecules is ignored. To convert mg/L of NH_3 to mg/L of NH_3-N, multiply by 0.822.

The equilibrium between the unionized form (NH_3) and the ionized form (NH_4^+) depends on pH, temperature, and, to a much lesser degree, on ionic strength (salinity or concentration of total dissolved solids; see Chapter 3).

- At 15°C and pH > 9.6, the fraction of NH_3 is greater than 0.5.
- At 15°C and pH < 9.6, the fraction of NH_4^+ is greater than 0.5.
- A temperature increase shifts the equilibrium of Equation 3.16 to the left, increasing the NH_3 concentration.
- A temperature decrease shifts the equilibrium of Equation 3.16 to the right, increasing the NH_4^+ concentration.
- An increase in ionic strength shifts the equilibrium of Equation 3.16 to the right, increasing the NH_4^+ concentration slightly. In waters with very high total dissolved solids (>10,000 mg/L), there will be a small but measurable decrease in the percentage of NH_3.
- Because pH and temperature can vary considerably along a stream or within a lake, the fraction of total ammonia that is unionized is also variable at different locations. Therefore, the amount of total ammonia is usually of regulatory concern, rather than only the unionized form.

Health Concerns

Total ammonia ($NH_3 + NH_4^+$) in drinking water is more of an esthetic than a health concern. The odor and taste of ammonia makes drinking water unpalatable at concentrations well below the appearance of any toxic effects to humans. The main health concern with ammonia is its potential oxidation to nitrite (NO_2^-) and nitrate (NO_3^-). Ingested nitrate and nitrite react with iron in blood hemoglobin to cause a blood oxygen deficiency disease called "methemoglobinemia," which is especially dangerous in infants (blue baby syndrome) because of their small total blood volume. There is no human or animal evidence of carcinogenicity.

Drinking Water Standards

The EPA has no primary or secondary drinking water standards for ammonia. However, the presence of NH_3 above 0.1 mg/L may raise the suspicion of recent pollution. The lifetime health advisory says that the concentration in drinking water that is not expected to cause any adverse effects over 70 years of exposure, with a margin of safety, is 30 mg/L.

   Some states have adopted ammonia limits for water that will receive treatment to produce drinking water. For example, Colorado's ammonia standard for water classified as domestic water supply is 0.05 mg/L–N, 30-day average, for total ammonia ($NH_3 + NH_4^+$).

Rules of Thumb

1. Only NH_3, the unionized form, has significant toxicity for aquatic life.
2. To convert mg/L of unionized or total ammonia to mg/L as nitrogen, multiply by 0.822. For example

$$17.4 \text{ mg/L } NH_3 = 0.822 \times 17.4 = 14.3 \text{ mg/L-N.}$$

3. Since pH 9.6 is higher than the pH of most natural waters, ammonia-nitrogen in natural waters usually is mostly in the less toxic ionized ammonium form (NH_4^+).
4. In high pH waters (pH > 9), the NH_3 fraction can reach levels toxic to aquatic life.
5. The ionized form is not volatile and cannot be removed by air stripping. The unionized form, NH_3, is volatile and can be removed by air stripping.

ANTIMONY (SB), CAS # 1440-36-0

Background

Antimony is a metalloid (having properties intermediate between metals and nonmetals) in the same chemical group (Group 5A) as arsenic, with which it has some chemical similarities, including toxicity. However, it is only about one tenth as abundant in the earth's crust and soils. The symbol Sb for the element is from stibium, the Latin name for antimony. In the environmental literature, antimony is often included with the metals because it is usually analyzed, along with other metals, by inductively coupled plasma (ICP) or atomic absorption (AA) techniques. Common sources of antimony in drinking water are discharges from petroleum refineries, fire retardants, ceramics, electronics, and solder. It is also found in batteries, pigments, ceramics, and glass.

   Antimony is usually adsorbed strongly to iron, manganese, and aluminum compounds in soils and sediments. Soil concentrations normally range between 1 mg/L and 9 mg/L. The amount commonly dissolved in rivers is small, less than 0.005 mg/L. There is no evidence of bioconcentration of most antimony compounds,

Health Concerns

Antimony is used in medicines for treating parasite infections. It is present in meats, vegetables, and seafood in an average concentration of about 0.2 ppb (μg/L) to 1.1 ppb. An average person ingests about 5 μg of antimony every day in food and drink. Short-term exposures above the MCL may cause nausea, vomiting, and diarrhea. Potential health effects from long-term exposure above the MCL are an increase in blood cholesterol and a decrease in blood glucose. There is insufficient evidence to state whether antimony has the potential to cause cancer.

Drinking Water Standards

Maximum contaminant level goal: 0.006 mg/L.
Maximum contaminant level: 0.006 mg/L.

Other Comments

Treatment/best available technologies: Coagulation and filtration, reverse osmosis.

ARSENIC (As), CAS # 7440-38-2

Background

Chemically, arsenic is classified as a *metalloid*, having properties intermediate between metals and nonmetals. In the environmental literature, it is often included with the metals because it is usually analyzed, along with other metals, by inductively coupled plasma (ICP) or atomic absorption (AA) techniques. Inorganic arsenic occurs naturally in many minerals, especially in ores of copper and lead. Smelting of these ores introduces arsenic to the atmosphere as dust particles.

In minerals, arsenic is combined mostly with oxygen, chlorine, and sulfur. Inorganic arsenic compounds are used mainly as wood preservatives, insecticides, and herbicides. Organic forms of arsenic found in plants and animals are combined with carbon and hydrogen. Organic arsenic is generally less toxic than inorganic arsenic. Arsenic is not abundant, with an average concentration in the lithosphere of about 1.5 mg/kg (ppm). Background levels in soils typically range from 1 to 95 mg/kg. Average levels in U.S. soils are around 5–7 mg/kg. It is widely distributed and is found naturally in many foods at levels of 20–140 ppb, subjecting most Americans to a constant low exposure, perhaps around 50 μg per day. Normal human blood contains 0.2–1.0 mg/L of arsenic; however, there is no evidence that arsenic is an essential nutrient.

Many arsenic compounds are water soluble and may be found in groundwater, especially in the western U.S. The average concentration for U.S. surface water is around 3 ppb. Groundwater levels average about 1–2 ppb, except in some western states where groundwater is in contact with volcanic rock and sulfide minerals high in arsenic. In western mining regions, arsenic levels as high as 48,000 ppb have been observed. Many people who are dependent on well water in the West ingest higher than average levels of inorganic arsenic through their drinking water supplies.

Health Concerns

High levels (>60 ppm) of arsenic in food or water can be fatal. Arsenic damages tissues in the nervous system, stomach, intestine, and skin. Breathing high levels can irritate lungs and throat. Lower levels can cause nausea, diarrhea, irregular heartbeat, blood vessel damage, reduction of red and white blood cells, and tingling sensations in hands and feet. Long term exposure to inorganic arsenic may cause darkening of the skin and the appearance of small warts on the palms, soles, and torso.

Inorganic arsenic was recognized as a possible carcinogen as early as 1879, when it was suggested that high rates of lung cancer in German miners might have been caused by inhaled arsenic. Arsenic is currently considered a carcinogen. Breathing inorganic arsenic increases the risk of lung cancer, and ingesting inorganic arsenic increases the risk of skin cancer and tumors of the bladder, kidney, liver, and lung.

A crisis of well-water contamination by arsenic was discovered in Bangladesh in 1992. The crisis was created through a well-intended effort by the United Nations Children's Fund (UNICEF) to provide Bangladesh with reliable water sources that are free of cholera and dysentery organisms. Millions of water wells were installed, and the water was tested for microbial contaminants but not for arsenic and other toxic metals. It is now estimated that 85% of Bangladesh's geographical area contains wells contaminated with inorganic arsenic. Tens of thousands of people now exhibit signs of arsenic poisoning. The World Bank, United Nations, and other sources have begun a multimillion-dollar-effort named the Bangladesh Arsenic Mitigation Water Supply Project to supply uncontaminated water to Bangladesh's 85,000 villages.

Drinking Water Standards

Maximum contaminant level goal: none.
Maximum contaminant level: 0.05 mg/L.

Other Comments

Treatment/best available technologies: Iron coprecipitation, activated alumina or carbon sorption, ion-exchange, reverse osmosis.

   A maximum concentration of 0.1 mg/L is recommended for irrigation water and for protection of aquatic plants.

Asbestos, CAS # 1332-21-4

Background

Asbestos is a generic term for different naturally formed fibrous silicate minerals that are classified into two groups, serpentine and amphibole, based on structure. Six minerals have been characterized as asbestos: chrysotile, crosidolite, anthophyllite, tremolite, actinolite, and andamosite. The most common form is chrysotile, which is a member of the serpentine group. The others belong to the amphibole group. These different forms of asbestos are composed of 40–60% silica, the remainder being oxides of iron, magnesium, and other metals. The EPA banned most uses of asbestos in the U.S. on July 12, 1989 because of potential adverse health effects in exposed persons.

   Although asbestos may be introduced into the environment by the dissolution of asbestos-containing minerals and from industrial effluents, the primary source is through the wear or breakdown of asbestos-containing materials. Because asbestos fibers are resistant to heat and most chemicals, they have been mined for use in over 3000 different products in the U.S., such as roofing materials, brake linings, asbestos-reinforced pipe, packing seals, gaskets, fire-resistant textiles, and floor tiles. The remaining currently allowed uses of asbestos include battery separators, sealant tape, asbestos thread, packing materials, and certain industrial uses of gaskets.

   Typical background levels in lakes and streams range from 1 to 10 million fibers/L. Asbestos is insoluble, nonvolatile, and nonbiodegradable and does not tend to adsorb to stream sediments. Asbestos fibers do not chemically decompose to other compounds in the environment and, therefore, can remain in the environment for decades or longer. Small asbestos fibers and fiber-containing particles may be carried for long distances by water currents before settling out. Larger fibers and particles tend to settle more quickly. Asbestos fibers do not pass through soils to groundwater.

   There are no data regarding the bioaccumulation of asbestos in aquatic organisms, but asbestos is not expected to bioaccumulate. Ordinary sand filtration removes about 90% of the fibers.

Health Concerns

There are no reliable data available on the acute toxic effects from short-term exposures to asbestos. Long-term inhalation has the potential to cause cancer of the lung and other internal organs. Long-term ingestion above the MCL increases the risk of developing benign intestinal polyps.

Drinking Water Standards

Maximum contaminant level goal: 7 million fibers per liter (MFL) for fibers > 10 microns in length.
Maximum contaminant level: 7 million fibers per liter (MFL)

Other Comments

Treatment/best available technologies: Coagulation and filtration, direct and diatomite filtration, corrosion control.

BARIUM (BA), CAS # 7440-39-3

Background

Barium is the sixth most abundant element in the lithosphere, averaging about 500 mg/kg. It exists mainly as the sulfate ($BaSO_4$, barite) and, to a lesser extent, the carbonate ($BaCO_3$, witherite). Traces of barium are found in most soils, natural waters, and foods. Background levels for soils range between 100 and 3000 mg/kg, with an average of 500 mg/kg. Although most groundwaters contain only a trace of barium, some geothermal groundwaters may contain as much as 10 mg/L.

Barium is released to water and soil in the disposal of drilling wastes, from copper smelting, and industrial waste streams. It is not very mobile in most soil systems. In water, the more toxic soluble salts are likely to precipitate as the less toxic insoluble sulfate and carbonate compounds. Background levels for soil range from 100 to 3000 ppm. Barium occurs naturally in almost all surface waters examined in concentrations of 2–340 µg/L, with an average of 43 µg/L. In surface water and most groundwater, only traces of the element are present. However, some wells may contain barium levels 10 times higher than the drinking water standard. Marine animals concentrate the element 7–100 times, and marine plants concentrate it 1000 times from seawater. Soybeans and tomatoes also accumulate soil barium 2–20 times.

Health Concerns

There is no evidence that barium is an essential nutrient. All soluble barium salts are considered toxic. Short-term exposure at levels above the MCL may cause gastrointestinal disturbances, muscular weakness, and liver, kidney, heart, and spleen damage. Long-term exposure above the MCL may cause hypertension. There is no evidence that barium can cause cancer. No health advisories have been established for short-term exposures.

Drinking Water Standards

Maximum contaminant level goal: 2 mg/L.
Maximum contaminant level: 2 mg/L.

Other Comments

Treatment/best available technologies: Ion-exchange, reverse osmosis, lime softening, electrodialysis.

BERYLLIUM (BE), CAS # 7440-41-7

Background

Beryllium is a metal found in natural deposits as ores containing other elements and in some precious stones such as emeralds and aquamarine. Beryllium is not likely to be found in natural waters above trace levels due to the insolubility of oxides and hydroxides at normal environmental pHs. It has been reported to occur in U.S. drinking water at 0.01 to 0.7 µg/L.

A major use of beryllium is as an alloy hardener. Its greatest use is in making metal alloys for nuclear reactors and the aerospace industry. It is also used as an alloy and oxide in electrical equipment and electronic components in military vehicle armor. The chloride is used as a catalyst and intermediate in chemical manufacture. The oxide is used in glass and ceramic manufacture. Beryllium enters the environment principally as dust from burning coal and oil and from the slag and ash dumps of coal combustion. Some tobacco leaves contain significant levels of beryllium, which can enter the lungs of those exposed to tobacco smoke. It is also found in discharges from other industrial and municipal operations. Rocket exhausts contain oxide, fluoride, and chloride compounds of beryllium.

Very little is known about what happens to beryllium compounds when released to the environment. Beryllium compounds of very low water solubility appear to predominate in soils. Leaching and transport through soils to groundwater is unlikely to be of concern. Erosion or runoff of beryllium compounds into surface waters is not likely, and it appears unlikely to leach to groundwater when released to land. Erosion and bulk transport of soil may carry beryllium sorbed to soils into surface waters, but most likely in particulate rather than dissolved form.

Health Effects

Beryllium is more toxic when inhaled as fine particles than when ingested orally. Short-term air exposure can cause inflammation (chemical pneumonitis) of the lungs when inhaled. Some people develop a sensitivity, or allergy, to inhaled beryllium, leading to chronic beryllium disease. Long-term ingestion in water above the MCL may lead to intestinal lesions. There is some evidence that beryllium may cause cancer from lifetime exposures at levels above the MCL.

Drinking Water Standards

Maximum contaminant level goal: 0.004 mg/L.
Maximum contaminant level: 0.004 mg/L.

Other Comments

Treatment/best available technologies: Activated alumina, coagulation and filtration, ion-exchange, lime softening, reverse osmosis.

BORON (B), CAS # 7440-42-8

Background

Boron is usually found in nature as the hydrated sodium borate salt kernite ($Na_2B_4O_7 \cdot 4H_2O$) or the calcium borate salt colemanite ($Ca_2B_6O_{11} \cdot 5H_2O$). Most environmentally important boron compounds are highly water-soluble. Natural weathering of boron-containing minerals is a major source of boron in certain geographical locations. In the U.S., the minerals richest in boron are found in the Mojave Desert region of California, where concentrations above 300 mg/L have been observed in boron-rich lakes. In other U.S. surface waters, an average boron level is around 100 µg/L, but concentrations vary widely (from around 0.02 to 0.3 mg/L), depending on local geologic and industrial conditions. Background soil levels in the U.S. range up to 300 mg/kg, with an average of around 26 mg/kg.

Sodium tetraborate (kernite) is also known as borax and finds use as an additive in detergents and other cleaning agents. A major use for boron is the manufacture of borosilicate glass which, because of its low coefficient of thermal expansion, is used in ovenware, laboratory glassware, piping, and sealed-beam headlights. Boric acid (H_3BO_3) is used as a weak antiseptic and eye-wash and as a "natural" insecticide. Other uses for boron compounds include fire retardants, leather tanning, pulp and paper whitening agents, and high-energy rocket fuels. Elemental boron is used for neutron absorption in nuclear reactors and in alloys with copper, aluminum, and steel. For these reasons, boron is common in sewage and industrial wastes. Effluent from municipal sewage treatment plants may contain up to 7 mg/L of boron, with an average of 1 mg/L in California.

Boron is essential to plant growth in very small amounts but may become toxic at higher amounts. For boron-sensitive plants, the toxic level may be as low as 1 mg/L. A maximum level of 0.75 mg/L in soil and irrigation water is generally accepted as protective for sensitive plants under long-term irrigation.

Boron is not known to be an essential nutrient for animals or humans. Boron mobility in water is greatest at pH < 7.5. Adsorption to soils and sediments is the main mechanism for removal from environmental waters. Sorption to oxide and hydroxide solids, particularly aluminum species, is enhanced above pH 7.5 and in the presence of Ca and Mg. There is no evidence that boron is bioconcentrated significantly by aquatic organisms, and naturally occurring levels of boron do not appear to have an adverse effect on aquatic life. It is sometimes suggested that boron concentrations in discharges to fresh waters be limited to 10 mg/L.

Health Concerns

Moderately high doses of boron compounds appear to have little detrimental health effects. The lethal dose of boric acid for adults varies from 15 to 20 g. Chronic ingestion may cause dry skin, skin eruptions, and gastric disturbances.

Drinking Water Standards

In general, boron in drinking water is not regarded as hazardous to human health, and there are no drinking primary or secondary drinking water standards.

Other Comments

Treatment/best available technologies: Because most boron compounds are highly water soluble, boron is not significantly removed by conventional wastewater treatment. Boron may be co-precipitated with aluminum, silicon, or iron solids.

CADMIUM (CD), CAS # 7440-43-9

Background

Cadmium is usually present in all soils and rocks. It occurs naturally in zinc, lead, and copper ores, in coal, and other fossil fuels and shales. It often is released during volcanic action. These deposits can serve as sources to groundwaters and surface waters, especially when they are in contact with soft, acidic waters. The adsorption of cadmium onto soils and silicon or aluminum oxides is strongly pH-dependent, increasing as conditions become more alkaline. When the pH is below 6–7, cadmium is desorbed from these materials. The oxide and sulfide compounds are relatively insoluble, while the chloride and sulfate salts are soluble. Soluble cadmium compounds have the potential to leach through soils to groundwater.

Average concentrations of cadmium in U.S. waters is about 0.001 mg/L. Cadmium concentrations in bed sediments are generally at least 10 times higher than in overlying water. Cadmium for industrial use is extracted during the production of other metals, chiefly zinc, lead, and copper. It is used for batteries, alloys, pigments, metal protective coatings, and as a stabilizer in plastics. It enters the environment mostly from industrial and domestic wastes, especially those associated with nonferrous mining, smelting, and municipal waste dumps. Because cadmium is chemically similar to zinc, an essential nutrient for plants and animals, it is readily assimilated into the food chain. Plants absorb cadmium from irrigation water. Low levels exist in all foods, highest in shellfish, liver, and kidney meats. Smoking can double the average daily intake; one cigarette typically contains 1 to 2 μg of cadmium. The recommended upper limit in irrigation water is 0.01 mg/L.

Health Concerns

Cadmium is acutely toxic; a lethal dose is about 1 g. Acute exposure can cause nausea, vomiting, diarrhea, muscle cramps, salivation, sensory disturbances, liver injury, convulsions, shock, and renal failure. It is eliminated from the body slowly and can bioaccumulate over many years of low exposure. Long-term exposure to low levels of cadmium in air, food, and water leads to a build-up of cadmium in the kidneys and may cause kidney disease. Other potential long-term effects are blood, liver, and lung damage, and fragile bones. There is no adequate evidence to state whether or not cadmium has the potential to cause cancer from lifetime exposures in drinking water.

Drinking Water Standards

Maximum contaminant level goal: 0.005 mg/L.
Maximum contaminant level: 0.005 mg/L.

Other Comments

Treatment/best available technologies: Coagulation and filtration, ion-exchange, lime softening, reverse osmosis.

CALCIUM (CA), CAS # 7440-70-2

Background

Calcium cations (Ca^{2+}) and calcium salts are among the most commonly encountered substances in water, arising mostly from dissolution of minerals. Calcium often is the most abundant cation in river water. Among the most common calcium minerals are the two crystalline forms of calcium carbonate, calcite and aragonite ($CaCO_3$, limestone is primarily calcite), calcium sulfate (the dehydrated form, $CaSO_4$, is anhydrite; the hydrated form, $CaSO_4 \cdot 2H_2O$, is gypsum), calcium magnesium carbonate ($CaMg(CO_3)_2$, dolomite), and, less often, calcium fluoride (CaF_2, fluorite). Water hardness is caused by the presence of dissolved calcium, magnesium, and sometimes iron (Fe^{2+}), all of which form insoluble precipitates with soap and are prone to precipitating in water pipes and fixtures as carbonates (see Chapter 3). Limestone ($CaCO_3$), lime (CaO), and hydrated lime ($Ca(OH)_2$) are heavily used in the treatment of wastewater and water supplies to raise the pH and precipitate metal pollutants.

Health Concerns

Calcium is an essential nutrient for plants and animals, essential for bone, nervous system, and cell development. The recommended daily intake for adults is between 800 and 1200 mg per day. Most of this is obtained in food. Drinking water typically accounts for 50 to 300 mg per day, depending on the water hardness and assuming ingestion of 2 L per day. Calcium in food and water is essentially nontoxic. A number of studies suggest that water hardness protects against cardiovascular disease. One possible adverse effect from ingesting high concentrations of calcium for long periods of time may be an increased risk of kidney stones. The presence of calcium in water decreases the toxicity of many metals to aquatic life. Stream standards for these metals are expressed as a function of hardness and pH. Thus, the presence of calcium in water is beneficial and no limits on calcium have been established for protection of human or aquatic health.

Drinking Water Standards

There are no upper limits for calcium concentrations. To the contrary, calcium in water is usually regarded as beneficial.

CHLORIDE (CL⁻), CAS # 7440-39-3

Background

Chlorides are widely distributed in nature, usually in the form of sodium, potassium, and calcium salts (NaCl, KCl, and $CaCl_2$), although many minerals contain small amounts of chloride as an impurity. Chloride in natural waters arises from weathering of chloride minerals, salting of roads for snow and ice control, seawater intrusion in coastal regions, irrigation drainage, and industrial wastewater. Chloride ion is extremely mobile. All chloride salts are very soluble except for chloride salts of lead ($PbCl_2$), silver (AgCl), and mercury (Hg_2Cl_2, $HgCl_2$). Chloride is not sorbed to soils and moves with water with little or no retardation. Consequently, it eventually moves to closed basins (as the Great Salt Lake in Utah) or to the oceans.

Concentrations in unpolluted surface waters and nongeothermal groundwaters are generally low, usually below 10 mg/L. Thus, chloride concentrations in the absence of pollution are normally less than those of sulfate or bicarbonate.

Health Concerns

Chloride is the most abundant anion in the human body and is essential to normal electrolyte balance of body fluids. A daily dietary intake for adults of about 9 mg of chloride per kilogram of body weight is considered essential for good health. Chlorides in water are more of a taste than a health concern, although high concentrations may be harmful to people with heart or kidney problems.

Drinking Water Standards

There are no primary drinking water standards for chloride. The EPA secondary standard for chloride is 250 mg/L.

Other Comments

Treatment/best available technologies: Conventional water treatment does not remove chloride ion. Reverse osmosis or nanofiltration is required.

CHROMIUM (CR), CAS # 7440-47-3

Background

Chromium occurs in minerals mostly as chrome iron ore or chromite ($FeCr_2O_2$), in which it is present as Cr(III) with oxidation number +3. Chromium in soils occurs mostly as insoluble chromium oxide (CrO_3), where it is present as Cr(VI) with an a oxidation number of +6. In natural waters, dissolved chromium exists as either Cr^{3+} cations or in anions such as chromate (CrO_4^{2-}) and dichromate ($Cr_2O_7^{2-}$), where it is hexavalent with oxidation number +6. Though widely distributed in soils and plants, it generally is present at low concentrations in natural waters. Background levels in water typically range between 0.2 and 20 µg/L, with an average of 1 µg/L.

As a positively charged ion, trivalent chromium (Cr^{3+}) readily sorbs to negatively charged soils and minerals. Unsorbed Cr^{3+} forms insoluble colloidal hydroxides in the pH range of natural surface waters (6.5 to 9). Thus, it is unlikely that dissolved trivalent chromium will be present in surface waters at levels of concern. Trivalent chromium is also not likely to migrate to groundwater, most of it being retained in the upper 5 to 10 cm of soil.

The hexavalent form of chromium, existing in negatively charged complexes, is not sorbed to any extent by soil or particulate matter and is much more mobile than Cr(III). However, Cr(VI) is a strong oxidant and reacts readily with any oxidizable organic material present, with the resultant formation of Cr(III). In the absence of organic matter, Cr(VI) can be stable for long periods of time, particularly under aerobic conditions. Under anaerobic conditions, Cr(VI) is quickly reduced to low mobility Cr(III). Thus, most of the chromium in surface waters will be present in particulate form as suspended and bed sediments.

Chromium has many industrial uses. Some major applications are in metal alloys, protective coatings on metal, magnetic tapes, paint pigments, cement, paper, rubber, and composition floor covering.

The main natural environmental source is weathering of rocks and soil. Major anthropomorphic sources include metal alloy production, metal plating, cement manufacturing, and incineration of municipal refuse and sewage sludge.

Health Concerns

Trivalent chromium is an essential trace nutrient and plays a role in prevention of diabetes and atherosclerosis. Trivalent chromium is essentially nontoxic. The harmful effects of chromium to human health are caused by hexavalent chromium. Since oxidants such as chlorine or ozone readily oxidize trivalent chromium to the toxic hexavalent form, water quality limits are usually written for total chromium concentrations.

The EPA has found chromium potentially to cause skin irritation or ulceration due to acute exposures at levels above the MCL. Chromium also has the potential to cause damage to the liver, kidney circulatory, and nerve tissues, and dermatitis due to long-term exposures at levels above the MCL. There is no evidence that chromium in drinking water has the potential to cause cancer from lifetime exposures.

Drinking Water Standards

Maximum contaminant level goal: 0.1 mg/L (total Cr).
Maximum contaminant level: 0.1 mg/L (total Cr).

These standards are based on the total concentration of the trivalent and hexavalent forms of dissolved chromium (Cr^{3+} and Cr^{6+}).

Other Comments

Treatment/best available technologies: Coagulation and filtration, ion-exchange, reverse osmosis, lime softening (for Cr(III) only).

COPPER (CU), CAS # 7440-50-8

Background

In nature, copper sometimes occurs as the pure metal but more often in the form of mineral ores that contain 2% or less of the metal. The most common copper-bearing ores are sulfides, arsenites, chlorides, and carbonates. Chalcopyrite ($CuFeS_2$) is the most abundant of the copper ores, accounting for about 50% of the world's copper deposits. The weathering of copper deposits is the main natural source of copper in the aquatic environment, but dissolved copper rarely occurs in unpolluted source water above 10 µg/L, limited by the solubility of copper hydroxide ($Cu(OH)_2$), co-precipitation with less soluble metal hydroxides, and adsorption. In some cases, copper salts may be added to reservoirs for the control of algae. Copper concentrations in acid mine drainage may reach several hundred mg/L, but, if the pH is raised to 7 or higher, most of the copper will precipitate. Smelting operations and municipal incineration may also introduce copper into surface waters.

Copper occurs in drinking water primarily due to corrosion of copper pipes and fittings, which are widely used for interior plumbing of residences and other buildings. This is the reason for an

EPA Action Level based on samples taken from distribution system taps, rather than an MCL. All water is corrosive to some degree toward copper, even water termed noncorrosive or water treated to make it less corrosive. Corrosivity toward copper metal increases with decreasing pH, especially below pH 6.5.

Health Concerns

Copper is an essential nutrient, but at high doses it has been shown to cause stomach and intestinal distress, liver and kidney damage, and anemia. Persons with Wilson's disease may be at a higher risk of health problems due to copper than the general public. There is inadequate evidence to state whether or not copper has the potential to cause cancer from a lifetime exposure in drinking water.

Drinking Water Standards

Maximum contaminant level goal: 1.3 mg/L.
Action Level: > 1.3 mg/L in 10% or more of tap water samples.

Other Comments

Treatment/best available technologies: For treating source water: Ion exchange, lime softening, reverse osmosis, coagulation, and filtration. For corrosion control: pH and alkalinity adjustment, calcium adjustment, silica- or phosphate-based corrosion inhibition.

CYANIDE (CN⁻), CAS # 57-12-5; HYDROGEN CYANIDE (HCN), CAS # 74-90-8

Background

Cyanide is a product of natural animal and vegetative decay processes and also is a component in many industrial waste streams. It is used extensively in mining to separate metals, particularly gold, from ores. In water, an equilibrium exists between the ionized (CN^-) and unionized (HCN) forms, the fraction of each depending on pH (see Equation 7.1 and Figure 7.1).

$$CN^- + H^+ \leftrightarrow HCN. \tag{7.1}$$

Below pH = 9, the predominant form is HCN. HCN is more toxic than CN^- and is the dominant form in most natural waters. HCN is volatile while CN^- is nonvolatile.

The most common industrially used form, hydrogen cyanide, is used in the production of nylon and other synthetic fibers and resins. Some cyanide compounds are used as herbicides. The major sources of cyanide releases to water are discharges from metal finishing industries — iron and steel mills and organic chemical industries. Disposal of cyanide wastes in landfills is a major source of releases to soil.

Cyanides are not persistent when released to water or soil and are not likely to accumulate in aquatic life. They rapidly evaporate and are broken down by microbes. They do not bind to soils and may leach to groundwater. (Cyanide-containing herbicides, such as Tabun, have moderate potential for leaching but are readily biodegraded; therefore, they are not expected to bioconcentrate.)

Soluble cyanide compounds, such as hydrogen and potassium cyanide, have low adsorption to soils with high pH, high carbonate, and low clay content. Soluble cyanide compounds are not exptected to bioconcentrate.

Insoluble cyanide compounds, such as the copper and silver salts, adsorb to soils and sediments and have the potential to bioconcentrate. Insoluble forms do not biodegrade to hydrogen cyanide.

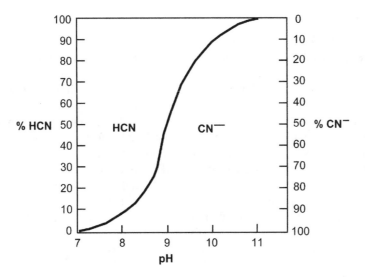

FIGURE 7.1   Distribution of cyanide between the HCN and CN⁻ forms, as a function of pH.

Health Concerns

Short-term exposure to cyanide compounds above the MCL may cause rapid breathing, tremors, and other neurological effects. Long-term exposure at levels above the MCL may cause weight loss, thyroid effects, and nerve damage. There is inadequate evidence for carcinogenicity from lifetime exposures in drinking water.

Drinking Water Standards

Maximum contaminant level goal: 0.2 mg/L.
Maximum contaminant level: 0.2 mg/L.

Other Comments

Treatment/best available technologies: Ion-exchange, reverse osmosis, chlorination.

FLUORIDE (F⁻), CAS # 16984-48-8

Background

Most soils and rocks contain trace amounts of fluoride. Much higher concentrations are found in areas of active or dormant volcanic activity. Common fluoride-containing minerals include fluorite (or fluorspar, CaF_2), cryolite (Na_3AlF_6), and fluorapatite ($Ca_5F(PO_4)_3$). Weathering of minerals is the main source of fluoride in unpolluted waters, where concentrations are usually less than 1 mg/L, but may sometimes exceed 50 mg/L. Where dissolved calcium is present, the formation of fluorite may limit fluoride concentrations. High fluoride concentrations are more likely in water with low calcium concentrations. Groundwater usually contains higher concentrations than surface water, and groundwater concentrations as high as 10 mg/L are common.

The formation of fluoride complexes may be important in solubilizing beryllium, aluminum, tin, and iron in natural waters. Addition of fluoride to drinking water and toothpaste for reducing dental caries, and its subsequent discharge in sewage, also contribute to aquatic fluoride. Discharges from aluminum, steel, and phosphate production are important industrial sources of fluoride in water.

Health Concerns

Small amounts of fluoride appear to be an essential nutrient. People in the U.S. ingest about 2 mg/day in water and food. A concentration of about 1 mg/L in drinking water effectively reduces dental caries without harmful effects on health. Dental fluorosis can result from exposure to concentrations above 2 mg/L in children up to about 8 years of age. In its mild form, fluorosis is characterized by white opaque mottled areas on tooth surfaces. Severe fluorosis causes brown to black stains and pitting. Although the matter is controversial, the EPA has determined that dental fluorosis is a cosmetic and not a toxic and/or an adverse health effect. Water hardness limits fluoride toxicity to humans and fish. The severity of fluorosis decreases in harder drinking water. Crippling skeletal fluorosis in adults requires the consumption of about 20 mg or more of fluoride per day over a 20-year period. In the U.S., cases of crippling skeletal fluorosis have been observed that are associated with the consumption of 2 L of water per day containing 4 mg/L of fluoride. The EPA has concluded that 0.12 mg/kg/day of fluoride can protect against crippling skeletal fluorosis. Fluoride therapy, where 20 mg/day is ingested for medical purposes, is sometimes used to strengthen bone, particularly spinal bones.

Drinking Water Standards

Maximum contaminant level goal: 4.0 mg/L.
Maximum contaminant level: 4.0 mg/L.

Other Comments

Treatment/best available technologies: Anion-exchange, nanofiltration, reverse osmosis.

IRON (FE), CAS # 7439-89-6

Background

Iron is naturally released into waters by weathering of pyritic ores containing iron sulfide (FeS_2) and other iron-bearing minerals in igneous, sedimentary, and metamorphic rocks. It also comes from many human sources: mineral processing, coke and coal burning, acid-mine drainage, iron and steel industry wastes, and corrosion of iron and steel. Because iron is an essential nutrient for animals and plants, it is present in organic matter, in soil, and in sewage and landfill leachate. Many microorganisms use iron as an energy source and play an important part in iron oxidation and reduction processes.

In the aquatic environment, iron is present in two oxidation states: ferrous (Fe^{2+}) and ferric (Fe^{3+}). The reduced ferrous state is highly soluble in the pH range of unpolluted surface waters, while the oxidized ferric state is associated with compounds of low solubility at pH values above 5. For example, Fe^{3+} reacts with water to form low solubility iron oxyhydroxides, which form yellow to red-brown precipitates often seen on rocks and sediments in surface waters with high iron concentrations. Iron oxyhydroxides often form as colloidal suspensions of gels or flocs. These have large surface areas and a strong adsorptive capacity for other dissolved ionic species. Co-precipitation with iron at elevated pHs has been developed as a treatment process for removing other dissolved metals. For example, removal of dissolved zinc by lime precipitation to a concentration below 0.1 mg/L requires co-precipitation with iron.

Since the ferrous state is easily oxidized to the ferric state — a process often enhanced by aerobic iron bacteria that leave slimy deposits of ferric iron — dissolved iron is mainly found under reducing conditions in groundwater or anaerobic surface waters. In well-aerated water above pH 5, dissolved iron concentrations are less than 30 μg/L, while groundwater concentrations may be as high as 50 mg/L. When concentrations in the milligram per liter are reported for aerated surface waters, the iron is generally associated with sediments. When sampling groundwater for dissolved iron, it is important to purge the well adequately, filter the water immediately on site to remove iron sediments, and acidify the filtrate to prevent further precipitation. It is not uncommon for dissolved Fe^{2+} in groundwater to enter a well and become oxidized to Fe^{3+} after exposure to oxygen in the well or the distribution system. The result is rust-colored water from precipitated ferric iron compounds and potential staining of plumbing fixtures and laundry. Such water often has objectionable taste and eventually may clog plumbing fixtures, reducing the flow of water.

Health Concerns

Iron is an essential nutrient in animal and plant metabolism. It is not normally considered a toxic substance. It is not regulated in drinking water except as a secondary standard for aesthetic reasons. Adults require between 10 and 20 mg of iron per day. Excessive iron ingestion may result in haemochromatosis, a condition of tissue damage from iron accumulation. This condition rarely occurs from dietary intake alone but has resulted from prolonged consumption of acidic foods cooked in iron utensils and from the ingestion of large quantities of iron tablets.

Drinking Water Standards

The EPA has no primary drinking water standard for iron. The EPA secondary drinking water standard (nonenforceable) is 0.3 mg/L as total iron.

Other Comments

Treatment/best available technologies: Lime precipitation, aeration, cation-exchange, microfiltration, reverse osmosis.

LEAD (PB), CAS # 7439-92-1

Background

Lead minerals are found mostly in igneous, metamorphic and sedimentary rocks. The most abundant lead mineral is galena (PbS). Oxide, carbonate, and sulfate minerals are lanarkite (PbO), cerrusite ($PbCO_3$), and anglesite ($Pb(SO_4)$), respectively. Commercial ores have concentrations of lead in the range 30–80 g/kg. Metallic lead and the common lead minerals have very low solubility. Most environmental lead (perhaps 85%) is associated with sediments; the rest is in dissolved form. Although some lead enters the environment from natural sources by weathering of minerals, particularly galena, anthropogenic sources are about 100 times greater. Mining, milling and smelting of lead and metals associated with lead, such as zinc, copper, silver, arsenic and antimony, are major sources, as are combustion of fossil fuels and municipal sewage. Commercial products that are major sources of lead pollution include lead-acid storage batteries, electroplating, construction materials, ceramics and dyes, radiation shielding, ammunition, paints, glassware, solder, piping, cable sheathing, roofing, and, until around 1980, gasoline additives such as tetramethyllead and tetraethyllead.

In areas away from mining and smelters, the use of leaded gasoline exceeded all other sources between 1940 and 1980. In 1970, new regulations in the Clean Air Act led to a reduction in lead additives to gasoline as well as tighter restrictions on industrial emissions. By 1985, these measures had resulted in an overall decrease in lead emissions of around 20%. Organic lead additives to automotive gasoline were completely eliminated in 1996, but soils and water bodies still carry the lead legacy from earlier years.

Levels of dissolved lead in natural surface waters are generally low. Lead sulfides, sulfates, oxides, carbonates, and hydroxides are almost insoluble. Because of their greater abundance, carbonates and hydroxides impose an upper limit on the concentrations of lead that can occur in lakes, rivers and groundwaters. The global mean lead concentration in lakes and rivers is estimated to be between 1.0 and 10.0 µg/L.

Pb^{2+} is the stable ionic species in most of the natural environment. Sorption is the dominant mechanism controlling the distribution of lead in the aquatic environment, where it forms complexes with organic ligands to yield soluble, colloidal, and particulate compounds that sorb to humic materials. At low lead concentrations typically found in the aquatic environment, most of the lead in the dissolved phase is likely to be in the form of organic ligand complexes. In the presence of clay suspensions at pH 5–7, most lead is precipitated and sorbed as sparingly soluble hydroxides. Soluble lead is removed from natural waters mainly by association with sediments and suspended particulates. Lead solubility is very low (<1 µg/L at pH 8.5–11) in water containing carbon dioxide and sulfate. At constant pH, the solubility of lead decreases with increasing alkalinity. Lead is bioaccumulated by aquatic organisms, including benthic bacteria, freshwater plants, invertebrates and fish.

Lead in drinking water results primarily from corrosion of materials containing lead and copper in distribution systems, and from lead and copper plumbing materials used to plumb publicly and privately owned buildings connected to the distribution system. Very little lead enters public distribution systems in water from treatment plants. Most public water systems serve at least some buildings with lead solder and/or lead service lines. About 20% of all public water systems have some lead service lines and connections within their distribution systems.

All water is corrosive to metal plumbing materials to some degree, even water termed noncorrosive or water treated to make it less corrosive. The corrosivity of water to lead is influenced by water pH, total alkalinity, dissolved inorganic carbonate, calcium, and hardness. Galvanic corrosion of lead into water also occurs with lead-soldered copper pipes, due to differences in the electrochemical potential of the two metals. Grounding of household electrical systems to plumbing can accelerate galvanic corrosion.

Lead is not very mobile under normal environmental conditions. It is retained in the upper 2–5 cm of soil, especially soils with at least 5% organic matter or a pH of 5 or above. Leaching is not important under normal conditions. It is expected to slowly undergo speciation to the more insoluble sulfate, sulfide, oxide, and phosphate salts.

Metallic lead can be dissolved by pure water in the presence of oxygen, but if the water contains carbonates and silicates, protective films are formed preventing further attack. The solubility of Pb is 10 µg/L above pH 8, while near pH 6.5 the solubility can exceed 100 µg/L. Lead is effectively removed from the water column by adsorption to organic matter and clay minerals, precipitation as insoluble salts, and reaction with hydrous iron and manganese oxides. Under appropriate conditions, dissolution due to anaerobic microbial action may be significant in subsurface environments. In an oxidizing environment, the least soluble common forms of lead are the carbonate, hydroxide, and hydroxycarbonate. In reducing conditions where sulfur is present, PbS is formed as an insoluble solid.

Health Concerns

Short-term exposure to lead at relatively low concentrations can cause interference with red-blood-cell chemistry, delays in normal physical and mental development in babies and young children, slight deficits in the attention span, hearing, and learning abilities of children, and slight increases in the blood pressure of some adults. It appears that some of these effects — particularly changes in the levels of certain blood enzymes and in aspects of children's neuro-behavioral development — may occur at blood lead levels so low as to be essentially without a threshold. Long-term exposure to lead has been linked to cerebrovascular and kidney disease in humans. Lead has the potential to cause cancer from a lifetime exposure at levels above the action level.

Drinking Water Standards

Maximum contaminant level goal: zero.
Maximum contaminant level: Because plumbing in homes and commercial buildings is the main source of lead in drinking water, the EPA has established a tap water action level rather than an MCL. Action Level: >0.015 mg/L in more than 10% of tap water samples.

Other Comments

Treatment/best available technologies: Ion-exchange, lime softening, reverse osmosis, coagulation and filtration.
Corrosion Control: pH and alkalinity adjustment, calcium adjustment, silica- or phosphate-based corrosion inhibition.

MAGNESIUM (MG), CAS # 7439-95-4

Background

Magnesium is used in the textile, tanning, and paper industries. Lightweight alloys of magnesium are used extensively in molds, die castings, extrusions, rolled sheets and plate forgings, mechanical handling equipment, portable tools, luggage, and general household goods. The carbonates, chlorides, hydroxides, oxides, and sulfates of magnesium are used in the production of magnesium metal, refractories, fertilizers, ceramics, explosives and medicinals.

Magnesium is abundant in the earth's crust and is a common constituent of natural water. Along with calcium, it is one of the main contributors to water hardness. The aqueous chemistry of magnesium is similar to that of calcium, such that carbonates and oxides are formed. Magnesium compounds are more soluble than their calcium counterparts. As a result, large amounts of magnesium are rarely precipitated. Magnesium carbonates and hydroxides precipitate at pH > 10. Magnesium concentrations can be extremely high in certain closed saline lakes. Natural sources contribute more magnesium to the environment than all anthropogenic sources. Magnesium is commonly found in magnesite, dolomite, olivine, serpentine, talc, and asbestos minerals. The principal sources of magnesium in natural water are ferromagnesium minerals in igneous rocks and magnesium carbonates in sedimentary rocks. Water in watersheds with magnesium-containing rocks may contain magnesium in the concentration range of 1–100 mg/L. The sulfates and chlorides of magnesium are very soluble, and water that comes in contact with such deposits may contain hundreds of milligrams of magnesium per liter.

Health Concerns

Magnesium is an essential nutrient for plants and animals used mainly for bone and cell development. It accumulates in calcareous tissues and is found in edible vegetables (700–5600 mg/kg), marine algae (6400–20,000 mg/kg), marine fish (1200 mg/kg), and mammalian muscle (900 mg/kg) and bone (700–1800 mg/kg). Magnesium is one of the principal cations of soft tissue. It is an essential part of the chlorophyll molecule. Recommended daily intake for adults is 400–450 mg per day, of which drinking water can supply from 12 to 250 mg per day, depending on the magnesium concentration and assuming ingestion of 2 L per day. Magnesium salts are used medicinally as cathartics and anticonvulsants. In general, the presence of magnesium in water is beneficial, and no limits on magnesium have been established for protection of human or aquatic health.

Drinking Water Standards

There are no primary or secondary drinking water standards for magnesium. Magnesium in drinking water may provide nutritional benefits for people with magnesium-deficient diets.

Manganese (Mn), CAS # 7439-96-5

Background

Manganese is an abundant, widely distributed metal. It does not occur in nature as the elemental metal but is found in various salts and minerals frequently along with iron compounds. Soils, sediments, and metamorphic and sedimentary rocks are significant natural sources of manganese. The most important manganese mineral is pyrolusite (MnO_2). Other manganese minerals are manganese carbonate ($MnCO_3$, rhodocrosite) and manganese silicate ($MnSiO_3$, rhodonite). Ferromanganese minerals, such as biotite mica ($K(Mg,Fe)_3(AlSi_3O_{10})(OH)_2$) and amphibole (($(Mg,Fe)_7Si_8O_{22}(OH)_2$), contain large amounts of manganese. The weathering of manganese deposits contributes small amounts of manganese to natural waters.

Manganese, its alloys and manganese compounds are commonly used in the steel industry for manufacturing metal alloys and dry cell batteries, and in the chemical industry for making paints, varnishes, inks, dyes, glass, ceramics, matches, fireworks, and fertilizers. The iron and steel industry and acid mine drainage release a large portion of the manganese found in the environment. Iron and steel plants also release manganese into the atmosphere, from which it is redistributed by atmospheric deposition.

Manganese seldom reaches concentrations of 1.0 mg/L in natural surface waters and is usually present in quantities of 0.2 mg/L or less. Concentrations higher than 0.2 mg/L may occur in groundwaters and deep stratified lakes and reservoirs under reducing conditions. Subsurface and acid mine waters may contain 10 mg/L. Manganese is similar to iron in its chemical behavior and is frequently found in association with iron. In the absence of dissolved oxygen, manganese normally is in the reduced manganous (Mn^{2+}) form, but it is readily oxidized to the manganic (Mn^{4+}) form. Permanganates (Mn^{7+}) are not persistent because they are strong oxidizers and rapidly are reduced in the process of oxidizing organic materials. Nitrate, sulfate, and chloride salts of manganese are quite soluble in water, whereas oxides, carbonates, phosphates, sulfides, and hydroxides are only sparingly soluble. In natural waters, a substantial fraction of manganese is present in suspended form. In surface waters, divalent manganese (Mn^{2+}) is rapidly oxidized to insoluble manganese dioxide (MnO_2), which then precipitates as a black solid often observed as black stains on rocks. In drinking water distribution systems, precipitation of MnO_2 may cause unsightly black staining of fixtures and laundry.

Health Concerns

Manganese is an essential trace element for microorganisms, plants, and animals and is therefore contained in all, or nearly all, organisms. Manganese is a ubiquitous element that is essential for normal physiologic functioning in all animal species. The total body load of manganese in an average adult is about 12 mg. Health problems in humans may arise from deficient and excessive intakes of manganese. Thus, any quantitative risk assessment for manganese must consider that, although manganese is an essential nutrient, excessive intake causes toxic symptoms. An average dietary intake in the U.S. ranges between 2 and 10 mg/day, with an average around 4 mg/day. Grains and cereals are the richest dietary sources of manganese, followed by fruits and vegetables. Meat, fish, and poultry contain little manganese. Drinking water supplies almost always contain less than the secondary standard of 0.05 mg/L and drinking water generally contributes no more than about 0.07 mg/day to an adult diet. A maximum adult dietary intake of 20 mg/day is recommended to avoid manganese toxicity. Manganese is not considered to be a cancer risk.

Drinking Water Standards

The EPA has no primary drinking water standard for manganese. The EPA secondary drinking water standard (nonenforceable) is 0.05 mg/L.

Other Comments

Treatment/best available technologies: Lime precipitation, aeration, cation-exchange, microfiltration, reverse osmosis.

MERCURY (HG), CAS # 7439-97-6

Background

Mercury is a liquid metal found in natural deposits of ores containing other elements. Mercury deposits occur in all types of rocks: igneous, sedimentary, and metamorphic. Although cinnabar (HgS) is the most common mercury ore, mercury is present in more than 30 common ore and gangue minerals. Mercury exists in the environment as the elemental metal, as monovalent and divalent salts, and as organic mercury compounds, the most important of which are methyl mercury ($HgCH_3^+$) and dimethyl mercury ($Hg(CH_3)_2$). Methyl and dimethyl mercury are formed from inorganic mercury by microorganisms found in bottom sediments and sewage sludge. There are other microorganisms that can demethylate mercury back to the inorganic form.

Mercury is noteworthy among environmental pollutants by virtue of its volatility and the ease by which inorganic mercury can be converted to organic forms by microbial processes. Its volatility accounts for the fact that mercury is present in the atmosphere as metallic mercury vapor and as volatilized organic mercury compounds. Terrestrial environments appear to be major sources of atmospheric mercury, with contributions from evapotranspiration of leaves, decaying vegetation, and degassing of soils. The major source of mercury movement in the environment is the natural degassing of the earth's crust, which may introduce between 25,000 and 150,000 tons of mercury (Hg) per year into the atmosphere. It is not unusual for atmospheric concentrations of mercury in an area to be up to 4 times the level in contaminated soils. Atmospheric mercury can enter terrestrial and aquatic habitats via particle deposition and precipitation. Measuring atmospheric concentrations of mercury from aircraft is a form of aerial prospecting for mineral formations of other metals associated with elemental mercury. Inorganic forms of mercury can be converted to soluble organic forms by anaerobic microbial action in the biosphere. In the atmosphere, 50% of volatile mercury is metallic mercury vapor. Twenty-five to fifty percent of mercury in water is organic. Mercury in the environment is deposited and revolatilized many times, with a residence time in the atmosphere of several days. In the volatile phase it can be transported hundreds of kilometers.

Twenty thousand tons of mercury per year are also released into the environment by human activities such as combustion of fossil fuels, operation of metal smelters, cement manufacture, and other industrial releases. Mercury is used in the chloralkali industry, where mercury is used as an electrode to produce chlorine, caustic soda (sodium hydroxide), and hydrogen by electrolysis of molten sodium chloride. It is also used to produce electrical products such as dry-cell batteries, fluorescent light bulbs, switches, and other control equipment. Electrical products account for 50% of mercury used. Aquatic pollution originates in sewage, metal refining operations, chloralkali plant wastes, industrial and domestic products such as thermometers and batteries, and from solid wastes in major urban areas, where electrical mercury switches account for a significant release of mercury to the environment.

In most unpolluted surface waters, mercuric hydroxide ($Hg(OH)_2$) and mercuric chloride ($HgCl_2$) are the predominant mercury species, with concentrations less than 0.001 mg/L. In polluted waters, concentrations up to 0.03 mg/L may occur. In aquatic systems, mercury binds to dissolved matter or fine particulates. In freshwater habitats, it is common for mercury compounds to be sorbed to particulate matter and sediments. Sediment binding capacity is related to organic content and is slightly affected by pH. Mercury tends to combine with sulfur in anaerobic bottom sediments. Organic methyl mercury bioconcentrates along aquatic food chains to the extent that fish in mildly polluted waters may become unsafe for food use.

Health Concerns

Mercury is highly toxic. Organic alkyl mercury compounds, such as ethylmercuric chloride (C_2H_5HgCl) which used to be used as fungicides, produce illness or death from the ingestion of only a few milligrams. Because inorganic forms of mercury can be converted to very toxic methyl and dimethyl mercury by anaerobic microorganisms, any form of mercury must be considered as potentially hazardous to the environment. Most human mercury exposure is due to consumption of fish. The EPA has found that short-term and long-term exposure to mercury at levels in drinking water above the MCL may cause kidney damage. There is inadequate evidence to state whether or not mercury has the potential to cause cancer from lifetime exposures in drinking water.

Drinking Water Standards

Maximum contaminant level goal: 0.002 mg/L.
Maximum contaminant level: 0.002 mg/L.

Other Comments

Treatment/best available technologies: Granular activated carbon for influent mercury concentrations above 10 µg/L, coagulation and filtration, lime softening, and reverse osmosis for influent mercury concentrations less than 10 µg/L.

MOLYBDENUM (MO), CAS # 7439-98-7

Background

Molybdenum is widely distributed in trace amounts in nature, occurring chiefly as insoluble molybdenite (MoS_2) and soluble molybdates (MoO_4^{2-}). Molybdenum is relatively mobile in the environment because soluble compounds predominate at pH > 5. The solubility of molybdenum increases as redox potential is lowered. Below pH 5, adsorption and coprecipitation of the molybdate anion by hydrous oxides of iron and aluminum are effective at removing dissolved molybdenum. The weathering of igneous and sedimentary rocks (especially shales) is the main natural source of molybdenum to the aquatic environment.

Molybdenum metal is used in the manufacture of special steel alloys and electronic apparatus. Molybdenum salts are used in the manufacture of glass, ceramics, pigments, and fertilizers. The use of fertilizers containing molybdenum is the single most important anthropogenic input to the aquatic environment. Other contributions to the aquatic environment come from mining and milling of molybdenum, the use of molybdenum products, the mining and milling of some uranium and copper ores, and the burning of fossil fuels. Fresh water usually contains less than 1 mg/L molybdenum. Concentrations ranging between 0.03 and 10 µg/L are typical of unpolluted waters. Levels as high as 1500 µg/L have been observed in rivers of industrial areas. The average concentration of molybdenum in finished drinking water is about 1 to 4 µg/L.

Health Concerns

Molybdenum is an essential trace nutrient for all plants and animals. It is considered nontoxic to humans, but excessive levels (0.14 mg/kg body weight; 10 mg/day for a 70 kg adult) may cause high uric acid levels and an increased chance of gout. The recommended daily intake is 70–250 µg/day for adults. Local concentrations may vary by a factor of 10 or more depending on regional geology, causing both deficient and excessive intake of molybdenum by plants and ruminants. Average adults contain about 5 mg of molybdenum in their body and ingest about 100 to 300 µg/day. Twenty enzymes in plants and animals are known to be built around molybdenum, including xanthine oxidase, which helps to produce uric acid, essential for eliminating excess nitrogen from the body.

Drinking Water Standards

There are no primary or secondary drinking water standards for molybdenum.

NICKEL (NI), CAS # 7440-02-0

Background

Nickel is found in many ores as sulfides, arsenides, antimonides, silicates, and oxides. Its average crustal concentration is about 75 mg/kg. Because nickel is an important industrial metal, industrial waste streams can be a major source of environmental nickel. Inadvertent formation of volatile nickel carbonyl can occur in various industrial processes that use nickel catalysts, such as coal gasification, petroleum refining, and hydrogenation of fats and oils. Nickel oxide is present in residual fuel oil and in atmospheric emissions from nickel refineries. The atmosphere is a major conduit for nickel as particulate matter. Contributions to atmospheric loading come from both natural sources and anthropogenic activity, with input from both stationary and mobile sources. Nickel particulates eventually precipitate from the atmosphere to soils and waters. Soil-borne nickel enters waters with surface runoff or by percolation of dissolved nickel into groundwater.

Nickel is one of the most mobile heavy metals in the aquatic environment. Its concentration in unpolluted water is controlled largely by co-precipitation and sorption with hydrous oxides of iron and manganese. In polluted environments, nickel forms soluble complexes with organic material. In reducing environments where sulfides are present, insoluble nickel sulfide is formed. Average concentrations in U.S. surface waters are typically between 10 µg/L and 100 µg/L, with concentrations as high as 11,000 µg/L in streams receiving mine discharges. In surface waters, sediments generally contain more nickel than the overlying water.

Health Concerns

Nickel appears to be an essential trace nutrient in animals and humans, but its exact role is not yet understood. Daily intake of nickel, mainly from food, is around 150 µg/day; the adult requirement is between 5 and 50 µg/day. Although a few nickel compounds — such as nickel carbonyl — are poisonous and carcinogenic, most nickel compounds are nontoxic. The toxicity of dissolved nickel that is ingested orally is low, comparable to zinc, chromium, and manganese, perhaps because only 2–3% of ingested nickel is absorbed.

The EPA has not found nickel to cause adverse human health effects from short-term exposures at levels above the former MCL. Long-term exposure above the former MCL can cause decreased body weight, heart and liver damage, and dermatitis. There is no evidence that nickel has the potential to cause cancer from lifetime exposures in drinking water.

Drinking Water Standards

The EPA remanded the drinking water standard for nickel in 1995. Prior to 1995, the maximum contaminant level goal (MCLG) and maximum contaminant level (MCL) both were 0.1 mg/L. Currently there are no drinking water standards for nickel.

Other Comments

Treatment/best available technologies: ion-exchange, lime softening, reverse osmosis.

Nitrate (NO_3^-), CAS # 14797-55-8; Nitrite (NO_2^-), CAS # 14797-65-0

Background (see Chapter 3 for a more detailed discussion).

Nitrate and nitrite are highly soluble in water. Due to their high solubility and weak retention by soil, nitrate and nitrite are very mobile, moving through soil at approximately the same rate as water. Thus, nitrate and nitrite have a high potential to migrate to groundwater. Because they are not volatile, nitrate and nitrite are likely to remain in water until consumed by plants or other organisms.

Nitrate is the oxidized form and nitrite is the reduced form. Aerated surface waters will contain mainly nitrate, and groundwaters, with lower levels of dissolved oxygen, will contain mostly nitrite. They readily convert between the oxidized and reduced forms depending on the redox potential. Nitrite in groundwater is converted to nitrate when brought to the surface or exposed to air in wells. Nitrate in surface water is converted to nitrite when it percolates through soil to oxygen-depleted groundwater.

The main inorganic sources of contamination of drinking water by nitrate are potassium nitrate and ammonium nitrate. Both salts are used mainly as fertilizers. Ammonium nitrate is also used in explosives and blasting agents. Because nitrogenous materials in natural waters tend to be converted to nitrate, all environmental nitrogen compounds — particularly organic nitrogen and ammonia — should be considered as potential nitrate sources. Primary sources of organic nitrates include human sewage and livestock manure — especially from feedlots.

Health Concerns

Nitrate is a normal dietary component. A typical adult ingests around 75 mg/day, mostly from the natural nitrate content of vegetables, particularly beets, celery, lettuce, and spinach. Short-term exposure to levels of nitrate in drinking water that are higher than the MCL can cause serious illness or death, particularly in infants. Nitrate is converted to nitrite in the body. Nitrite oxidizes Fe^{2+} in blood hemoglobin to Fe^{3+}, rendering the blood unable to transport oxygen. Infants are much more sensitive than adults to this problem because of their small total blood supply. Symptoms include shortness of breath and blueness of the skin. This can be an acute condition in which health deteriorates rapidly over a period of days.

Long-term exposure to levels of nitrate and/or nitrite in excess of the MCL may cause diuresis, increased starchy deposits, and hemorrhaging of the spleen. There is inadequate evidence to state whether or not nitrates or nitrites have the potential to cause cancer from lifetime exposures in drinking water.

Drinking Water Standards

Nitrate: *Maximum contaminant level goal:* 10 mg/L.
          Maximum contaminant level: 10 mg/L.

Nitrite: *Maximum contaminant level goal:* 1.0 mg/L.
          Maximum contaminant level: 1.0 mg/L.

Total (Nitrate + Nitrite): *Maximum contaminant level goal:* 10 mg/L.
                            Maximum contaminant level: 10 mg/L.

Selenium (Se), CAS # 7782-49-2

Background

Selenium is widely distributed in the earth's crust at concentrations averaging 0.09 mg/kg. It occurs in igneous rocks, with sulfides in volcanic sulfur deposits, in hydrothermal deposits, and in porphyry copper deposits. The major source of selenium in the environment is the weathering of rocks and soils. In addition, volcanic activity contributes to its natural occurrence in waters in trace amounts. Volcanic activity is an important source of selenium in regions with high soil concentrations.

Most selenium for industrial and commercial purposes is produced from electrolytic copper-refining shines and from flue dusts from copper and lead smelters. Anthropogenic sources of selenium in water bodies include effluents from copper and lead refineries, municipal sewage, and fallout of emissions from fossil fuel combustion. Selenium in surface waters can range between 0.1 μg/L and 2700 μg/L, with most values between 0.2 μg/L and 20 μg/L.

Selenate is more mobile under oxidizing conditions than under reducing conditions and can be reduced by bacteria in anaerobic environments to form methylated selenium compounds, which are volatile. Dissolved selenium exists mostly as the selenite (SeO_3^{2-}) and selenate (SeO_4^{2-}) anions. Ferric selenite, however, is insoluble and offers a treatment for removing dissolved selenium. Alkaline and oxidizing conditions favor the formation of soluble selenates, which also are the biologically available forms for plants and animals. Acidic and reducing conditions readily reduce selenates and selenites to insoluble elemental selenium, which precipitates from the water column.

Health Concerns

Selenium is a nutritionally essential trace element for all vertebrates and most plants. Human blood contains about 0.2 mg/L, 1000 times higher than typical surface waters, demonstrating that selenium is bioaccumulated. The adult daily requirement is between 20 and 200 μg/day. Drinking water seldom contains a few micrograms per liter of selenium, not enough for it to be a significant dietary source. Depending on the food eaten, daily intake is between 6 and 200, with levels near 150 μg/day being typical. Cereals, nuts, and seafood are good sources of selenium. Ingestion of quantities above the recommended maximum daily intake of 450 μg/day increases the risk of selenium poisoning, the most obvious symptom of which is bad breath and body odor caused by volatile methyl selenium produced by the body to eliminate excess selenium.

The EPA has found that short-term exposure to selenium at levels above the MCL may cause hair and fingernail changes, damage to the peripheral nervous system, fatigue, and irritability. Long-term exposure above the MCL may cause hair and fingernail loss, damage to kidney and liver tissue, and damage to the nervous and circulatory systems. There is no evidence that selenium has the potential to cause cancer from lifetime exposures in drinking water.

Drinking Water Standards

Maximum contaminant level goal: 0.05 mg/L.
Maximum contaminant level: 0.05 mg/L.

Other Comments

Treatment/best available technologies: Activated alumina, coagulation and filtration, lime softening, reverse osmosis, electrodialysis.

SILVER (AG), CAS # 7440-22-4

Background

Silver is a white, lustrous, ductile metal that occurs naturally in its pure, elemental form and in ores, mostly as argentite (Ag_2S). Other silver ores include cerargyrite ($AgCl$), proustite (($AgS)_3 \cdot As_2S_3$), and pyrargyrite (($Ag_2S)_3 \cdot Sb_2S_3$). Silver is also found associated with lead, gold, copper, and zinc ores. Silver is among the less common but most widely distributed elements in the earth's crust. Its concentration in normal soil averages around 0.3 mg/kg.

A large portion of silver consumption is for photographic materials. Also, because silver has the highest known electrical and thermal conductivities of all metals, it finds extensive use in electrical and electronic products such as batteries, switch contacts, and conductors. Other major uses include sterling and plated metalwork, jewelry, coins and medallions, brazing alloys and solders, catalysts, mirrors, fungicides, and dental and medical supplies.

Natural processes, such as weathering and volcanic activity, release silver to the environment. Silver has been found associated with sulfides, sulfates, chlorides, and ammonia salts in deposits and discharges of hot springs and volcanic materials. Anthropogenic sources of silver include discharges from landfills and waste lagoons, fallout from incineration and industrial emissions, and direct waste discharge to water. Some home water treatment devices use silver as an antibacterial agent and may represent a contamination source. Surface waters in nonindustrial regions average around 0.2 to 0.3 µg/L of silver, while the streambed sediments range between 140 to 600 µg/kg of silver. In industrial areas, silver concentrations in surface waters may reach 40 µg/L and stream sediment concentrations 1500 µg/kg. Finished drinking water seldom contains more than 1 µg/L of silver.

Metallic silver is stable over much of the pH and redox range found in natural waters but has very low water solubility. Insoluble silver compounds, such as $AgCl$, Ag_2S, Ag_2Se, and Ag_3AsS_3, may be present in aquatic systems in colloidal form, adsorbed to various humic substances, or incorporated with sediments. At pH <7.5 under aerobic conditions, the Ag^+ cation is soluble and mobile. Around pH 7.5–8.0, aquatic Ag^+ reacts with water to form the insoluble oxide. Silver is dispersed through the aquatic environment as dissolved and colloidal species, but it eventually resides in the bottom sediments. Sorption, particularly by manganese dioxide, and precipitation of silver halides, particularly silver chloride, are the main processes that remove dissolved silver from the water column. These processes, along with the low crustal abundance of silver, account for its low observed concentrations in the aqueous phase.

Health Concerns

There is no evidence that silver is an essential nutrient. Metallic silver is not considered toxic, but most of its salts exhibit toxic properties. Large oral doses of silver nitrate can cause severe gastrointestinal irritation and ingestion of 10 g is likely to be fatal. Chronic human exposure to silver in drinking water seems only to cause argyria, a discoloration of the skin resulting from the deposition of metallic silver in tissues. The EPA considers this condition to be mainly cosmetic. The minimum adult cumulative dose of silver for inducing argyria is about 1000 mg, an amount likely to be encountered only in industrial environments. The accumulation of 1000 mg of silver over a lifetime (70 years) would require the retention of 40 µg/day. Since around 90% of the silver intake is excreted, the required daily intake for inducing argyria would be more like 400 µg/day. An average daily diet may contain 20 to 80 µg of silver and drinking 2 L of water may contribute an additional 2 µg. Thus, food and drinking water are not likely to deliver toxic quantities of silver.

Drinking Water Standards

The EPA has no primary drinking water standard for silver. The EPA secondary drinking water standard (nonenforceable) is 0.10 mg/L.

Other Comments

Treatment/best available technologies: Lime softening at pH 11, sand filtration followed by activated carbon, ion-exchange, nano- and ultrafiltration.

Sulfate (SO_4^{2-}), CAS # 14808-79-8

Background

Sulfate minerals are widely distributed in nature, and the sulfate anion (SO_4^{2-}) is a common constituent of unpolluted water. Sulfate anion is the stable, oxidized form of sulfur and is readily soluble in water. Lead, barium, and strontium sulfate compounds, however, are insoluble. Under anaerobic conditions, sulfates serve as an oxygen source for bacteria that reduce dissolved sulfate to sulfide, which then may be volatilized to the atmosphere as H_2S, precipitated as insoluble salts, or incorporated in living organisms. These processes are common in the anaerobic regions of wetlands and lakes fed by surface and groundwaters with high sulfate levels. Oxidation of sulfides returns the sulfur to the sulfate form.

Sulfates may be leached from most sedimentary rocks, including shales, with the most appreciable contributions from such sulfate deposits as gypsum ($CaSO_4\cdot2H_2O$) and anhydrite ($CaSO_4$). The oxidation of sulfur-bearing organic materials can contribute sulfates to waters. Industrial discharges are another significant source of sulfates. It is estimated that about one half of the river sulfate load arises from mineral weathering and volcanism, the other half from biochemical and anthropogenic sources. Tanneries, sulfite-pulp mills, textile plants, sulfuric acid production, metal-working industries, and mine drainage wastes are all sources of sulfate polluted water. Air emissions from industrial fuel combustion and the roasting of sulfur-containing ores, carry large amounts of sulfur dioxide and sulfur trioxide into the atmosphere, adding sulfates to surface waters through precipitation. Sulfate concentrations normally vary between 10 mg/L and 80 mg/L in most surface waters, although they may reach thousands of milligrams per liter near industrial discharges. High sulfate concentrations are also present in areas of acid mine drainage and in well waters and surface waters in arid regions where sulfate minerals are present, particularly where irrigation runoff drains soils containing sulfate minerals.

Health Concerns

The EPA has established a secondary drinking water standard for sulfate of 250 mg/L based on aesthetic effects such as taste and odor. The EPA estimates that about 3% of the public drinking water systems in the U.S. may have sulfate levels of 250 mg/L or greater. The main health concern regarding sulfate in drinking water is that diarrhea may be associated with ingesting water that contains high concentrations of sulfate. In the 1996 amendments to the Safe Drinking Water Act, Congress mandated that the EPA must determine by August 2001 whether to regulate sulfate in drinking water. The EPA has published the results of a study, *Health Effects from Exposure to High Levels of Sulfate in Drinking Water Study,* EPA 815-R-99-001, January 1999. As a supplement to the study, the EPA convened a workshop to review the results of the study and to discuss the relevant scientific literature (*Health Effects from Exposure to Sulfate in Drinking Water Workshop,* EPA 815-99-002, January 1999). The EPA is currently receiving public comments on the conclusions of the study and workshop. If the EPA decides to regulate sulfate, they must publish a proposed MCL by August 2003 and issue a final standard by February 2005.

The conclusions of the study and workshop are

- In the study of adults, there was no statistically significant association between the concentration of sulfate in drinking water and the frequency of diarrhea. In a study of infants in South Dakota, there was no significant association between sulfate intake and the incidence of diarrhea.
- There is insufficient data to identify the level of sulfate in drinking water below which adverse human effects are unlikely — for example, to establish an MCL. It was found that some people experience a laxative effect when consuming tap water containing 1000 mg/L to 1200 mg/L of sulfate (as sodium sulfate). However, no studies found an increase in diarrhea, dehydration, or weight loss.

- Based on limited data, people seem to acclimate to the presence of sulfate in drinking water.
- Not enough data could be obtained for completing a dose-response study in infants.
- There is not enough scientific evidence on which to base a regulation, but workshop panelists favored a health advisory in places where drinking water has sulfate levels ≥500 mg/L.

Drinking Water Standards

The EPA has no primary drinking water standard for sulfate. The EPA secondary drinking water standard (nonenforceable) is 250 mg/L, based on aesthetic effects.

Other Comments

Treatment/best available technologies: Anion-exchange, reverse osmosis, nanofiltration, anaerobic precipitation as metal sulfide.

HYDROGEN SULFIDE (H_2S), CAS # 7783-06-4

Background (see Chapters 3 and 6, and Sections 3.10 and 6.8, for more detailed discussions).

Natural waters acquire sulfur compounds mainly from geochemical weathering of sulfide and sulfate minerals, fertilizers, decomposition of organic matter, and atmospheric deposition from industrial fuel combustion. During the decomposition of organic matter, sulfur is released largely as hydrogen sulfide (H_2S), which oxidizes rapidly to sulfate under aerobic conditions. Therefore, under aerobic conditions in aquatic systems, sulfate is the predominant form of sulfur and concentrations of hydrogen sulfide are very low.

Under anaerobic conditions, sulfate and organic sulfur compounds are reduced to the sulfide anion (S^{2-}) by bacterial reduction. The presence of sulfate-reducing bacteria in drinking water distribution systems can be a major cause of taste and odor problems. Sulfide anion is commonly found under aquatic anaerobic conditions wherever sulfur is present, such as in domestic and industrial wastewater and sludges, the hypolimnion of stratified lakes, and the bottoms of wetlands. Sulfide anion reacts with water to form hydrogen sulfide (H_2S) (see Chapter 3). H_2S is a flammable, poisonous gas with a characteristic strong odor of rotten eggs. Hydrogen sulfide and the sulfide salts of the alkali and alkaline metals (Groups 1A and 2A of the Periodic Table) are soluble in water. Soluble sulfide salts dissociate in water, forming the sulfide anion, which then reacts with water to form H_2S.

If transition metal cations are present — particularly iron and manganese — metal sulfides of low solubility are formed and precipitated at neutral and alkaline pH values. Thus, anaerobic zones of lakes and wetlands, where high levels of dissolved metals are present along with sulfide, are likely to contain metal sulfides in the sediments and very little dissolved sulfide anion. Only after most of the dissolved metals have been precipitated can H_2S accumulate in the water. Black sediments of eutrophic lakes and wetlands consist largely of precipitated iron and manganese sulfides, S^{2-} dissolved in interstitial water, acid-soluble sulfide compounds, elemental sulfur, organic sulfur, and sulfates. H_2S also is added in large quantities to the atmosphere from volcanic gases, industrial sources, and biochemical activity in water and soil.

The human nose is very sensitive to the rotten-egg odor of hydrogen sulfide. Most people can detect the smell of H_2S in water containing as little as 1 µg/L. Typical concentrations of H_2S in unpolluted surface water are <0.25 µg/L. H_2S > 2.0 µg/L constitutes a chronic hazard to aquatic life. Groundwater usually contains little or no sulfide because contact with metal-bearing minerals results in the formation of metal sulfides with low solubility. Note, however, that sulfides may be found in wells, arising from sulfate-reducing bacteria present in the well. Some brines, especially those associated with petroleum deposits, may contain several hundred mg/L of dissolved hydrogen sulfide.

Health Concerns

Hydrogen sulfide is acutely toxic to humans. It is a leading cause of death in the workplace, most often by accidental inhalation of high concentrations (>1000 ppm). However, the EPA has no drinking water standards for hydrogen sulfide because its disagreeable taste and odor make water unpalatable at concentrations much lower than the toxic levels. A guideline value is that the presence of H_2S should not be detectable by consumers.

Drinking Water Standards

The EPA has no drinking water standards for hydrogen sulfide.

Other Comments

Treatment/best available technologies: See Chapter 3, Section 3.10.

THALLIUM (TL), CAS # 7440-28-0

Background

Thallium is a soft, lead-like metal that is widely distributed in trace amounts. It may be found naturally as the pure metal or associated with potassium and rubidium in copper, gold, zinc, and cadmium ores. Thallium compounds are nonvolatile, although many are water soluble. Thallium is generally present in trace amounts in fresh water. Unpolluted soil levels range between 0.1 and 0.8 mg/kg, with an average around 0.2 mg/kg.

Thallium compounds are used mainly in the electronics industry and, to a limited extent, in the manufacture of pharmaceuticals, alloys, and high refractive index glass. Manmade sources of thallium pollution include gaseous emission of cement factories, coal burning power plants, and metal sewers. Small amounts of thallium in fallout from these sources frequently contaminate food crops in nearby farms and gardens, where it is readily absorbed by plant roots. The leaching of thallium from ore processing operations is the major source of elevated thallium concentrations in water. Thallium is a trace metal associated with copper, gold, zinc, and cadmium.

Most thallium is released into the environment by weathering of minerals. Human sources of thallium are wastes from the production of other metals, for example, from the roasting of pyrite during the production of sulfuric acid, and in mining and smelting operations of copper, gold, zinc, lead, and cadmium. Waste streams of these industries may contain as much as 90 µg/L.

In the aquatic environment, thallium is transported as soluble complexes with humic materials (above pH 7), sorption to clay minerals, and bioaccumulation. In reducing environments, thallium may be precipitated as elemental metal or, in the presence of sulfur, as the insoluble sulfide. In waters of high oxygen content, Tl^+ is the dominant oxidation state, forming soluble chloride, carbonate and hydroxy salts. Thallium sorption to sediments is pH dependent. Thallium is strongly sorbed by montmorillonite clay at pH 8 but only slightly at pH 4. In a study of heavy metal cycling in a lake in southwestern Michigan, thallium was detected only in the sediments. Since thallium is soluble in most aquatic systems, it is readily available to aquatic organisms and is quickly bioaccumulated by fish and plants.

Health Concerns

Thallium is a toxic metal with no known nutritional value. On the contrary, it is notorious for its use by murderers as a poison; the lethal dose for an adult is around 800 mg. Environmental exposure is mainly through contaminated foods, which are estimated to contain, on average, about 2 ppb of thallium. The adult total body burden of thallium is 0.1–0.5 mg thallium — the greatest part of which is carried by muscle tissue. Short-term exposures to thallium at levels above the MCL can cause gastrointestinal irritation, numbness of toes and fingers, the sensation of burning feet, and muscle cramps. Long-term exposure to thallium at levels above the MCL can cause damage to liver, kidney, intestinal, and testicular tissues, as well as changes in blood chemistry and hair loss. There is no evidence that thallium has the potential to cause cancer from lifetime exposures in drinking water.

Drinking Water Standards

Maximum contaminant level goal: 0.0005 mg/L.
Maximum contaminant level: 0.002 mg/L.

Other Comments

Treatment/best available technologies: Activated alumina, ion-exchange, reverse osmosis, nano-filtration.

Vanadium (V), CAS # 7440-62-2

Background

Vanadium (V) is widely dispersed in the earth's crust at an average concentration of about 150 ppm. Deposits of ore-grade mineable V are rare. There are more than 65 known vanadium-bearing minerals. In the U.S., vanadium occurs in uranium-bearing ores and in phosphate shales and rocks in parts of the western states. Vanadium is also present in coal and crude oil. The bulk of commercial vanadium is obtained as a byproduct or coproduct from the processing of iron, titanium, and uranium ores, and to a lesser extent, from phosphate, bauxite, and chromium ores, and the ash or coke from burning or refining petroleum. The main use for vanadium is as an alloy additive.

Vanadium enters the aquatic environment mainly by surface erosion and natural seepage from carbon-rich deposits such as tar and oil sands, atmospheric deposition, weathering of vanadium-rich ores and clays, and leaching of coal mine wastes. Vanadium concentrations in fresh water typically range between <0.3 µg/L to around 200 µg/L. Groundwater concentrations of vanadium are typically less than 1 µg/L.

Health Concerns

There are no particular health or safety hazards associated with vanadium and its compounds. Dust and fine powders present a moderate fire hazard. Vanadium compounds — of which the most common is vanadium pentoxide — may irritate the conjunctivae and respiratory tract. Toxic effects have been observed from airborne concentrations of vanadium compounds of several milligrams or more per cubic meter of air. OSHA threshold limits in the workplace are 0.5 mg/m^3 for dust and 0.05 mg/m^3 for fumes. Oral toxicity in humans is minimal.

As an environmental pollutant, vanadium is of concern mainly because of its high levels in residual fuel oils and its subsequent contribution to atmospheric particulate levels from the combustion of these fuels in urban areas.

Drinking Water Standards

The EPA has no drinking water standards for vanadium.

ZINC (ZN), CAS # 7440-66-6

Background

Zinc is a common contaminant in surface and groundwaters, stormwater runoff, and industrial waste streams. Its average concentration in the earth's crust is around 70 mg/kg. Because it is very chemically reactive, zinc is not found free in nature but is always associated with one of the 55 known zinc minerals (mainly sulfides, oxides, carbonates, and silicates). Ninety percent of industrial zinc production is from the minerals sphalerite (also called zincblende) and wurtzite, which are different crystalline forms of zinc sulfide (ZnS). Zinc minerals tend to be associated with minerals of other metals, particularly lead, copper, cadmium, mercury, and silver. The most important chemical property of zinc is its high reduction potential, which places it above iron in the electromotive series. Because of this, zinc coatings are used to protect iron and steel from corrosion. When iron or steel coated with zinc is exposed to corrosive conditions, zinc displaces iron from solution and prevents dissolution of the iron. In the process of dissolving, zinc reduces dissolved iron, and many other dissolved metals, to the metallic state.

Zinc occurs in natural waters in both suspended and dissolved forms. The dissolved form is the divalent cation, Zn^{2+}. Dissolved zinc is readily sorbed to or occluded in mineral clays and humic colloids. In water of low alkalinity and below pH 7, Zn^{2+} is the dominant form. As pH and alkalinity increase, the potential for sorption and humic complex formation also increase. Above pH 7, sorption to metal oxides, clays, and apatite (calcium phosphate minerals, $Ca_5(F,Cl,OH,0.5CO_3)(PO_4)_3$) can bind over 90% of dissolved zinc. Sorption is generally not significant below pH 6. At low pH or under anaerobic conditions, metal oxides can redissolve and release bound zinc cation into solution. However, if the redox potential rises and aerobic conditions are established, zinc sulfide is oxidized to soluble zinc sulfate ($ZnSO_4$), also releasing Zn^{2+} into solution.

Zinc concentrations in unpolluted surface waters typically range between about 5 μg/L and 50 μg/L. Streams draining from mined areas often exceed 100 μg/L. Industries with waste streams containing significant levels of zinc include steel works with galvanizing operations, zinc and brass metal works, zinc and brass plating, and production of viscose rayon yarn, ground wood pulp, and newsprint paper. Reported concentrations of zinc in industrial waste streams reach 48,000 mg/L. More representative values are between 10 and 200 mg/L.

Health Concerns

Zinc is an essential nutrient and is not toxic to humans. About 1 g/day may be ingested without ill effects. Recommended dietary allowance is 15 mg/day for adults.

Drinking Water Standards

The EPA has no primary drinking water standard for zinc. The EPA secondary drinking water standard (nonenforceable) is 5 mg/L, based on a metallic taste detectable by many people above that level.

Other Comments

Treatment/best available technologies: Chemical precipitation, ion-exchange, evaporative recovery of salts, reverse osmosis, electrolytic plating.

Appendix A

Drinking Water Standards

(Adapted from: http://www.epa.gov/ogwdw000/wot/appa.html, July 1999.)

TABLE 1
National Primary Drinking Water Regulations

Inorganic Chemicals	MCLG[1] (mg/L)[4]	MCL[2] or TT[3] (mg/L)[4]	Potential Health Effects from Ingestion of Water	Sources of Contaminant in Drinking Water
Antimony	0.006	0.006	Increase in blood cholesterol, decrease in blood glucose.	Discharge from petroleum refineries, fire retardants, ceramics, electronics, solder.
Arsenic	none[5]	0.05	Skin damage, circulatory system problems, increased risk of cancer.	Discharge from semiconductor production; petroleum refining, wood preservatives, animal feed additives, herbicides, erosion of natural deposits.
Asbestos (fibers >10 μm)	7 million fibers/L (MFL)	7 MFL	Increased risk of developing benign intestinal polyps.	Decay of asbestos cement in water mains, erosion of natural deposits.
Barium	2	2	Increase in blood pressure.	Discharge of drilling wastes and from metal refineries, erosion of natural deposits.
Beryllium	0.004	0.004	Intestinal lesions.	Discharge from metal refineries and coal burning factories, discharge from electrical, aerospace, and defense industries.
Cadmium	0.005	0.005	Kidney damage.	Corrosion of galvanized pipes, erosion of natural deposits, discharge from metal refineries, runoff from waste batteries and paints.
Chromium (total)	0.1	0.1	Some people who use water containing chromium well in excess of the MCL for many years could experience allergic dermatitis.	Discharge from steel and pulp mills, erosion of natural deposits.
Copper	1.3	Action Level = 1.3; TT[6]	Short-term exposure: gastrointestinal distress. Long-term exposure: liver or kidney damage. Those with Wilson's disease should consult their doctor if their water systems exceed the action level.	Corrosion of household plumbing, erosion of natural deposits, leaching from wood preservatives.
Cyanide (as free cyanide)	0.2	0.2	Nerve damage or thyroid problems.	Discharge from steel/metal factories, discharge from plastic and fertilizer factories.

TABLE 1 (continued)
National Primary Drinking Water Regulations

Inorganic Chemicals	MCLG[1] (mg/L)[4]	MCL[2] or TT[3] (mg/L)[4]	Potential Health Effects from Ingestion of Water	Sources of Contaminant in Drinking Water
Fluoride	4.0	4.0	Bone disease (pain and tenderness of the bones), children may get mottled teeth.	Water additive which promotes strong teeth, erosion of natural deposits, discharge from fertilizer and aluminum factories.
Lead	zero	Action Level = 0.015, TT[6]	Infants and children: delays in physical or mental development. Adults: kidney problems, high blood pressure.	Corrosion of household plumbing systems, erosion of natural deposits. Banned as a component of plumbing materials in August 1998.
Inorganic Mercury	0.002	0.002	Kidney damage.	Erosion of natural deposits; discharge from refineries and factories, runoff from landfills and cropland.
Nickel	Remanded in 1995.[7] Prior values: MCLG = 0.1 MCL = 0.1		No health effects found at acute exposures above 0.1 mg/L. Chronic exposures above 0.1 mg/L may cause decreased body weight, dermatitis, heart and liver damage.	Industrial atmospheric releases, metal plating wastes, mineral formations, oil refining, organic chemical production, steel production.
Nitrate (measured as Nitrogen)	10	10	"Blue Baby Syndrome" in infants under 6 months — life threatening without immediate medical attention. Symptoms: infant looks blue and has shortness of breath.	Runoff from fertilizer use, leaching from septic tanks, sewage, erosion of natural deposits.
Nitrite (measured as Nitrogen)	1	1	"Blue Baby Syndrome" in infants under 6 months — life threatening without immediate medical attention. Symptoms: Infant looks blue and has shortness of breath.	Runoff from fertilizer use, leaching from septic tanks, sewage, erosion of natural deposits.
Selenium	0.05	0.05	Hair or fingernail loss, numbness in fingers or toes, circulatory problems.	Discharge from petroleum refineries, erosion of natural deposits, discharge from mines.
Sulfate	Proposed as of 1999: MCL = 500 MCLG = 500		Laxative effect.	Erosion of natural deposits.
Thallium	0.0005	0.002	Hair loss, changes in blood, kidney, intestine, or liver problems.	Leaching from ore-processing sites, discharge from electronics, glass, and pharmaceutical companies.

TABLE 1 (continued)
National Primary Drinking Water Regulations

Organic Chemicals	MCLG[1] (mg/L)[4]	MCL[2] or TT[3] (mg/L)[4]	Potential Health Effects from Ingestion of Water	Sources of Contaminant in Drinking Water
Acrylamide	zero	TT[8]	Nervous system or blood problems, increased risk of cancer.	Added to water during sewage/wastewater treatment.
Alachlor	zero	0.002	Eye, liver, kidney or spleen problems, anemia, increased risk of cancer.	Runoff from herbicide used on row crops.
Atrazine	0.003	0.003	Cardiovascular system problems, reproductive difficulties.	Runoff from herbicide used on row crops.
Benzene	zero	0.005	Anemia, decrease in blood platelets, increased risk of cancer.	Discharge from factories, leaching from gas storage, tanks and landfills.
Benzo(a)pyrene	zero	0.0002	Reproductive difficulties, increased risk of cancer.	Leaching from linings of water storage tanks and distribution lines.
Carbofuran	0.04	0.04	Problems with blood or nervous system, reproductive difficulties.	Leaching of soil fumigant used on rice and alfalfa.
Carbon tetrachloride	zero	0.005	Liver problems, increased risk of cancer.	Discharge from chemical plants and other industrial activities.
Chlordane	zero	0.002	Liver or nervous system problems, increased risk of cancer.	Residue of banned termiticide.
Chlorobenzene	0.1	0.1	Liver or kidney problems.	Discharged from chemical and agricultural chemical factories.
2,4-D	0.07	0.07	Kidney, liver, or adrenal gland problems.	Runoff from herbicide used on row crops.
Dalapon	0.2	0.2	Minor kidney changes.	Runoff from herbicide used on rights of way.
1,2-Dibromo-3-chloropropane (DBCP)	zero	0.0002	Reproductive difficulties, increased risk of cancer.	Runoff/leaching from soil fumigant used on soybeans, cotton, pineapples, and orchards.
o-Dichlorobenzene	0.6	0.6	Liver, kidney, or circulatory system problems.	Discharge from industrial chemical factories.
p-Dichlorobenzene	0.075	0.075	Anemia, liver, kidney or spleen damage, changes in blood.	Discharge from industrial chemical factories.
1,2-Dichloroethane	zero	0.005	Increased risk of cancer.	Discharge from industrial chemical factories.
1-1-Dichloroethylene	0.007	0.007	Liver problems.	Discharge from industrial chemical factories.
cis-1,2-Dichloro-ethylene	0.07	0.07	Liver problems.	Discharge from industrial chemical factories.
trans-1,2-Dichloroethylene	0.1	0.1	Liver problems.	Discharge from industrial chemical factories.

TABLE 1 (continued)
National Primary Drinking Water Regulations

Organic Chemicals	MCLG[1] (mg/L)[4]	MCL[2] or TT[3] (mg/L)[4]	Potential Health Effects from Ingestion of Water	Sources of Contaminant in Drinking Water
Dichloromethane	zero	0.005	Liver problems, increased risk of cancer.	Discharge from pharmaceutical and chemical factories.
1-2-Dichloropropane	zero	0.005	Increased risk of cancer.	Discharge from industrial chemical factories.
Di-(2-ethylhexyl) adipate	0.4	0.4	General toxic effects or reproductive difficulties.	Leaching from PVC plumbing systems, discharge from chemical factories.
Di-(2-ethylhexyl) phthalate	zero	0.006	Reproductive difficulties, liver problems, increased risk of cancer.	Discharge from rubber and chemical factories.
Dinoseb	0.007	0.007	Reproductive difficulties.	Runoff from herbicide used on soybeans and vegetables.
Dioxin (2,3,7,8-TCDD)	zero	0.00000003	Reproductive difficulties, increased risk of cancer.	Emissions from waste incineration and other combustion, discharge from chemical factories.
Diquat	0.02	0.02	Cataracts.	Runoff from herbicide use.
Endothall	0.1	0.1	Stomach and intestinal problems.	Runoff from herbicide use.
Endrin	0.002	0.002	Nervous system effects.	Residue of banned insecticide.
Epichlorohydrin	zero	TT[8]	Stomach problems, reproductive difficulties, increased risk of cancer.	Discharge from industrial chemical factories, added to water during treatment process.
Ethylbenzene	0.7	0.7	Liver or kidney problems.	Discharge from petroleum refineries.
Ethylene dibromide	zero	0.00005	Stomach problems, reproductive difficulties, increased risk of cancer.	Discharge from petroleum refineries.
Glyphosate	0.7	0.7	Kidney problems, reproductive difficulties.	Runoff from herbicide use.
Heptachlor	zero	0.0004	Liver damage, increased risk of cancer.	Residue of banned termiticide.
Heptachlor epoxide	zero	0.0002	Liver damage, increased risk of cancer.	Breakdown of hepatachlor.
Hexachloro-benzene	zero	0.001	Liver or kidney problems, reproductive difficulties, increased risk of cancer.	Discharge from metal refineries and agricultural chemical factories.
Hexachloro-cyclopentadiene	0.05	0.05	Kidney or stomach problems.	Discharge from chemical factories.
Lindane	0.0002	0.0002	Liver or kidney problems.	Runoff/leaching from insecticide used on cattle, lumber, gardens.
Methoxychlor	0.04	0.04	Reproductive difficulties.	Runoff/leaching from insecticide used on fruits, vegetables, alfalfa, livestock.

TABLE 1 (continued)
National Primary Drinking Water Regulations

Organic Chemicals	MCLG[1] (mg/L)[4]	MCL[2] or TT[3] (mg/L)[4]	Potential Health Effects from Ingestion of Water	Sources of Contaminant in Drinking Water
Oxamyl (Vydate)	0.2	0.2	Slight nervous system effects.	Runoff/leaching from insecticide used on apples, potatoes, and tomatoes.
Polychlorinated biphenyls (PCBs)	zero	0.0005	Skin changes, thymus gland problems, immune deficiencies, reproductive or nervous system difficulties, increased risk of cancer.	Runoff from landfills, discharge of waste chemicals.
Pentachlorophenol	zero	0.001	Liver or kidney problems, increased risk of cancer.	Discharge from wood preserving factories.
Picloram	0.5	0.5	Liver problems.	Herbicide runoff.
Simazine	0.004	0.004	Problems with blood.	Herbicide runoff.
Styrene	0.1	0.1	Liver, kidney, and circulatory problems.	Discharge from rubber and plastic factories, leaching from landfills.
Tetrachloro-ethylene (PCE)	zero	0.005	Liver problems, increased risk of cancer.	Leaching from PVC pipes, discharge from factories and dry cleaners.
Toluene	1	1	Nervous system, kidney, or liver problems.	Discharge from petroleum factories.
Total Trihalomethanes (TTHMs)	N/A[14]	0.08	Liver, kidney or central nervous system problems, increased risk of cancer.	Byproduct of drinking water disinfection.
Toxaphene	zero	0.003	Kidney, liver, or thyroid problems, increased risk of cancer.	Runoff/leaching from insecticide used on cotton and cattle.
2,4,5-TP (Silvex)	0.05	0.05	Liver problems.	Residue of banned herbicide.
1,2,4-Trichloro-benzene	0.07	0.07	Changes in adrenal glands.	Discharge from textile finishing factories.
1,1,1-Trichloro-ethane	0.20	0.2	Liver, nervous system, or circulatory problems.	Discharge from metal degreasing sites and other factories.
1,1,2-Trichloro-ethane	0.003	0.005	Liver, kidney, or immune system problems.	Discharge from industrial chemical factories.
Trichloroethylene (TCE)	zero	0.005	Liver problems, increased risk of cancer.	Discharge from petroleum refineries.
Vinyl chloride	zero	0.002	Increased risk of cancer.	Leaching from PVC pipes, discharge from plastic factories.
Xylenes (total)	10	10	Nervous system damage.	Discharge from petroleum factories and chemical factories.

TABLE 1 (continued)
National Primary Drinking Water Regulations

Radionuclides	MCLG[1] (mg/L)[4]	MCL[2] or TT[3] (mg/L)[4]	Potential Health Effects from Ingestion of Water	Sources of Contaminant in Drinking Water
Beta particles and photon emitters	none[5]	4 millirems per year	Increased risk of cancer.	Decay of natural and man-made deposits.
Gross alpha particle activity	none[5]	15 pico-curies per Liter (pCi/L)	Increased risk of cancer.	Erosion of natural deposits.
Radium 226 and Radium 228 (combined)	none[5]	5 pCi/L	Increased risk of cancer.	Erosion of natural deposits.
Radon	Proposed as of 1999: MCL = 300 pCi/L MCLG = zero		Increased risk of cancer.	Decay product of radium.
Uranium	Proposed as of 1999: MCL = 20 µg/L (equiv. to 30 pCi/L) MCLG = zero		Kidney effects, increased risk of cancer.	Erosion of natural deposits.

Microorganisms	MCLG[1] (mg/L)[4]	MCL[2] or TT[3] (mg/L)[4]	Potential Health Effects from Ingestion of Water	Sources of Contaminant in Drinking Water
Giardia lamblia	zero	TT[9]	Giardiasis, a gastroenteric disease.	Human and animal fecal waste.
Heterotrophic plate count	N/A	TT[9]	HPC has no health effects but can indicate how effective treatment is at controlling microorganisms.	N/A*
Legionella	zero	TT[9]	Legionnaire's Disease, commonly known as pneumonia.	Found naturally in water; multiplies in heating systems.
Total coliforms (including fecal coliform and *E. Coli*)	zero	5.0%[10]	Used as an indicator that other potentially harmful bacteria may be present.[11]	Human and animal fecal waste.
Turbidity	N/A	TT[9]	Turbidity has no health effects but can interfere with disinfection and provide a medium for microbial growth. It may indicate the presence of microbes.	Storm runoff, discharges into source water, and soil erosion.
Viruses (enteric)	zero	TT[9]	Gastroenteric disease.	Human and animal fecal waste.
Cryptosporidium		Proposed as of 1999: TT[3]	Diarrhea, nausea, stomach cramps.	Parasite common in lakes and rivers contaminated with sewage and animal wastes.

Note: National primary drinking water regulations (NPDWRs, or primary standards) are legally enforceable standards that apply to public water systems. Primary standards protect the quality of drinking water by limiting the levels of specific contaminants that can adversely affect public health, and that are known or anticipated to occur in public water systems. Table 1 divides these contaminants into inorganic chemicals, organic chemicals, radionuclides, and microorganisms.

TABLE 2
National Secondary Drinking Water Regulations

Contaminant	Secondary Standard
Aluminum	0.05 to 0.2 mg/L
Chloride	250 mg/L
Color	15 (color units)
Copper	1.0 mg/L
Corrosivity	noncorrosive
Fluoride	2.0 mg/L
Foaming agents	0.5 mg/L
Iron	0.3 mg/L
Manganese	0.05 mg/L
Odor	3 threshold odor number
pH	6.5–8.5
Silver	0.10 mg/L
Sulfate	250 mg/L
Total dissolved solids	500 mg/L
Zinc	5 mg/L

Note: National secondary drinking water regulations (NSDWRs or secondary standards) are nonenforceable guidelines regulating contaminants that may cause cosmetic effects (such as skin or tooth discoloration) or aesthetic effects (such as taste, odor, or color) in drinking water. The EPA recommends secondary standards to water systems but does not require systems to comply; however, states may choose to adopt them as enforceable standards.

TABLE 3
Regulations for Disinfectants and Disinfectant Byproducts

Disinfectant Residual	MRDLG[12] (mg/L)	MRDL[13] (mg/L)	Compliance Based on
Chlorine	4 (as Cl_2)	4.0 (as Cl_2)	Annual average
Chloramine	4 (as Cl_2)	4.0 (as Cl_2)	Annual average
Chlorine dioxide	0.8 (as Cl_2)	0.8 (as Cl_2)	Daily samples

TABLE 4
Regulations for Disinfectant Byproducts

Disinfection Byproducts	MCLG[1] (mg/L)	MCL[2] (mg/L)	Potential Health Effects from Ingestion of Water	Sources of Contaminant in Drinking Water
Total trihalomethanes[14]	N/A[15]	0.080	Cancer and other effects.	Drinking water chlorination and chloramination byproduct.
Chloroform	0	—	Cancer, liver, kidney, reproductive effects.	Drinking water chlorination and chloramination byproduct.
Bromodichloromethane	0	—	Cancer, liver, kidney, reproductive effects.	Drinking water chlorination and chloramination byproduct.
Dibromochloromethane	0.06	—	Nervous system, liver, kidney, reproductive effects.	Drinking water chlorination and chloramination byproduct.
Bromoform	0	—	Cancer, nervous system, liver, kidney.	Drinking water ozonation, chlorination and chloramination byproduct.
Haloacetic acids (five) (HAA5)[16]	N/A[15]	0.060	Cancer and other effects.	Drinking water chlorination and chloramination byproduct.
Dichloroacetic acid	0	—	Cancer and other effects.	Drinking water chlorination and chloramination byproduct.
Trichloroacetic acid	0.3	—	Cancer and other effects.	Drinking water chlorination and chloramination byproduct.
Chlorite	0.8	1.0	Hemolytic anemia.	Chlorine dioxide disinfection byproduct.
Bromate	0	0.010	Cancer.	Ozonation byproduct.

NOTES FOR TABLES 1, 2, AND 3

*N/A = not applicable

[1] Maximum contaminant level goal (MCLG) — The maximum level of a contaminant in drinking water at which a person drinking 2 L per day for 70 years would experience no known or anticipated adverse health effect, and which allows for an adequate margin of safety. MCLGs are set at zero for contaminants believed by the EPA to cause cancer, assuming that any exposure — no matter how small — poses some risk of cancer. MCLGs are nonenforceable health goals for public water systems.

[2] Maximum contaminant level (MCL) — The maximum permissible level of a contaminant in water that is delivered to any user of a public water system. MCLs are enforceable standards set as close to the MCLGs as technically and economically possible. Except for contaminants regulated as carcinogens, most MCLs and MCLGs are the same. Even when MCLs are less strict than MCLGs, the margins of safety ensure that MCLs provide substantial public health protection.

[3] Treatment technique — An enforceable procedure or level of technical performance that public water systems must follow to ensure control of a contaminant.

[4] Units are in milligrams per liter (mg/L) unless otherwise noted.

[5] MCLGs were not established before the 1986 Amendments to the Safe Drinking Water Act. Therefore, there is no MCLG for this contaminant.

[6] Lead and copper are regulated in a "treatment technique" which requires systems to take tap water samples at sites with lead pipes or copper pipes that have lead solder and/or are served by lead service lines. The action level — which triggers water systems into taking treatment steps if exceeded in more than 10% of tap water samples — for copper is 1.3 mg/L, and for lead is 0.015mg/L.

[7] The MCL and MCLG for nickel were remanded on February 9, 1995 (both had been equal to 0.1 mg/L). There currently is no EPA legal limit on the amount of nickel in drinking water. The EPA is currently reconsidering the limit for nickel.

[8] Each water system must certify, in writing, to the state (using third-party or manufacturer's certification) that when acrylamide and epichlorohydrin are used in drinking water systems, the combination (or product) of dose and monomer level does not exceed the levels specified, as follows:

- Acrylamide = 0.05% dosed at 1 mg/L (or equivalent)
- Epichlorohydrin = 0.01% dosed at 20 mg/L (or equivalent)

[9] The Surface Water Treatment Rule requires systems using surface water or groundwater under the direct influence of surface water to (1) disinfect their water and (2) filter their water or meet criteria for avoiding filtration, so that the following contaminants are controlled at the following levels:

- *Giardia lamblia*: 99.9% killed/inactivated
- Viruses: 99.99% killed/inactivated
- *Legionella*: No limit, but the EPA believes that if *Giardia* and viruses are inactivated, *Legionella* will also be controlled.
- Turbidity: At no time can turbidity (cloudiness of water) go above 5 nephelolometric turbidity units (NTU); systems that filter must ensure that the turbidity go no higher than 1 NTU (0.5 NTU for conventional or direct filtration) in at least 95% of the daily samples in any month.
- HPC: No more than 500 bacterial colonies per milliliter.

[10] No more than 5.0% samples total coliform-positive in a month. (For water systems that collect fewer than 40 routine samples per month, no more than one sample can be total coliform-positive). Every sample that has total coliforms must be analyzed for fecal coliforms. There cannot be any fecal coliforms.

[11] Fecal coliform and *E. coli* are bacteria whose presence indicates that the water may be contaminated with human animal wastes. Microbes in these wastes can cause diarrhea, cramps, nausea, headaches, or other symptoms.

[12] Maximum residual disinfectant level goal (MRDLG) — A residual level at which no known or anticipated adverse effects on the health of people would occur, and which allows for an adequate margin of safety. MRDLGs are nonenforceable health goals for public water systems, based only on health effects and exposure information. They do not reflect the benefits of controlling waterborne microbial contaminants.

[13] Maximum residual disinfectant level (MRDL) — MRDLs are similar to MCLs. MRDLs are enforceable standards, analogous to MCLs, that recognize the benefits of adding a disinfectant to water on a continuous basis and of addressing emergency situations such as distribution system pipe ruptures. As with MCLs, the EPA has set the MRDLs as close to the MRDLGs as possible.

[14] Total trihalomethanes is the sum of the concentrations of chloroform, bromodichloromethane, dibromochloromethane, and bromoform.

[15] N/A — not applicable because there are MCLGs for individual THMs and HAAs.

[16] Haloacetic acids (five) (HAA5) is the sum of the concentrations of mono-, di-, and trichloroacetic acids and mono- and dibromoacetic acids.

Appendix B

National Recommended Water Quality Criteria

(Adapted from: EPA 822-Z-99-001, April 1999.)

Section 304(a)(1) of the Clean Water Act requires the EPA to develop, publish, and revise water quality criteria so that they accurately reflect the latest scientific knowledge. These criteria are based solely on a scientific interpretation of data concerning the relation between pollutant concentrations and environmental and human health effects. Considerations of technological feasibility or economic impacts of attaining the recommended criteria are not included in the criteria development process. Section 304(a) criteria are intended to provide guidance to states and tribes for adopting water quality standards and discharge limitations that are protective of designated water uses within their jurisdictions. States and tribes have three options (40 CFR 131.11):

1. Adopt the EPA recommended criteria.
2. Adopt the EPA criteria modified to reflect site-specific conditions.
3. Adopt different criteria derived using scientifically defensible methods.

Options 2 and 3 must result in criteria that are sufficient to protect designated water uses.

The EPA-recommended water quality criteria are not regulations. They are the product of compiling and interpreting a vast quantity of scientific data that the EPA recommends for states and tribes to use as guidance in adopting water quality standards pursuant to Section 303(c) of the Clean Water Act and the implementing of federal regulations in 40 CFR 131. The EPA water quality criteria do not represent legally binding requirements on EPA, states, tribes, or the public.

The following tables list all priority toxic pollutants and some nonpriority toxic pollutants, with criteria for environmental and human health effects and for organoleptic effects. If some criteria are not listed for a pollutant, the EPA has no national recommendation for those criteria. Also listed are the chemical abstracts service (CAS) registry numbers, which provide a unique identification number for each chemical.

Criterion maximum concentration (CMC) — an estimation of the highest concentration of a pollutant in surface water to which an aquatic community can be exposed briefly without resulting in an unacceptable effect; equivalent to an acute water quality standard.

Criterion continuous concentration (CCC) — an estimation of the highest concentration of a pollutant in surface water to which an aquatic community can be exposed indefinitely without resulting in an unacceptable effect; equivalent to a chronic water quality standard.

TABLE 1
Water Quality Criteria: Priority Toxic Pollutants

	Priority Pollutant	CAS Number	Fresh Water CMC (µg/L)	Fresh Water CCC (µg/L)	Salt Water CMC (µg/L)	Salt Water CCC (µg/L)	Human Health for Consumption of Water + Organism (µg/L)	Human Health for Consumption of Organism Only (µg/L)
1	Antimony	7440360	—	—	—	—	14 B,Z	4300 B
2	Arsenic	7440382	340 A,D,K	150 A,D,K	69 A,D,BB	36 A,D,BB	0.018 C,M,S	0.14 C,M,S
3	Beryllium	7440417	—	—	—	—	J,Z	J
4	Cadmium	7440439	4.3 D,E,K	2.2 D,E,K	42 D,BB	9.3 D,BB	J,Z	J
5a	Chromium III	16065831	570 D,E,K	74 D,E,K	—	—	J,Z Total	J
5b	Chromium VI	18540299	16 D,K	11 D,K	1100 D,BB	50 D,BB	J,Z Total	J
6	Copper	7440508	13 D,E,K,CC	9.0 D,E,K,CC	4.8 D,CC,FF	3.1 D,CC,FF	1300 U	—
7	Lead	7439921	65 D,E,BB,GG	2.5 D,E,BB,GG	210 D,BB	8.1 D,BB	J	J
8	Mercury	7439976	1.4 D,K,HH	0.77 D,K,HH	1.8 D,EE,FF	0.94 D,EE,HH	0.050 B	0.051 B
9	Nickel	7440020	470 D,E,K	52 D,E,K	74 D,BB	8.2 D,BB	610 B	4600 B
10	Selenium	7782492	L,R,T	50 T	290 D,BB,DD	71 D,BB,DD	170 Z	11,000
11	Silver	7440223	3.4 D,E,G	—	1.9 D,G	—	—	—
12	Thallium	7440280	—	—	—	—	1.7 B	6.3 B
13	Zinc	7440666	120 D,E,K	120 D,E,K	90 D,BB	81 D,BB	91,000 U	69,000 U
14	Cyanide	57125	22 K,Q	5.2 K,Q	1 Q,BB	1 Q,BB	700 B,Z	220,000 B,H
15	Asbestos	1332214	—	—	—	—	7×10^6 fibers/L I	—
16	2,3,7,8-TCDD Dioxin	1746016	—	—	—	—	1.3E-8 C	1.4E-8 C
17	Acrolein	107028	—	—	—	—	320	780
18	Acrylonitrile	107131	—	—	—	—	0.059 B,C	0.66 B,C
19	Benzene	71432	—	—	—	—	1.2 B,C	71 B,C
20	Bromoform	75252	—	—	—	—	4.3 B,C	360 B,C
21	Carbon Tetrachloride	56235	—	—	—	—	0.25 B,C	4.4 B,C
22	Chlorobenzene	108907	—	—	—	—	680 B,Z	21,000 B,H
23	Chlorodibromomethane	124481	—	—	—	—	0.41 B,C	34 B,C
24	Chloroethane	75003	—	—	—	—	—	—

TABLE 1 (continued)
Water Quality Criteria: Priority Toxic Pollutants

	Priority Pollutant	CAS Number	Fresh Water CMC (μg/L)	Fresh Water CCC (μg/L)	Salt Water CMC (μg/L)	Salt Water CCC (μg/L)	Human Health for Consumption of Water + Organism (μg/L)	Human Health for Consumption of Organism Only (μg/L)
25	2-Chloroethylvinyl Ether	110758	—	—	—	—	—	—
26	Chloroform	67663	—	—	—	—	5.7 B,C	470 B,C
27	Dichlorobromomethane	75274	—	—	—	—	0.56 B,C	46 B,C
28	1,1-Dichloroethane	75343	—	—	—	—	—	—
29	1,2-Dichloroethane	107062	—	—	—	—	0.38 B,C	99 B,C
30	1,1-Dichloroethylene	75354	—	—	—	—	0.057 B,C	3.2 B,C
31	1,2-Dichloropropane	78875	—	—	—	—	0.52 B,C	39 B,C
32	1,3-Dichloropropene	542756	—	—	—	—	10 B	1700 B
33	Ethylbenzene	100414	—	—	—	—	3100 B,Z	29,000 B
34	Methyl Bromide	74839	—	—	—	—	48 B	4,000 B
35	Methyl Chloride	74873	—	—	—	—	J	J
36	Methylene Chloride	75092	—	—	—	—	4.7 B,C	1600 B,C
37	1,1,2,2-Tetrachloroethane	79345	—	—	—	—	0.17 B,C	11 B,C
38	Tetrachloroethylene	127184	—	—	—	—	0.8 C	8.85 C
39	Toluene	108883	—	—	—	—	6800 B,Z	200,000 B
40	1,2-Trans-dichloroethylene	156605	—	—	—	—	700 B,Z	140,000 B
41	1,1,1-Trichloroethane	71556	—	—	—	—	J,Z	J
42	1,1,2-Trichloroethane	79005	—	—	—	—	0.60 B,C	42 B,C
43	Trichloroethylene	79016	—	—	—	—	2.7 C	81 C
44	Vinyl Chloride	75014	—	—	—	—	2.0 C	525 C
45	2-Chlorophenol	95578	—	—	—	—	120 B,U	400 B,U
46	2,4-Dichlorophenol	120832	—	—	—	—	93 B,U	790 B,U
47	2,4-Dimethylphenol	105679	—	—	—	—	540 B,U	2300 B,U
48	2-Methyl-4,6-dinitrophenol	534521	—	—	—	—	13.4	765
49	2,4-Dinitrophenol	51285	—	—	—	—	70 B	14,000 B

TABLE 1 (continued)
Water Quality Criteria: Priority Toxic Pollutants

	Priority Pollutant	CAS Number	Fresh Water CMC (µg/L)	Fresh Water CCC (µg/L)	Salt Water CMC (µg/L)	Salt Water CCC (µg/L)	Human Health for Consumption of Water + Organism (µg/L)	Human Health for Consumption of Organism Only (µg/L)
50	2-Nitrophenol	88755	—	—	—	—	—	—
51	4-Nitrophenol	100027	—	—	—	—	—	—
52	3-Methyl-4-chlorophenol	59507	—	—	—	—	U	U
53	Pentachlorophenol	87865	19 F,K	15 F,K	13 BB	7.9 BB	0.28 B,C	8.2 B,C,H
54	Phenol	108952	—	—	—	—	21,000 B,U	4,600,000 B,H,U
55	2,4,6-Trichlorophenol	88062	—	—	—	—	2.1 B,C,U	6.5 B,C
56	Acenaphthene	83329	—	—	—	—	1200 B,U	2700 B,U
57	Acenaphthalene	208968	—	—	—	—	—	—
58	Anthracene	120127	—	—	—	—	9600 B	110,000 B
59	Benzidine	92875	—	—	—	—	0.00012 B,C	0.00054 B,C
60	Benzo(a)anthracene	56553	—	—	—	—	0.0044 B,C	0.049 B,C
61	Benzo(a)pyrene	50328	—	—	—	—	0.0044 B,C	0.049 B,C
62	Benzo(b)fluoranthene	205992	—	—	—	—	0.0044 B,C	0.049 B,C
63	Benzo(g,h,i)perylene	191242	—	—	—	—	—	—
64	Benzo(k)fluoranthene	207089	—	—	—	—	0.0044 B,C	0.049 B,C
65	Bis(2-chloroethoxy)methane	111911	—	—	—	—	—	—
66	Bis(2-chloroethyl)ether	111444	—	—	—	—	0.31 B,C	1.4 B,C
67	Bis(2-chloroisopropyl)ether	39638329	—	—	—	—	1400 B	170,000 B
68	Bis(2-ethylhexyl)phthalate[x]	117817	—	—	—	—	1.8 B,C	5.9 B,C
69	4-Bromophenyl Phenylether	101553	—	—	—	—	—	—
70	Butylbenzyl phthalate[w]	85687	—	—	—	—	3000 B	5200 B
71	2-Chloronaphthalene	91587	—	—	—	—	1700 B	4300 B
72	4-Chlorophenyl Phenylether	7005723	—	—	—	—	—	—
73	Chrysene	218019	—	—	—	—	0.0044 B,C	0.049 B,C
74	Dibenzo(a,h,)anthracene	53703	—	—	—	—	0.0044 B,C	0.049 B,C

TABLE 1 (continued)
Water Quality Criteria: Priority Toxic Pollutants

Priority Pollutant	CAS Number	Fresh Water CMC (µg/L)	Fresh Water CCC (µg/L)	Salt Water CMC (µg/L)	Salt Water CCC (µg/L)	Human Health for Consumption of Water + Organism (µg/L)	Human Health for Consumption of Organism Only (µg/L)
75	1,2-Dichlorobenzene	95501	—	—	—	—	2,700 B,Z	17,000 B,Z
76	1,3-Dichlorobenzene	541731	—	—	—	—	400	2600
77	1,4-Dichlorobenzene	106467	—	—	—	—	400 Z	2600
78	3,3'-Dichlorobenzidine	91941	—	—	—	—	0.04 B,C	0.077 B,C
79	Diethylphthalate^w	84662	—	—	—	—	23,000 B	120,000 B
80	Dimethylphthalate	131113	—	—	—	—	313,000	2,900,000
81	Di-n-Butylphthalate	84742	—	—	—	—	2700 B	12,000 B
82	2,4-Dinitrotoluene	121142	—	—	—	—	—	—
83	2,6-Dinitrotoluene	606202	—	—	—	—	—	—
84	Di-n-Octylphthalate	117840	—	—	—	—	—	—
85	1,2-Diphenylhydrazine	122667	—	—	—	—	0.040 B,C	0.54 B,C
86	Fluoranthene	206440	—	—	—	—	300 B	370 B
87	Fluorene	86737	—	—	—	—	1300 B	14,000 B
88	Hexachlorobenzene	118741	—	—	—	—	0.00075 B,C	0.00077 B,C
89	Hexachlorobutadiene	87683	—	—	—	—	0.44 B,C	50 B,C
90	Hexachlorocyclopentadiene	77474	—	—	—	—	240 B,U,Z	17,000 B,H,U
91	Hexachloroethane	67721	—	—	—	—	1.9 B,C	8.9 B,C
92	Ideno(1,2,3-cd)pyrene	193395	—	—	—	—	0.0044 B,C	0.049 B,C
93	Isophorone	78591	—	—	—	—	36 B,C	2600 B,C
94	Naphthalene	91203	—	—	—	—	—	—
95	Nitrobenzene	98953	—	—	—	—	17 B	1900 B,H,U
96	N-Nitrosodimethylamine	62759	—	—	—	—	0.00069 B,C	8.1 B,C
97	N-Nitrosodi-n-propylamine	621647	—	—	—	—	0.005 B,C	1.4 B,C
98	N-Nitrosodiphenylamine	86306	—	—	—	—	5.0 B,C	16 B,C
99	Phenanthrene	85018	—	—	—	—	—	—

TABLE 1 (continued)
Water Quality Criteria: Priority Toxic Pollutants

	Priority Pollutant	CAS Number	Fresh Water CMC (µg/L)	Fresh Water CCC (µg/L)	Salt Water CMC (µg/L)	Salt Water CCC (µg/L)	Human Health for Consumption of Water + Organism (µg/L)	Organism Only (µg/L)
100	Pyrene	129000	—	—	—	—	960 B	11,000 B
101	1,2,4-Trichlorobenzene	120821	—	—	—	—	260 Z	940
102	Aldrin	309002	3.0 G	—	1.3 G	—	0.00013 B,C	0.00014 B,C
103	alpha-BHC	319846	—	—	—	—	0.0039 B,C	0.013 B,C
104	beta-BHC	319857	—	—	—	—	0.014 B,C	0.046 B,C
105	gamma-BHC (Lindane)	58899	—	—	—	—	0.019 C	0.063 C
106	delta-BHC	319868	—	—	—	—	—	—
107	Chlordane	57749	2.4 G	0.0043 G,AA	0.09 G	0.004 G,AA	0.0021 B,C	0.0022 B,C
108	4,4'-DDT	50293	1.1 G	0.001 G,AA	0.13 G	0.001 G,AA	0.00059 B,C	0.00059 B,C
109	4,4'-DDE	72559	—	—	—	—	0.00059 B,C	0.00059 B,C
110	4,4'-DDD	72548	—	—	—	—	0.00083 B,C	0.00084 B,C
111	Dieldrin	60571	0.24 K	0.056 K,O	0.71 G	0.0019 G,AA	0.00014 B,C	0.00014 B,C
112	alpha-Endosulfan	959988	0.22 G,Y	0.056 G,Y	0.034 G,Y	0.0087 G,Y	110 B	240 B
113	beta-Endosulfan	33213659	0.22 G,Y	0.056 G,Y	0.034 G,Y	0.0087 G,Y	110 B	240 B
114	Endosulfan sulfate	1031078	—	—	—	—	110 B	240 B
115	Endrin	72208	0.086 K	0.036 K,O	0.037 G	0.0023 G,AA	0.76 B	0.81 B
116	Endrinaldehyde	7421934	—	—	—	—	0.76 B	0.81 B
117	Heptachlor	76448	0.52 G	0.0038 G,AA	0.053 G	0.0036 G,AA	0.00021 B,C	0.00021 B,C
118	Heptachlor epoxide	1024573	0.52 G,V	0.0038 G,AA	0.053 G,V	0.0036 G,V,AA	0.00010 B,C	0.00011 B,C
119	Polychlorinated biphenyls (PCBs)	—	—	0.014 N,AA	—	0.03 N,AA	0.00017 B,C,P	0.00017 B,C,P
120	Toxaphene	8001352	0.73	0.0002 AA	0.21	0.0002 AA	0.00073 B,C	0.00075 B,C

A This recommended water quality criterion was derived from data for arsenic (III) but is applied here to total arsenic, which might imply that arsenic (III) and arsenic (V) are equally toxic to aquatic life and that their toxicities are additive. In the arsenic criteria document (EPA 440/5-84-033, January 1985), Species Mean Acute Values are given for both arsenic (III) and arsenic (V) for five species, and the ratios of the SMAVs for each species range from 0.6 to 1.7. Chronic values are available for both arsenic (III) and arsenic (V) for one species; for the fathead minnow, the chronic value for arsenic (V) is 0.29 times the chronic value for arsenic (III). No data are available concerning whether the toxicities of the forms of arsenic to aquatic organisms are additive.

TABLE 1 (continued)
Water Quality Criteria: Priority Toxic Pollutants

B This criterion has been revised to reflect The Environmental Protection Agency's q1* or RfD, as contained in the Integrated Risk Information System (IRIS) as of April 8, 1998. The fish tissue bioconcentration factor (BCF) from the 1980 Ambient Water Quality Criteria document was retained in each case.

C This criterion is based on carcinogenicity of 10^{-6} risk. Alternative risk levels may be obtained by moving the decimal point (for example, for a risk level of 10^{-5}, move the decimal point in the recommended criterion one place to the right).

D Freshwater and saltwater criteria for metals are expressed in terms of the dissolved metal in the water column. The recommended water quality criteria value was calculated by using the previous 304(a) aquatic life criteria expressed in terms of total recoverable metal, and multiplying it by a conversion factor (CF). The term CF represents the recommended conversion factor for converting a metal criterion expressed as the total recoverable fraction in the water column to a criterion expressed as the dissolved fraction in the water column. (Conversion factors for saltwater CCCs are not currently available. Conversion factors derived for saltwater CMCs have been used for both saltwater CMCs and CCCs.) See "Office of Water Policy and Technical Guidance on Interpretation and Implementation of Aquatic Life Metals Criteria," October 1, 1993, by Martha G. Prothro, Acting Assistant Administrator for Water, available from the Water Resource Center, USEPA, 401 M St., SW, mail code RC4100, Washington, DC 20460; and 40CFR §131.36(b)(1). Conversion factors applied in the table can be found in Appendix A to the Preamble-Conversion Factors for Dissolved Metals.

E The freshwater criterion for this metal is expressed as a function of hardness (mg/L) in the water column. The value given here corresponds to a hardness of 100 mg/L. Criteria values for other hardness may be calculated from the following: CMC (dissolved) = $\exp\{m_A [\ln(hardness)] + b_A\}(CF)$, or CCC (dissolved) = $\exp\{m_C [\ln(hardness)] + b_C\}(CF)$ and the parameters specified in Appendix B to the Preamble-Parameters for Calculating Freshwater Dissolved Metals Criteria that Are Hardness-Dependent.

F Freshwater aquatic life values for pentachlorophenol are expressed as a function of pH and are calculated as follows: CMC = $\exp(1.005(pH) - 4.869)$; CCC = $\exp(1.005(pH) - 5.134)$. Values displayed in the table correspond to a pH of 7.8.

G This criterion is based on 304(a) aquatic life criterion issued in 1980 and was issued in one of the following documents: Aldrin/Dieldrin (EPA 440/5-80-019), Chlordane (EPA 440/5-80-027), DDT (EPA 440/5-80-038), Endosulfan (EPA 440/5-80-046), Endrin (EPA 440/5-80-047), Heptachlor (440/5-80-052), Hexachlorocyclohexane (EPA 440/5-80-054), Silver (EPA 440/5-80-071). The minimum data requirements and derivation procedures were different in the 1980 guidelines than in the 1985 guidelines. For example, a "CMC" derived using the 1980 Guidelines was derived to be used as an instantaneous maximum. If assessment is to be done using an averaging period, the values given should be divided by 2 to obtain a value that is more comparable to a CMC derived using the 1985 guidelines.

H No criterion for protection of human health from consumption of aquatic organisms excluding water was presented in the 1980 criteria document or in the 1986 Quality Criteria for Water. Nevertheless, sufficient information was presented in the 1980 document to allow the calculation of a criterion, even though the results of such a calculation were not shown in the document.

I This criterion for asbestos is the maximum contaminant level (MCL) developed under the Safe Drinking Water Act (SDWA).

TABLE 1 (continued)
Water Quality Criteria: Priority Toxic Pollutants

J EPA has not calculated a human health criterion for this contaminant. However, permit authorities should address this contaminant in NPDES permit actions using the state's existing narrative criteria for toxics.

K This recommended criterion is based on a 304(a) aquatic life criterion that was issued in the *1995 Updates: Water Quality Criteria Documents for the Protection of Aquatic Life in Ambient Water*, (EPA 820-B-96-001, September 1996). This value was derived using the GLI Guidelines (60FR15393-15399, March 23, 1995; 40CFR132 Appendix A); the difference between the 1985 guidelines and the GLI guidelines are explained on page iv of the 1995 updates. None of the decisions concerning the derivation of this criterion were affected by any considerations that are specific to the Great Lakes.

L The CMC = 1/[(f1/CMC1) + (f2/CMC2)] where f1 and f2 are the fractions of total selenium that are treated as selenite and selenate, respectively, and CMC1 and CMC2 are 185.9 µg/L and 12.83 µg/L, respectively.

M EPA is currently reassessing the criteria for arsenic. Upon completion of the reassessment the agency will publish revised criteria as appropriate.

N PCBs are a class of chemicals which includes aroclors 1242, 1254, 1221, 1232, 1248, 1260, and 1016 (CAS numbers 53469219, 11097691, 11104282, 11141165, 12672296, 11096825 and 12674112 respectively). The aquatic life criteria apply to this set of PCBs.

O The derivation of the CCC for this pollutant did not consider exposure through the diet, which is probably important for aquatic life occupying upper trophic levels.

P This criterion applies to total PCBs, i.e., the sum of all congener or isomer analyses.

Q This recommended water quality criterion is expressed as µg free cyanide (as CN)/L.

R This value was announced (61FR58444-58449, November 14, 1996) as a proposed GLI 303(c) aquatic life criterion. The EPA is currently working on this criterion; therefore, this value might change substantially in the near future.

S This recommended water quality criterion refers to the inorganic form only.

T This recommended water quality criterion is expressed in terms of total recoverable metal in the water column. It is scientifically acceptable to use the conversion factor of 0.922 that was used in the GLI to convert this to a value that is expressed in terms of dissolved metal.

U The organoleptic effect criterion is more stringent than the value for priority toxic pollutants.

V This value was derived from data for heptachlor, and the criteria document provides insufficient data to estimate the relative toxicities of heptachlor and heptachlor epoxide.

W Although the EPA has not published a final criteria document for this compound, it is the EPA's understanding that sufficient data exist to allow calculation of aquatic criteria. It is anticipated that the industry intends to publish in the peer reviewed literature draft aquatic life criteria generated in accordance with EPA guidelines. The EPA will review such criteria for possible issuance as national WQC.

TABLE 1 (continued)
Water Quality Criteria: Priority Toxic Pollutants

X There is a full set of aquatic life toxicity data that show that DEHP is not toxic to aquatic organisms at or below its solubility limit.

Y This value was derived from data for endosulfan and is most appropriately applied to the sum of alpha-endosulfan and beta-endosulfan.

Z A more stringent MCL has been issued by the EPA. Refer to drinking water regulations (40 CFR 141) or safe drinking water hotline (1-800-426-4791) for values.

AA This CCC is based on the Final Residue Value procedure in the 1985 guidelines. Since the publication of the Great Lakes Aquatic Life Criteria Guidelines in 1995 (60FR15393-15399, March 23, 1995), the Agency no longer uses the Final Residue Value procedure for deriving CCCs for new or revised 304(a) aquatic life criteria.

BB This water quality criterion is based on a 304(a) aquatic life criterion that was derived using the 1985 guidelines (*Guidelines for Deriving Numerical National Water Quality Criteria for the Protection of Aquatic Organisms and Their Uses*, PB85-227049, January 1985) and was issued in one of the following criteria documents: Arsenic (EPA 440/5-84-033), Cadmium (EPA 440/5-84-032), Chromium (EPA 440/5-84-029), Copper (EPA 440/5-84-031), Cyanide (EPA 440/5-84-028), Lead (EPA 440/5-84-027), Nickel (EPA 440/5-86-004), Pentachlorophenol (EPA 440/5-86-009), Toxaphene, (EPA 440/5-87- 003).

CC When the concentration of dissolved organic carbon is elevated, copper is substantially less toxic, and use of Water-Effect Ratios might be appropriate.

DD The selenium criteria document (EPA 440/5-87-006, September 1987) provides that, if selenium is as toxic to saltwater fishes in the field as it is to freshwater fishes in the field, the status of the fish community should be monitored whenever the concentration of selenium exceeds 5.0 μg/L in salt water because the saltwater CCC does not take into account uptake via the food chain.

EE This recommended water quality criterion was derived on page 43 of the mercury criteria document (EPA 440/5-84-026, January 1985). The saltwater CCC of 0.025 μg/L given on page 23 of the criteria document is based on the Final Residue Value procedure in the 1985 guidelines. Since the publication of the Great Lakes Aquatic Life Criteria Guidelines in 1995 (60FR15393-15399, March 23, 1995), the Agency no longer uses the Final Residue Value procedure for deriving CCCs for new or revised 304(a) aquatic life criteria.

FF This recommended water quality criterion was derived in *Ambient Water Quality Criteria Saltwater Copper Addendum* (Draft, April 14, 1995) and was promulgated in the Interim Final National Toxics Rule (60FR22228-222237, May 4, 1995).

GG EPA is actively working on this criterion; therefore, this recommended water quality criterion may change substantially in the near future.

HH This recommended water quality criterion was derived from data for inorganic mercury (II) but is applied here to total mercury. If a substantial portion of the mercury in the water column is methylmercury, this criterion will probably be under-protective. In addition, even though inorganic mercury is converted to methylmercury and methylmercury bioaccumulates to a great extent, this criterion does not account for uptake via the food chain because sufficient data were not available when the criterion was derived.

TABLE 2
Water Quality Criteria: Nonpriority Toxic Pollutants

#	Nonpriority Pollutant	CAS Number	Fresh Water CMC (µg/L)	Fresh Water CCC (µg/L)	Salt Water CMC (µg/L)	Salt Water CCC (µg/L)	Human Health Water + Organism (µg/L)	Human Health Organism Only (µg/L)
1	Alkalinity	—	—	20,000 F	—	—	—	—
2	Aluminum (pH 6.5–9.0)	7429905	750 G,I	87 G,I,L	—	—	—	—
3	Ammonia	7664417	Freshwater criteria are pH dependent, see EPA 822-R-98-008.		Saltwater criteria are pH and temperature dependent, see EPA 440/5-88-004.		D	—
4	Aesthetic qualities	—	Narrative statement, see *Gold Book.**					
5	Bacteria	—	For primary recreation and shellfish uses, see *Gold Book.**					
6	Barium	7440393	—	—	—	—	1,000 A	—
7	Boron	7440428			Narrative statement, see *Gold Book.**		—	—
8	Chloride	16887006	860,000 G	230,000 G	—	—	—	—
9	Chlorine	7782505	19	11	13	7.5	C	—
10	Chlorophenoxy herbicide 2,4,5-TP	93721	—	—	—	—	10 A	—
11	Chlorophenoxy herbicide 2,4-D	94757	—	—	—	—	100 A,C	—
12	Chloropyrifos	2921882	0.083 G	0.041 G	0.011 G	0.0056 G	—	—
13	Color	—		Narrative statement, see *Gold Book.**			F	—
14	Demeton	8065483	—	0.1 F	—	0.1 F	—	—
15	bis(chloromethyl)Ether	542881	—	—	—	—	0.00013 E	0.00078 E
16	Gases, total dissolved	—		Narrative statement, see *Gold Book.**			F	—
17	Guthion	86500	—	0.01 F	—	0.01 F	—	—
18	Hardness	—		Narrative statement, see *Gold Book.**			—	—
19	Hexachlorocyclohexane (technical)	319868	—	—	—	—	0.0123	0.0414
20	Iron	7439896	—	—	—	—	300 A	—
21	Malathion	121755	—	0.1 F	—	0.1 F	—	—

TABLE 2 (continued)
Water Quality Criteria: Nonpriority Toxic Pollutants

	Nonpriority Pollutant	CAS Number	Fresh Water CMC (µg/L)	Fresh Water CCC (µg/L)	Salt Water CMC (µg/L)	Salt Water CCC (µg/L)	Human Health for Consumption of Water + Organism (µg/L)	Human Health for Consumption of Organism Only (µg/L)
22	Manganese	7439965	—	—	—	—	50 A	100 A
23	Methoxychlor	72435	—	0.03 F	—	0.03 F	100 A,C	—
24	Mirex	2385855	—	0.001 F	—	0.001 F	—	—
25	Nitrates	14797558	—	—	—	—	10,000 A	—
26	Nitrosamines	—	—	—	—	—	0.0008	1.24
27	Dinitrophenols	25550587	—	—	—	—	70	14,000
28	N-Nitrosodibutylamine	924163	—	—	—	—	0.0064 A	0.587 A
29	N-Nitrosodiethylamine	55185	—	—	—	—	0.0008 A	1.24 A
30	N-Nitrosopyrrolidine	930552	—	—	—	—	0.016	91.9
31	Oil and Grease	—	Narrative statement, see *Gold Book.**				F	
32	Oxygen, dissolved	7782447	Warmwater and coldwater matrix, see *Gold Book.**				O	
33	Parathion	56382	0.065 J	0.013 J	—	—		
34	Pentachlorobenzene	608935	—	—	—	—	3.5 E	4.1 E
35	pH		—	6.5–9 F	—	6.5–8.5 F,K	5–9	—
36	Phosphorus, elemental	7723140	—	—	—	0.1 F,K		
37	Phosphorus, phosphate	—	Narrative statement, see *Gold Book.**					
38	Solids, dissolved (and salinity)	—	Species dependent criteria, see *Gold Book.**				250,000 A	
39	Solids, suspended (and turbidity)	—	Species dependent criteria, see *Gold Book.**				F	
40	Sulfide, hydrogen sulfide	7783064	—	2.0 F	—	2.0 F	—	—
41	Tainting substances	—	Narrative statement, see *Gold Book.**					
42	Temperature	—	Species dependent criteria, see *Gold Book.**					
43	1,2,4,5-Tetrachlorobenzene	95943	—	—	—	—	2.3 E	2.9 E
44	Tributyltin (TBT)	—	0.46 N	0.063 N	0.37 N	0.010 N	—	—
45	2,4,5-Trichlorophenol	95954	—	—	—	—	2,600 B,E	9,800 B,E

TABLE 2 (continued)
Water Quality Criteria: Nonpriority Toxic Pollutants

A This human health criterion is the same as originally published in the *Red Book* which predates the 1980 methodology and did not utilize the fish ingestion BCF approach. This same criterion value is now published in the *Gold Book*.*

B The organoleptic effect criterion is more stringent than the value presented in the nonpriority pollutants table.

C A more stringent maximum contaminant level (MCL) has been issued by the EPA under the Safe Drinking Water Act. Refer to drinking water regulations 40CFR141 or the safe drinking water hotline (1-800-426-4791) for values.

D According to the procedures described in the *Guidelines for Deriving Numerical National Water Quality Criteria for the Protection of Aquatic Organisms and Their Uses* — except possibly where a very sensitive species is important at a site — freshwater aquatic life should be protected if both conditions specified in Appendix C to the Preamble-Calculation of Freshwater Ammonia Criterion are satisfied.

E This criterion has been revised to reflect The Environmental Protection Agency's q1* or RfD, as contained in the Integrated Risk Information System (IRIS) as of April 8, 1998. The fish tissue bioconcentration factor (BCF) used to derive the original criterion was retained in each case.

F The derivation of this value is presented in the *Red Book* (EPA 440/9-76-023, July, 1976).

G This value is based on a 304(a) aquatic life criterion that was derived using the 1985 guidelines (*Guidelines for Deriving Numerical National Water Quality Criteria for the Protection of Aquatic Organisms and Their Uses*, PB85-227049, January 1985) and was issued in one of the following criteria documents: Aluminum (EPA 440/5-86-008); Chloride (EPA 440/5-88-001); Chloropyrifos (EPA 440/5-86-005).

I This value is expressed in terms of total recoverable metal in the water column.

J This value is based on a 304(a) aquatic life criterion that was issued in the *1995 Updates: Water Quality Criteria Documents for the Protection of Aquatic Life in Ambient Water* (EPA-820-B-96-001). This value was derived using the GLI Guidelines (60FR15393-15399, March 23, 1995; 40CFR132 Appendix A); the differences between the 1985 guidelines and the GLI guidelines are explained on page iv of the 1995 updates. No decision concerning this criterion was affected by any considerations that are specific to the Great Lakes.

K According to page 181 of the *Red Book*,

"For open ocean waters where the depth is substantially greater than the euphotic zone, the pH should not be changed more than 0.2 units from the naturally occurring variation or any case outside the range of 6.5 to 8.5. For shallow, highly productive coastal and estuarine areas where naturally occurring pH variations approach the lethal limits of some species, changes in pH should be avoided but in any case should not exceed the limits established for fresh water, i.e., 6.5–9.0."

TABLE 2 (continued)
Water Quality Criteria: Nonpriority Toxic Pollutants

L There are three major reasons why the use of Water-Effect Ratios might be appropriate. (1) The value of 87 µg/L is based on a toxicity test with the striped bass in water with pH = 6.5-6.6 and hardness <10 mg/L. Data in *Aluminum Water-Effect Ratio for the 3M Plant Effluent Discharge, Middleway, West Virginia* (May 1994) indicate that aluminum is substantially less toxic at higher pH and hardness, but the effects of pH and hardness are not well-quantified at this time. (2) In tests with the brook trout at low pH and hardness, effects increased with increasing concentrations of total aluminum even though the concentration of dissolved aluminum was constant, indicating that total recoverable is a more appropriate measurement than dissolved, at least when particulate aluminum is primarily aluminum hydroxide particles. In surface waters, however, the total recoverable procedure might measure aluminum associated with clay particles, which might be less toxic than aluminum associated with aluminum hydroxide. (3) The EPA is aware of field data indicating that many high quality waters in the U.S. contain more than 87 µg aluminum/L, when either total recoverable or dissolved is measured.

M U.S. EPA. 1973, Water Quality Criteria 1972. EPA-R3-73-033. National Technical Information Service, Springfield, VA; U.S. EPA. 1977. Temperature Criteria for Freshwater Fish: Protocol and Procedures. EPA-600/3-77-061. National Technical Information Service, Springfield, VA.

N This value was announced (62FR42554, August 7, 1997) as a proposed 304(a) aquatic life criterion. Although the EPA has not responded to public comment, it is publishing this as a 304(a) criterion as guidance for states and tribes to consider when adopting water quality criteria.

O U.S. EPA, 1986, Ambient Water Quality Criteria for Dissolved Oxygen. EPA 440/5-86-003, National Technical Information Service, Springfield, VA.

TABLE 3
Water Quality Criteria: Organoleptic Effects (Taste and Odor)

	Pollutant	CAS Number	Organoleptic Effect Criteria (μg/L)
1	Acenaphthene	83329	20
2	Chlorobenzene (monochlorobenzene)	108907	20
3	3-Chlorophenol	—	0.1
4	4-Chlorophenol	106489	0.1
5	2,3-Dichlorophenol	—	0.04
6	2,5-Dichlorophenol	—	0.5
7	2,6-Dichlorophenol	87650	0.2
8	3,4-Dichlorophenol	—	0.3
9	2,4,5-Trichlorophenol	95954	1
10	2,4,6-Trichlorophenol	88062	2
11	2,3,4,6-Tetrachlorophenol	58902	1
12	2-Methyl-4-chlorophenol	—	1800
13	3-Methyl-4-chlorophenol	59507	3000
14	3-Methyl-6-chlorophenol	—	20
15	2-Chlorophenol	95578	0.1
16	Copper	7440508	1000
17	2,4-Dichlorophenol	120832	0.3
18	2,4-Dimethylphenol	105679	400
19	Hexachlorocyclopentadiene	77474	1
20	Nitrobenzene	98953	30
21	Pentachlorophenol	87865	30
22	Phenol	108952	300
23	Zinc	7440666	5000

Note: The reference for pollutants 1 through 22 is the EPA *Gold Book*.* The reference for pollutant 23 is 45 FR79341.

The compilation contains 304(a) criteria for pollutants with toxicity-based criteria as well as nontoxicity-based criteria. The basis for the nontoxicity-based criteria is organoleptic effects (e.g., taste and odor) that would make water and edible aquatic life unpalatable but not toxic to humans.

NATIONAL RECOMMENDED WATER QUALITY CRITERIA — ADDITIONAL NOTES FOR TABLES 1, 2, AND 3

1. **Criteria maximum concentration and criterion continuous concentration.** The criteria maximum concentration (CMC) is an estimate of the highest concentration of a material in surface water to which an aquatic community can be exposed briefly without resulting in an unacceptable effect. The criterion continuous concentration (CCC) is an estimate of the highest concentration of a material in surface water to which an aquatic community can be exposed indefinitely without resulting in an unacceptable effect. The CMC and CCC are just two of the six parts of a aquatic life criterion; the other four parts are the acute averaging period, chronic averaging period, acute frequency of allowed exceedence, and chronic frequency of allowed exceedence. Because 304(a) aquatic life criteria are national guidance, they are intended to be protective of the vast majority of the aquatic communities in the U.S.

2. **Criteria recommendations for priority pollutants, nonpriority pollutants and organoleptic effects.** This compilation lists all priority toxic pollutants, some nonpriority toxic pollutants, and both human health effect and organoleptic effect criteria issued pursuant to CWA §304(a). Blank spaces indicate

that EPA has no CWA §304(a) criteria recommendations. For a number of nonpriority toxic pollutants not listed, CWA §304(a) "water + organism" human health criteria are not available, but the EPA has published MCLs under the SDWA that may be used in establishing water quality standards to protect water supply designated uses. Because of variations in chemical nomenclature systems, this listing of toxic pollutants does not duplicate the listing in Appendix A of 40 CFR Part 423. Also listed are the chemical abstracts service (CAS) registry numbers, which provide a unique identification for each chemical.

3. **Human health risk.** The human health criteria for the priority and nonpriority pollutants are based on carcinogenicity of 10^{-6} risk. Alternate risk levels may be obtained by moving the decimal point (e.g., for a risk level of 10^{-5}, move the decimal point in the recommended criterion one place to the right).

4. **Water quality criteria published pursuant to Section 304(a) or Section 303(c) of the CWA.** Many of the values in the compilation were published in the proposed California Toxics Rule (CTR, 62FR42160). Although such values were published pursuant to Section 303(c) of the CWA, they represent the Agency's most recent calculation of water quality criteria and thus are published today as the Agency's 304(a) criteria. Water quality criteria published in the proposed CTR may be revised when the EPA takes final action on the CTR.

5. **Calculation of dissolved metals criteria.** The 304(a) criteria for metals, shown as dissolved metals, are calculated in one of two ways. For freshwater metals criteria that are hardness-dependent, the dissolved metal criteria were calculated using a hardness of 100 mg/l as $CaCO_3$ for illustrative purposes only. The criteria for nonhardness-dependent saltwater and freshwater metals are calculated by multiplying the total recoverable criteria before rounding by the appropriate conversion factors. The final dissolved metals' criteria in the table are rounded to two significant figures. Information regarding the calculation of hardness-dependent conversion factors is included in the footnotes.

6. **Correction of chemical abstract services number.** The chemical abstract services number (CAS) for Bis(2-Chloroisopropyl)ether has been corrected in the table. The correct CAS number for this chemical is 39638-32-9. Previous EPA publications listed 108-60-1 as the CAS number for this chemical.

7. **Maximum contaminant levels.** The compilation includes footnotes for pollutants with maximum contaminant levels (MCLs) more stringent than the recommended water quality criteria in the compilation. MCLs for these pollutants are not included in the compilation but can be found in the appropriate drinking water regulations (40 CFR 141.11-16 and 141.60-63) through the safe drinking water hotline (800-426-4791), or the Internet (http://www.epa.gov/ost/tools/dwstds-s.html).

8. **Category criteria.** In the 1980 criteria documents, certain recommended water quality criteria were published for categories of pollutants rather than for individual pollutants within that category. Subsequently, in a series of separate actions, the agency derived criteria for specific pollutants within a category. Therefore, in this compilation EPA is replacing criteria representing categories with individual pollutant criteria (e.g., 1,3-dichlorobenzene, 1,4-dichlorobenzene and 1,2-dichlorobenzene).

9. **Specific chemical calculations.**
   A. Selenium.
      a. Human health: In the 1980 Selenium document, a criterion for the protection of human health from consumption of water and organisms was calculated based on a BCF of 6.0 L/kg and a maximum water-related contribution of 35 µg Se/day. Subsequently, the EPA Office of Health and Environmental Assessment issued an erratum notice (February 23, 1982), revising the BCF for selenium to 4.8 L/kg. In 1988, the EPA issued an addendum (ECAO-CIN-668) revising the human health criteria for selenium. Later in the final National Toxic Rule (NTR, 57 FR 60848), the EPA withdrew previously published selenium human health criteria, pending agency review of new epidemiological data. This compilation includes human health criteria for selenium, calculated using a BCF of 4.8 L/kg along with the current IRIS RfD of 0.005 mg/kg/day. The EPA included these recommended water quality criteria in the compilation because the data necessary for calculating a criterion in accordance with EPA's 1980 human health methodology are available.
      b. Aquatic life: This compilation contains aquatic life criteria for selenium that are the same as those published in the proposed CTR. In the CTR, the EPA proposed an acute criterion for selenium based on the criterion proposed for selenium in the Water Quality Guidance for the Great Lakes

System (61 FR 58444). The GLI and CTR proposals take into account data showing that selenium's two most prevalent oxidation states, selenite and selenate, present differing potentials for aquatic toxicity, as well as new data indicating that various forms of selenium are additive. The new approach produces a different selenium acute criterion concentration, or CMC, depending upon the relative proportions of selenite, selenate, and other forms of selenium that are present. The EPA notes that it is currently undertaking a reassessment of selenium and expects the 304(a) criteria for selenium will be revised based on the final reassessment (63FR261 86). However, until such time as revised water quality criteria for selenium are published by the agency, the recommended water quality criteria in this compilation are EPA's current 304(a) criteria.

B. 1,2,4-trichlorobenzene and zinc. Human health criteria for 1,2,4-trichlorobenzene and zinc have not been previously published. Sufficient information is now available for calculating water quality criteria for the protection of human health from the consumption of aquatic organisms and the consumption of aquatic organisms and water for both these compounds. Therefore, the EPA is publishing criteria for these pollutants in this compilation.

C. Chromium (III). The recommended aquatic life water quality criteria for chromium (III) included in the compilation are based on the values presented in the document titled *1995 Updates: Water Quality Criteria Documents for the Protection of Aquatic Life in Ambient Water*; however, this document contains criteria based on the total recoverable fraction. The chromium (III) criteria in this compilation were calculated by applying the conversion factors used in the Final Water Quality Guidance for the Great Lakes System (60 FR 15366) to the 1995 update document values.

D. Bis(Chloromethyl)ether, pentachlorobenzene, 1,2,4,5-tetrachlorobenzene, trichlorophenol. Human health criteria for these pollutants were last published in the EPA's *Quality Criteria for Water 1986*, EPA 440/5-86-001, or *Gold Book*. Some of these criteria were calculated using acceptable daily intake (ADIs) rather than RfDs. Updated q1*s and RfDs are now available in IRIS for bis(chloromethyl)ether pentachlorobenzene, 1,2,4,5-tetrachlorobenzene, and trichlorophenol and were used to revise the water quality criteria for these compounds. The recommended water quality criteria for bis(chloromethyl)ether were revised using an updated q1*, while criteria for pentachlorobenzene, 1,2,4,5-tetrachlorobenzene, and trichlorophenol were derived using an updated RfD value.

E. PCBs. In this compilation the EPA is publishing aquatic life and human health criteria based on total PCBs rather than individual arochlors. These criteria replace the previous criteria for the seven individual arochlors. Thus, there are criteria for a total of 102 of the 126 priority pollutants.

Appendix C

Sampling Containers, Minimum Sample Size, Preservation Procedures, and Storage Times

(Adapted from 40 CFR Parts 100-149, July 1, 1992.)

Measured Parameter	Container	Minimum Sample Size (mL)	Preservation	Maximum Storage Time (Recommended/Regulatory)
Acidity.	P, G(B)	100	Cool at 4°C.	14 d.
Alkalinity.	P, G	200	Cool at 4°C.	14 d.
BOD.	P, G	1000	Cool at 4°C.	48 h.
Boron.	P	100	None required.	6 months.
Bromide.	P, G	100	None required.	28 d.
Carbon, organic, total.	G	100	Analyze immediately, or add H_3PO_4 or H_2SO_4 to pH < 2; cool at 4°C.	28 d.
COD.	P, G	100	Analyze as soon as possible, or add H_2SO_4 to pH < 2, cool at 4°C.	28 d.
Chlorine, residual.	P, G	500	Analyze immediately.	Analyze immediately.
Color.	P, G	500	Cool at 4°C.	48 h.
Conductivity.	P, G	500	Cool at 4°C.	28 d.
Cyanide: total.	P, G	500	Add NaOH to pH >12, cool at 4°C in dark.	14 d or 24 h if sulfide present.
Cyanide: amenable to chlorination.	P, G	500	Add 100 mg $Na_2S_2O_3$/L.	14 d or 24 h if sulfide present.
Fluoride.	P	300	None required.	28 d.
Hardness.	P, G	100	Add HNO_3 to pH <2.	6 months.
Metals, general.	P(A), G(A)	500	For dissolved metals filter immediately, add HNO_2 to pH <2.	6 months.
Chromium VI.	P(A), G(A)	300	cool at 4°C.	24 h.
Mercury.	P(A), G(A)	500	Add HNO_2 to pH <2, 4°C, cool at 4°C.	28 d.
Nitrogen: ammonia.	P, G	500	Analyze as soon as possible or add H_2SO_4 to pH <2, cool at 4°C.	28 d.
Nitrate.	P, G	100	Analyze as soon as possible, or cool at 4°C.	48 h (28 d for chlorinated samples).
Nitrate + nitrite.	P, G	200	Add H_2SO_4 to pH <2, cool at 4°C.	48 h.
Nitrite.	P, G	100	Analyze as soon as possible, or cool at 4°C.	48 h.
Organic, Kjeldahl.	P, G	500	Cool at 4°C, add H_2SO_4 to pH <2.	28 d.
Oil and grease.	G, wide-mouth calibrated.	1000	Add H_2SO_4 to pH <2, cool at 4°C.	28 d/28 d.

Organic compounds

Measured Parameter	Container	Minimum Sample Size (mL)	Preservation	Maximum Storage Time (Recommended/Regulatory)
MBAS.	P, G	250	Cool at 4°C.	48 h.
Pesticides.	G(S), TFE-lined cap.		Cool at 4°C, 1000 mg ascorbic acid/L if residual chlorine present.	7 d/7 d until extraction, 40 d after extraction.
Phenols.	P, G	500	Cool at 4°C, add H_2SO_4 to pH <2.	*/28 d.
Purgeables by purge and trap.	G, TFE-lined cap.	50	Cool at 4°C, add HCl to pH <2, 1000 mg ascorbic acid/L if residual chlorine present.	7 d/14 d.
Oxygen, dissolved: electrode winkler.	G, BOD bottle	300	Analyze immediately, titration may be delayed after acidification.	Analyze immediately, 8 h/8 h.
Ozone.	G	1000	Analyze immediately.	Analyze immediately.
pH.	P, G	50	Analyze immediately.	Analyze immediately.

Measured Parameter	Container	Minimum Sample Size (mL)	Preservation	Maximum Storage Time (Recommended/Regulatory)
Organic compounds				
Phosphate.	G(A)	100	For dissolved phosphate, filter immediately, cool at 4°C.	48 h.
Silica.	P	200	Cool at 4°C, do not freeze.	28 d.
Solids.	P, G	200	Cool at 4°C.	7 d/2-7 d.**
Sulfate.	P, G	100	Cool at 4°C.	28 d.
Sulfide.	P, G	100	Cool at 4°C, add 4 drops 2N zinc acetate/l00 mL, add NaOH to pH >9.	7 d.
Temperature.	P, G		Analyze immediately.	Analyze immediately.
Turbidity.	P, G		Cool at 4°C.	48 h.

* See text in 40 CFR Parts 100–149, July 1, 1992 for additional details. For determinations not listed, use glass or plastic containers; preferably refrigerate during storage and analyze as soon as possible. Refrigerate = storage at 4°C, in the dark. P = plastic (polyethylene or equivalent); G = glass; G(A) or P(A) = rinsed with 1 + 1 HNO_3; G(B) = glass, borosilicate; G(S) = glass, rinsed with organic solvents; N.S. = not stated in cited reference; stat = no storage allowed, analyze immediately.
** Environmental Protection Agency, Rules and Regulations. Federal Register 49; No. 209, October 26, 1984. See this citation for possible differences regarding container and preservation requirements.

Index

A

AA technique, *see* Atomic absorption technique
ABP, *see* Acid-base potential
Acenaphthene, 93
Acid-base
  neutrality, 31
  potential (ABP), 158
  reactions, 33
Acidity, phenolphthalein, 39
Acid mine drainage, 1
Activated charcoal, sorption to, 195
Advanced Oxidation Process (AOP), 178
AEC, *see* Anion exchange capacity
Aeration mechanisms, 48
Aerobic biodegradation, 13
Aerobic respiration, 110
Air-stripping, removing ammonia by, 194
Air-water partition
  behavior, estimating, 79
  coefficient, 82
Alcohols
  double, 28
  of low molecular weight, 27
  straight chain, 28
  triple, 28
Algae, during daytime photosynthesis, 53
Alkalinity
  calculating, 40, 41
  effect of dissolved metal on, 65
  importance of to fish, 40
  measurement of, 66
  methyl orange, 39
  pH buffering effect of, 66
  phenolphthalein, 39
Alkanes, normal, 26
Aluminum, 58, 206–208
Ammonia, 54, 193, 208–209
  air-stripping, 56
  breakpoint chlorination for removing, 159
  calculating chlorine needed to remove, 162
  continued oxidation of, 208
  criteria and standards for, 54
  ionized, 53
  removal of, 53, 194
Ammunition, 221
Anaerobic biodegradation, 13
Anion exchange capacity (AEC), 183
Antimony, 209–210
Antiskid materials, 163

AOP, *see* Advanced Oxidation Process
Aquatic life, 2
  toxicity to, 34
  water quality standards, 46
Aquifer
  sediments, 115
  water pressure in, 134
Arenes, 123
Aromatic hydrocarbons, 123
Arsenic, 210–211
  poisoning, 211
  well-water contamination by, 210
Asbestos, 211–212
Asiatic clam, 179
Assimilable organic carbon, 169
Atomic absorption (AA) technique, 209, 210
Atrazine, 13, 76
Autotrophs, 74

B

Bacteria, 180
  benthic, 222
  denitrifying, 114
  killing, 169
  total coliform, 187
Barium, 212
BCF, *see* Bioconcentration factor
Bedrock, 74, 78
Benthic bacteria, 222
Benzene, 14, 93, 151
  partitioning data, for soil and water in equilibrium, 86
  spill, 91
  toluene, ethylbenzene, and xylene (BTEX), 105, 109,
    123
Benzyl mercaptan, 193
Beryllium, 212–213
Bioconcentration factor (BCF), 91
Biodegradable organic carbon, 169
Biodegradation, 13
  basic requirements of, 104
  field test for using alkalinity as indicator of, 119
  processes, 106, 107
  reactions, normal sequence of, 112
Biological oxygen demand (BOD), 49, 82, 206
Bioremediation, *in situ*, 108
Blue baby syndrome, 55
BOD, *see* Biological oxygen demand
Bond character, 20